[redacted]

				1 H Hydrogen 1	2 He Helium 4
5 B Boron 11	6 C Carbon 12	7 N Nitrogen 14	8 O Oxygen 16	9 F Fluorine 19	10 Ne Neon 20
13 Al Aluminium 27	14 Si Silicon 28	15 P Phosphorus 31	16 S Sulphur 32	17 Cl Chlorine 35.5	18 Ar Argon 40
31 Ga Gallium 70	32 Ge Germanium 73	33 As Arsenic 75	34 Se Selenium 79	35 Br Bromine 80	36 Kr Krypton 84
49 In Indium 115	50 Sn Tin 119	51 Sb Antimony 122	52 Te Tellurium 128	53 I Iodine 127	54 Xe Xenon 131
81 Tl Thallium 203	82 Pb Lead 207	83 Bi Bismuth 209	84 Po Polonium 210	85 At Astatine 210	86 Rn Radon 222

Cr romium 52	25 Mn Manganese 55	26 Fe Iron 56	27 Co Cobalt 59	28 Ni Nickel 59	29 Cu Copper 64	30 Zn Zinc 65
lo bdenum 96	43 Tc Technetium 99	44 Ru Ruthenium 101	45 Rh Rhodium 103	46 Pd Palladium 106	47 Ag Silver 108	48 Cd Cadmium 112
W ngsten 184	75 Re Rhenium 186	76 Os Osmium 190	77 Ir Iridium 192	78 Pt Platinum 195	79 Au Gold 197	80 Hg Mercury 201

[redacted]

Chemistry by Concept

Chemistry by Concept

A Comprehensive Account of the Principles of Chemistry for GCE 'O' Level and CSE Courses

Antony Spiers, B.Sc.(Econ.)
Maria Grey College;
Formerly Head of Science,
Twyford Comprehensive School,
Acton

Derek Stebbens, M.A., B.Sc.
Head of Chemistry,
Westminster School

Heinemann Educational Books
London

Heinemann Educational Books Ltd
LONDON EDINBURGH MELBOURNE AUCKLAND TORONTO
HONG KONG SINGAPORE KUALA LUMPUR
IBADAN NAIROBI JOHANNESBURG NEW DELHI

ISBN 0 435 64830 6

First published 1973

The 3-dimensional model on the cover shows the pattern of successive ionization energies for the elements hydrogen to magnesium in the Periodic Table.

Published by
Heinemann Educational Books Ltd
48 Charles Street, London W1X 8AH

Artwork drawn by John Brennan at Oxford Illustrators Ltd

Filmset by Keyspools Ltd, Golborne, Lancs
Printed in Great Britain by
BAS Printers Limited, Over Wallop, Stockbridge, Hants

Preface

This book is a departure from the usual textbook of chemistry. Modern syllabuses for first examinations are coming more and more to stress pattern rather than facts and this book is designed to reflect that trend. The earlier chapters, however, introduce a certain amount of data by way of 'hats to hang on the pegs', and some later chapters summarize factual information.

The book is intended for all students of chemistry to first examinations at 16+ whether they follow a Nuffield Syllabus or the normal syllabus laid down by examining boards.

A textbook may have two functions: it may lead the student to generalizations after instruction on how to examine evidence; and it may be a tool for revision. In the latter function the experience acquired by the student in experimental situations is reinforced and developed in a short treatment of the key ideas. Many books fulfil the first function by employing open-ended techniques. Stimulating though these may be, they often leave those with insufficient time or laboratory accommodation in some confusion.

In this book the emphasis is on summarizing ideas after fairly brief reviews of evidence. No experimental instructions are included other than the experimental summary diagrams. We feel that the student needs the principles of chemistry to be clearly charted since we are sceptical that they could be arrived at without a very long and carefully-guided series of experiments.

1973

Antony Spiers
Derek Stebbens

Contents

	Page
PREFACE	v
1. Heat and Chemical Change	1
2. Matter and Electricity	8
3. Air and Water	14
4. Atoms and Molecules 1: Mass and formula	23
5. Atoms and Molecules 2: Size and movement	32
6. Periodic Classification 1: Physical properties	42
7. Periodic Classification 2: Chemical properties	52
8. Atomic Structure	60
9. Structure and Bonding 1: Ionic compounds	70
10. Structure and Bonding 2: Non-ionic bonding	86
11. Moles of Molecules and Avogadro's Law	107
12. Investigating Chemical Equations	114
13. Energy Changes 1: Heat energy and electrical energy	121
14. Energy Changes 2: Combustion, solution and neutralization	128
15. Rates of Reaction	138
16. Dynamic Equilibrium	146
17. Electron Transfer	155
18. The Electrochemical Series	161
19. Electrolysis	170
20. Extraction of Elements 1: Electrolytic methods	178
21. Extraction of Elements 2: Chemical methods	186
22. Group Properties of Normal Elements	196
23. Elements in the Period Sodium to Argon	205
24. Transition Metals	213
25. Carbon Chemistry	226
26. Analysis	236
QUESTIONS	241
INDEX	271

1 Heat and Chemical Change

Introduction

Gold coins and ornaments have lain in sea water for hundreds of years without tarnishing; diamonds and other jewels have been buried for thousands of years without losing their sparkle. But most substances are not like these: they change with time. Iron rusts, snow melts to water in the thaw, copper roofs go green with age, and fresh milk turns sour. Scientists have studied these and many other changes and found that some are brought about by changes in temperature alone.

If ice is heated it melts. Further heating raises the temperature of the water until it boils and turns to steam. When the steam is cooled it turns back to water and the water can be further cooled until it changes to ice. Compare this kind of change with the rusting of iron.

Ice can be frozen by warming it, but iron cannot be as easily 'unrusted'. This is because rust is a different substance from the iron and has different properties.

Table 1.1

Some properties of iron	*Some properties of rust*
Hard	Soft and scaly
Bends without snapping	Brittle
Magnetic	Non-magnetic
Conducts electricity	Does not conduct electricity
Colour is silvery-grey	Colour is reddish-brown
1 cm^3 has a mass of about 8 g	1 cm^3 has a mass of about 5 g

Table 1.1 lists only some of the properties of iron and rust. There are many more but these serve to distinguish between the two. Rust can be made from iron but it is a different substance. In changing from iron to rust the iron has undergone a **chemical change** or a **chemical reaction.** To change the rust back to iron another chemical reaction would have to be carried out. The rust would disappear and its place would be taken by iron, which has different properties from the rust.

Exothermic change

The rusting of iron is a rather slow change. An example of a very fast change is the explosion of dynamite. The burning of gas in a Bunsen burner takes place at a rate between the two. The different substances formed when gas burns are carbon dioxide (which is also a gas at ordinary temperatures, say 20–25 °C) and steam. The steam can be cooled to water on a cold surface held above the flame. Energy in the form of heat is given to the surroundings when gas burns. This is called an **exothermic** change (giving heat **out**).

Table 1.2
Exothermic Changes*

Substance burned	*Appearance of substance*	*Product*	*Appearance of product*
Town gas or natural gas	Colourless gas	Carbon dioxide and water	Colourless gas and colourless liquid
Magnesium	Silvery metal	Magnesium oxide	White solid
Alcohol	Colourless liquid	Carbon dioxide and water	Colourless gas and colourless liquid
Petrol	Colourless liquid	Carbon dioxide and water	Colourless gas and colourless liquid
Sulphur	Yellow solid	Sulphur dioxide	Colourless gas
Phosphorus	Yellowish or dark red solid	Phosphorus oxide	White smoke

*Substances and products are described at 20–25 °C

Endothermic change

Not all chemical changes are exothermic. Some only take place when energy is put in and are called **endothermic** changes. For example, trilead tetroxide (red lead) changes to a yellow solid and oxygen when it is heated. If the heating is stopped before the change is complete the reaction stops also. This reaction can be investigated with the apparatus shown in Figure 1.1. Table 1.3 summarizes

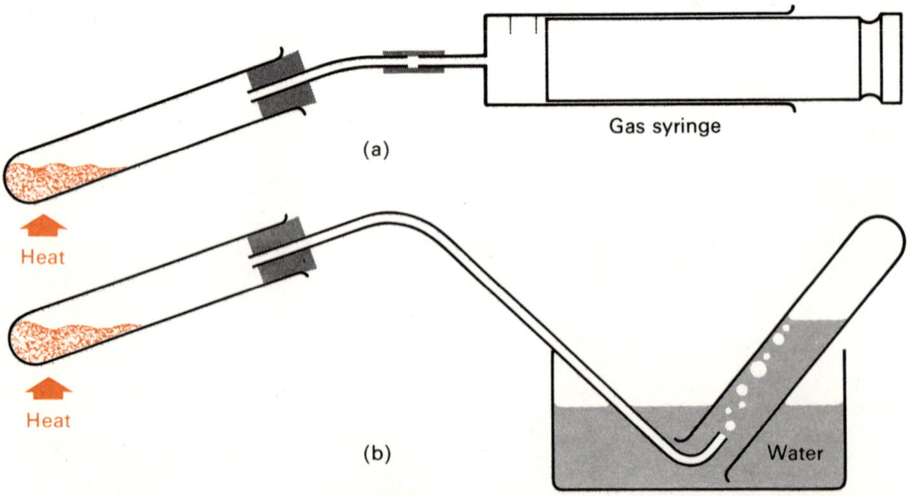

Figure 1.1 Two ways of collecting gas given off by a heated solid

several endothermic reactions. In all cases the products have different properties from the starting material.

Table 1.3
Endothermic Changes*

Substance heated	*Appearance of substance*	*Product*	*Appearance of product*
Trilead tetroxide	Orange-red powder	Lead monoxide and oxygen	Yellow solid and colourless gas
Basic copper carbonate	Green powder	Copper oxide, carbon dioxide, and water	Black solid, colourless gas, colourless liquid
Potassium permanganate	Violet or purple crystals	Potassium manganate, manganese dioxide, oxygen	Blackish/dark green solid, colourless gas
Lead nitrate	White crystals	Lead monoxide, nitrogen dioxide, oxygen	Yellow solid, brown gas, colourless gas
Copper sulphate crystals	Blue crystals	Anhydrous copper sulphate, water	White solid, colourless liquid

*Substances and products are described at 20–25 °C

The endothermic change from copper sulphate crystals to anhydrous ('without water') copper sulphate can easily be reversed. When the white powder has cooled a drop of water turns it back to blue. This is a useful test for the presence of water since no other substance causes the change. Heat is given **out** when the water rejoins the copper sulphate. So this is exothermic and is the reverse of the endothermic change which decomposed the copper sulphate crystals.

Table 1.4*
Properties of Substances Summarized

Substance	*Properties*	
Water	M.p. 0 °C B.p. 100 °C Density 1 g/cm^3 Turns anhydrous copper sulphate blue Reacts slowly with magnesium power at 20 °C to give hydrogen. Faster reaction at higher temperature	
	Dissolves	*Does not dissolve*
	potassium permanganate	manganese dioxide
	lead nitrate	lead monoxide
	copper sulphate	trilead tetroxide
	potassium manganate	basic copper carbonate

* Table 1.4 continues on pages 4 and 5.

Table 1.4—*continued*

Substance	*Properties*	
	Dissolves	*Does not dissolve*
	magnesium oxide (slightly) oxygen slightly carbon dioxide nitrogen dioxide sulphur dioxide phosphorus oxide alcohol	copper oxide iron iron rust sulphur petrol red phosphorus
Iron	*See* page 1, Table 1.1 M.p. 1540 °C B.p. 2900 °C Density 7.9 g/cm³ Rusts when put with air and water	
Iron rust (iron(III) oxide hydrate)	*See* page 1, Table 1.1 Gives off steam when heated. The resulting brown powder (iron(III) oxide) melts at 1565 °C Does not dissolve in water	
Magnesium	Soft, silvery metal B.p. 1100 °C Density 1.7 g/cm³ M.p. 650 °C Burns with brilliant flame Reacts slowly in powder form with water to give hydrogen	
Magnesium oxide	White powder M.p. 2800 °C Density 3.7 g/cm³ Dissolves slightly in water. Solution turns litmus blue (because it is alkaline)	
Alcohol	Colourless liquid M.p. about minus 100 °C B.p. about 80 °C Density about 0.8 g/cm³ Burns with blue flame Dissolves in water	
Carbon dioxide	Colourless gas denser than air B.p. minus 79 °C Density of solid 2.0 g/cm³ Turns lime water cloudy	
Sulphur	Yellow solid M.p. 119 °C B.p. 444 °C Density 2.1 g/cm³ Burns with blue flame Insoluble in water. Soluble in toluene	
Sulphur dioxide	Colourless, choking gas denser than air M.p. minus 76 °C B.p. minus 10 °C Density of solid 2.9 g/cm³ Dissolves in water. Solution turns litmus red (because it is acidic) Decolourizes potassium permanganate solution	

Table 1.4—*continued*

Substance	*Properties*
Phosphorus (white)	Yellowish, waxy solid M.p. 44 °C B.p. 280 °C Density 1.8 g/cm³ Bursts into flame in air above 40 °C Poisonous
Phosphorus (red)	Dark red solid M.p. 590 °C Density 2.3 g/cm³ Insoluble in water. Non-poisonous
Phosphorus pentoxide	White solid M.p. 560 °C Density 2.4 g/cm³ Dissolves in water. Solution turns litmus red (because it is acidic)
Trilead tetroxide (red lead)	Orange-red solid M.p. 500 °C Density 9.1 g/cm³ Insoluble in water Decomposes when heated. Products are lead monoxide and oxygen
Lead monoxide	Yellow solid M.p. 860 °C Density 9.6 g/cm³ Insoluble in water
Oxygen	Colourless gas M.p. minus 220 °C B.p. minus 183 °C Density of solid 1.1 g/cm³ Slightly soluble in water Assists burning: a glowing split bursts into flame in the pure gas
Basic copper carbonate	Green solid Decomposes on heating: products are black copper oxide, carbon dioxide and water Insoluble in water
Lead nitrate	White solid M.p. 360 °C Density 4.5 g/cm³ When heated, crystals shatter with crackling sound and decompose: products are lead monoxide, oxygen and nitrogen dioxide
Nitrogen dioxide	Brown gas M.p. minus 11 °C B.p. 22 °C Density of solid 1.5 g/cm³ Gas denser than air Poisonous. Soluble in water to form acidic solution
Potassium permanganate	Purple crystals Density 2.7 g/cm³ Soluble in water. Solution decolourized by sulphur dioxide When heated it decomposes easily to give dark green-black solid and oxygen

*All volumes are measured at about 20 °C.

Changes in mass of solid due to decomposition

When trilead tetroxide is heated it loses mass due to the oxygen given off but there is **no change of total mass** because

Mass of trilead tetroxide	=	Mass of lead monoxide	+	Mass of oxygen

This is true of any chemical reaction:

Total mass of reactant(s) = Total mass of product(s)

When trilead tetroxide is heated the mass of oxygen formed is small. 1 g loses only 23 mg oxygen. If the apparatus is allowed to cool back to room temperature the volume occupied by 23 mg oxygen is about 18 cm^3. Other substances which give oxygen on heating are listed in Table 1.5.

Table 1.5

Masses of Oxygen Lost when 1 g of Each Solid is Heated*

One gramme (1000 *mg*) *solid*	*Mass of oxygen given off*	*Volume of oxygen*
Trilead tetroxide	23 mg	18 cm^3
Potassium permanganate	101 mg	76 cm^3
Potassium nitrate	158 mg	119 cm^3
Potassium chlorate	392 mg	295 cm^3

* These changes are endothermic

In some cases there are more than two products. Lead nitrate gives lead monoxide, nitrogen dioxide and oxygen. Nitrogen dioxide is completely absorbed by a solution of sodium hydroxide, whereas oxygen is only slightly soluble in sodium hydroxide. Suggest a method of measuring the mass of each gas given off when lead nitrate is heated.

Changes in mass when solids burn in air

When magnesium burns it increases in mass. This is because it combines with oxygen from the air. Since there can be no change in total mass,

Mass of magnesium oxide	=	Mass of magnesium	+	Mass of oxygen

When 1 g magnesium is burned the mass increase is 667 mg. The volume of the oxygen used is 500 cm^3. Suitable apparatus for this and similar reactions is shown in Figure 1.2.

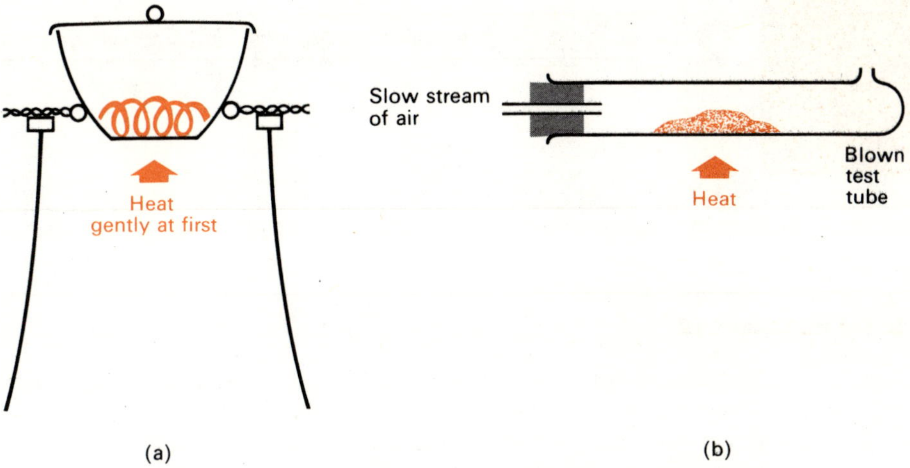

Figure 1.2 (a) Weighed coil of magnesium ribbon being heated in a porcelain crucible. (b) Copper or iron powder being heated in air

Table 1.6
Masses of Oxygen Gained by Burning 1 g of Each Solid*

One gramme (1000 *mg*) *solid*	*Mass of oxygen gained*	*Total mass of product*
Magnesium	667 mg	1.667 g
Copper	252 mg	1.252 g
Iron	429 mg	1.429 g

100 mg oxygen occupies about 75 cm^3 at room temperature

* These changes are exothermic

Summary

1. Chemical changes are recognized by the appearance of a substance different from the starting material.
2. The new substance has different properties.
3. Some chemical changes give heat to the surroundings and are called exothermic.
4. Some chemical changes absorb heat from the surroundings and are called endothermic.
5. In all reactions the total mass of reactant(s) is equal to the total mass of product(s).

2 Matter and Electricity

Solid substances

Some solid substances allow electricity to pass through them. Some do not. One way to investigate this is to make each substance in turn part of an electrical circuit and find out whether current flows. Provided the sample of substance under test is not more than a millimetre or so thick the expensive meter can be replaced by a torch bulb. Observations might be tabulated as in Table 2.1.

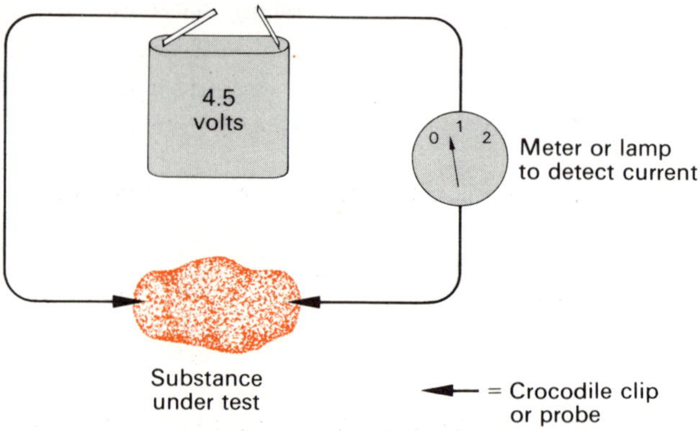

Figure 2.1 Testing substances for ability to conduct electricity

Table 2.1

Conductors	*Non-conductors*
Iron	Sulphur
Lead	Phosphorus
Copper	Glass
Magnesium	Magnesium oxide
Solder	Copper sulphate crystals
Brass	Sugar
	Candle wax
	Ice

Conductors such as copper are very good: they allow electricity to pass easily. Other conductors are not as good: carbon is an example. If a copper wire the length of a pencil is put between the two probes the lamp lights brightly, but if pencil 'lead' (which is carbon mixed with clay) is used, the lamp lights only dimly.

A sharp division between conductors and non-conductors cannot be made. Non-conductors are just very poor conductors. The best non-conductors are called **insulators.**

Solid conductors are usually elements. Metals are the best conductors but such non-metals as carbon (graphite form, **not** diamond) and silicon also conduct. Brass, solder and pewter are alloys, which are mixtures of metallic elements.

Table 2.2

Some metallic elements	*Some non-metallic elements*
Aluminium	Argon
Bismuth	Boron
Cadmium	Bromine
Calcium	Carbon
Copper	Chlorine
Gold	Hydrogen
Iron	Iodine
Lead	Phosphorus
Magnesium	Silicon
Mercury	Sulphur
Platinum	
Silver	
Tin	
Zinc	

Since alloys are mixtures they have the properties of the metals of which they are composed. Compounds are substances made by combining elements. The elements undergo chemical change when they combine and the compound does not have the properties of the elements of which it is composed. **Compounds never conduct electricity in the solid state.**

Liquid substances

Substances which conduct in the solid state also conduct when molten. Mercury, molten bismuth, molten copper and molten lead are all conductors.

Some compounds become good conductors when they melt. (*See* Figure 2.2 on page 10.)

Elements which conduct electricity are not changed chemically by the passage of current: copper can be used indefinitely as a conductor. Compounds are **decomposed** by passing an electric current. The elements are formed at the electrodes and the compound is used up. Molten compounds which conduct electricity are called **electrolytes.** They are always **salts** or **metal oxides and hydroxides.**

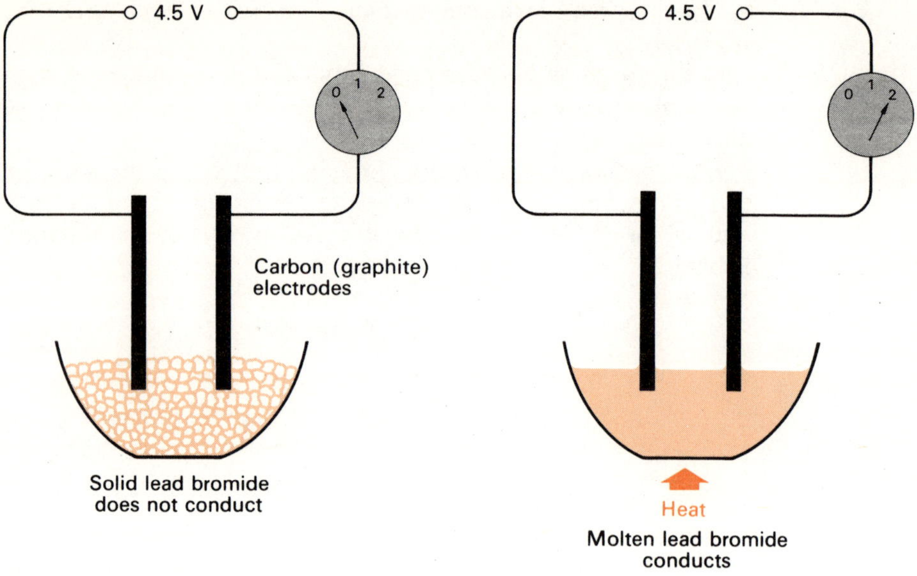

Figure 2.2

Table 2.3

Molten compound	*Product at positive (+) electrode*	*Product at negative (−) electrode**
Lead bromide (m.p. 450 °C)	Bromine	Lead
Potassium iodide (m.p. 720 °C)	Iodine	Potassium
Aluminium oxide (m.p. 2050 °C)	Oxygen	Aluminium

*Metals produced at a hot negative electrode must be protected from air

When compounds are electrolysed, electrical energy (equivalent to heat energy) is being put in to break up the compound. Therefore electrolysis reactions are **endothermic** and are the reverse of the exothermic changes which take place when the elements combine.

Compounds which do not conduct electricity in the molten state cannot undergo chemical change by electrolysis. They are called **non-electrolytes.** Alcohol is an example. Pure water seems on first investigation to be a non-conductor, but further experimentation shows it to be a **very poor** electrolyte.

Conduction of electricity by solutions

Dilute acids and solutions of electrolytes (salts and metal oxides and hydroxides) in water conduct electricity. Reaction takes place at each electrode. However, the reactions are complicated by the presence of water and study of the electrolysis of solutions will be left until later (Chapter 19). Table 2.4 lists products of electrolysis of various electrolyte solutions.

Table 2.4

Electrolyte solution in water	*Product at positive (+) electrode*	*Product at negative (−) electrode**
Copper sulphate	Oxygen	Copper
Lead acetate	Oxygen	Lead
Silver sulphate	Oxygen	Silver
Nickel sulphate	Oxygen	Nickel
Sodium nitrate	Oxygen	Hydrogen
Dilute sulphuric acid	Oxygen	Hydrogen
Sodium chloride	Chlorine	Hydrogen
Sodium bromide	Bromine	Hydrogen
Calcium hydroxide	Oxygen	Hydrogen

Tables summarizing the electrical conduction of solids and liquids appear on pages 12 and 13.

Summary of electrical conduction by liquids

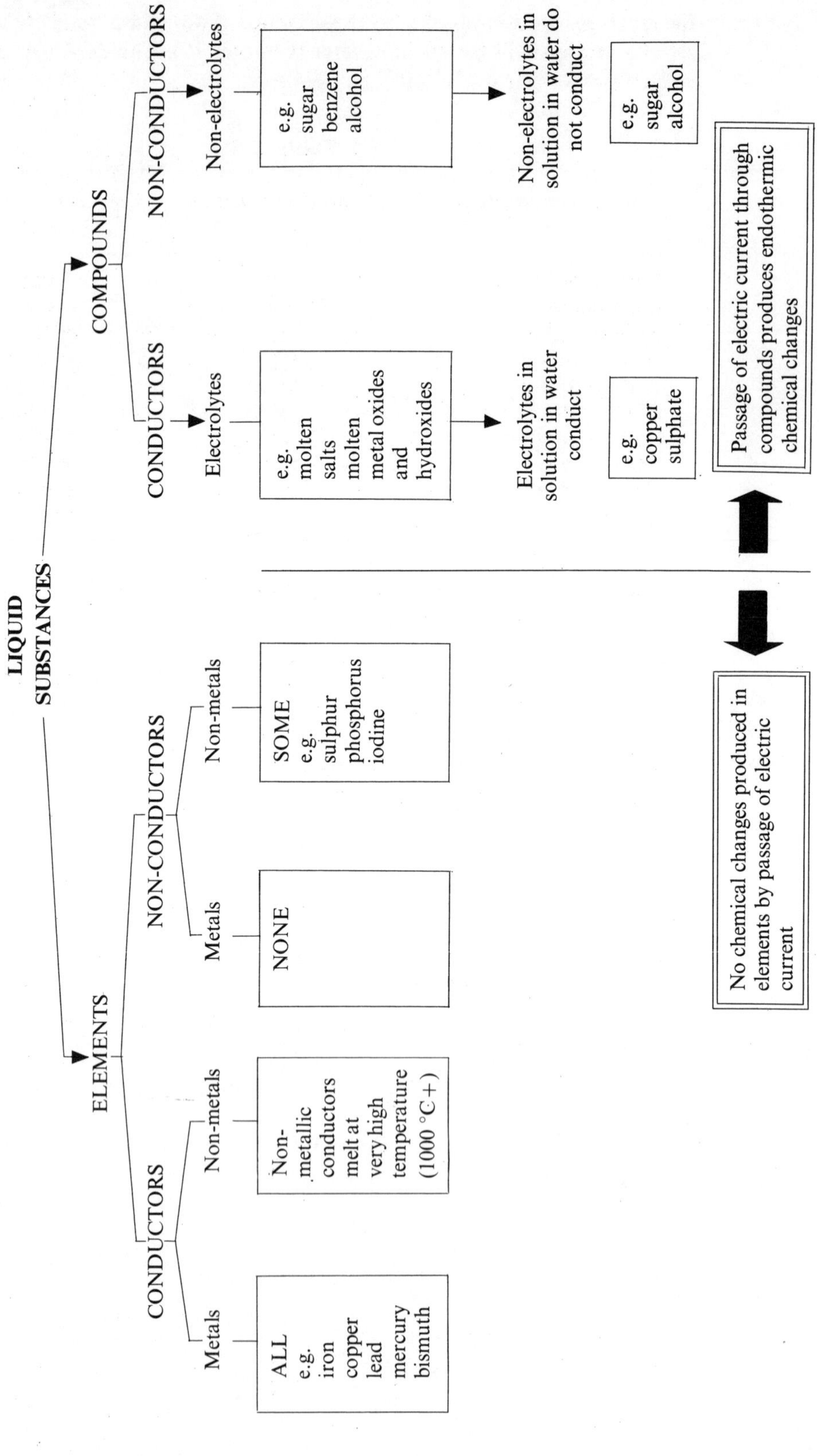

Summary of electrical conduction by solids

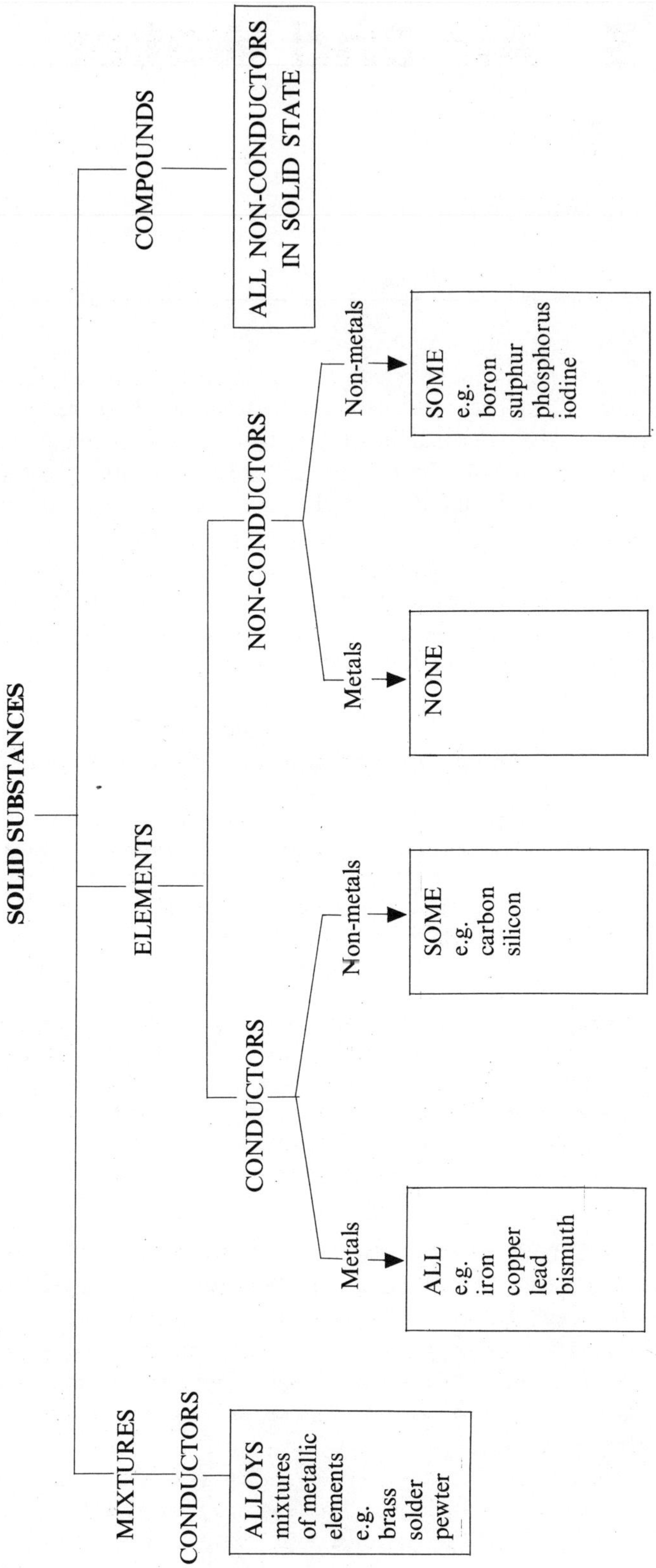

3 Air and Water

Oxygen and air

Nearly every element reacts with oxygen to form a compound, usually called an oxide. The process is called **oxidation**. Often the relative amounts of oxygen and the other element in the compound are indicated by the name: carbon monoxide, carbon dioxide, trilead tetroxide. This is not always so. Hydrogen oxide is called water, not dihydrogen oxide which would be its strict name, and dialuminium trioxide is called simply aluminium oxide.

Rocks and soil and other materials of the Earth's crust, including water, are made of compounds containing oxygen. It is not surprising that oxygen is the most plentiful of all elements in the world we see around us. Table 3.1 gives the abundance of the ten most common elements.

Table 3.1
Abundance of the Most Common Elements by Mass

Crust		*Sea water*		*Whole Earth*	
element	*per cent*	*element*	*per cent*	*element*	*per cent*
Oxygen	49.4	Oxygen	91	Iron	40–50
Silicon	25.8	Hydrogen	5.7	Oxygen	22–28
Aluminium	7.5	Chlorine	2	Silicon	11–15
Iron	4.7	Sodium	1	Magnesium	9
Calcium	3.4	Magnesium	0.1	Nickel	3–6
Sodium	2.6	Sulphur	0.08	Calcium	1–2
Potassium	2.4	Calcium	0.04	Aluminium	1–2
Magnesium	2.0	Potassium	0.04		
Hydrogen	0.9	Bromine	0.01		
Titanium	0.5	Carbon	0.003		

The Earth's crust is between 6 and 60 km thick. When the core is included the most abundant element is not oxygen but iron. However, in the region of Earth most familiar to us—the crust, the oceans, and the atmosphere—oxygen, silicon, aluminium, and hydrogen account for a high percentage.

Oxygen has many important uses. These are the main ones:

- In the making of steel it is used to burn impurities out of the iron.
- For welding and cutting steel and other metals a high temperature is necessary. It is achieved by burning acetylene gas in oxygen. Like other flammable substances, the acetylene burns very rapidly when mixed with oxygen and the required high temperature is reached.

- Oxygen is used in hospitals to help the breathing of premature babies and patients with pneumonia.

Thousands of tons of oxygen are used every year. It is obtained from air in the following way:

1. Water vapour and carbon dioxide are first removed. Both these substances would freeze to solids in the next stage and clog the apparatus.
2. The air is then compressed, cooled, allowed to expand, compressed again and cooled until it finally liquefies. At this stage the liquid contains oxygen, nitrogen, argon and tiny amounts of krypton and xenon.
3. The liquid air, at a temperature of about minus 200 °C, is allowed to warm up until the nitrogen boils. The boiling point of nitrogen is minus 196 °C. When all the nitrogen has boiled off the oxygen and argon remain. Argon may be separated from oxygen by a similar process of **fractional distillation**. Oxygen boils at minus 183 °C.

Oxygen is transported in bulk in liquid form. Special lorries having a large, spherical container covered in frost may sometimes be seen.

For such uses as welding and medical purposes the oxygen is stored as compressed gas in cylinders. A valve controls the release of gas from the cylinder.

At present, and no doubt for many years to come, Man's source of raw materials must be the crust, the oceans, and the atmosphere. About ninety elements can be found, either combined with others in compounds or native (uncombined). Six of the ninety are the noble gases:

helium	neon	argon
krypton	xenon	radon

With the exception of radon they are all found in air. The noble gases either do not combine with other elements or do so only with difficulty (*see* page 203). Fourteen of the other elements are non-metals:

boron	carbon	nitrogen	oxygen
fluorine	silicon	phosphorus	sulphur
chlorine	arsenic	selenium	bromine
tellurium	iodine		

This leaves about seventy elements which are metals.

Refer to Table 3.1: oxygen, silicon, aluminium, and hydrogen between them make up over 80 per cent of the crust, oceans and atmosphere. Nitrogen is a non-metal and is the main gas in air, but it does not appear amongst the ten most abundant elements. Only a small percentage of the crust and oceans is nitrogen and the mass of the atmosphere is about one millionth of the total mass of the Earth.

Air is a mixture of gases. Its composition is regulated to a large extent by plants and animals, which are the main controllers of the amounts of oxygen and carbon dioxide, through photosynthesis and respiration. Normal air contains water vapour, dust and traces of other gases. If air is dried and filtered to remove dust its approximate composition by volume is shown in the table on page 16.

Nitrogen	78%
Oxygen	21%
Noble gases (mainly argon)	0.93%
Carbon dioxide	0.03%
Other gases	traces only

Since air is a mixture these percentages vary slightly from place to place and under different climatic conditions.

Oxygen combines with metals to form solid oxides. In this way oxygen can be removed from a sample of air and the fraction of it present in the sample can be measured. This can be done either by weighing or by volumetric means. The solid oxide which is formed takes up very little more space than the metal from which it was made. Several methods of doing this are illustrated in Figures 3.1, 3.2 and 3.3

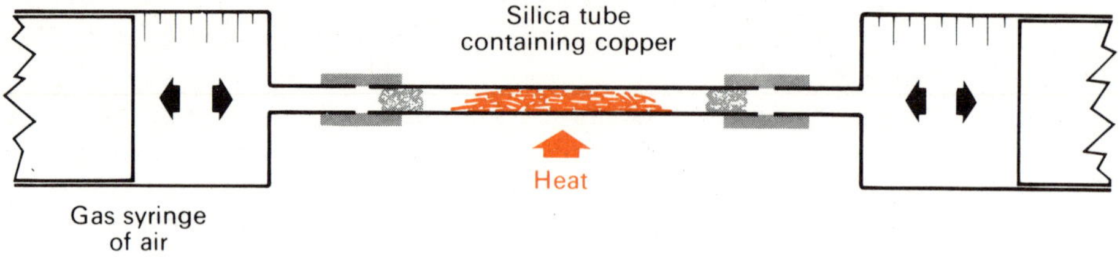

Figure 3.1 Air is pumped to and fro over the hot copper. The amount of oxygen used up is measured by the contraction after the apparatus cools

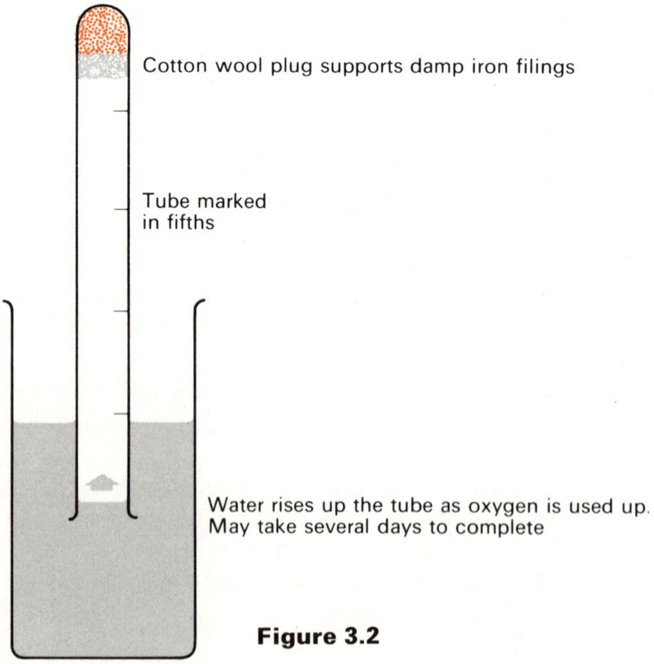

Figure 3.2

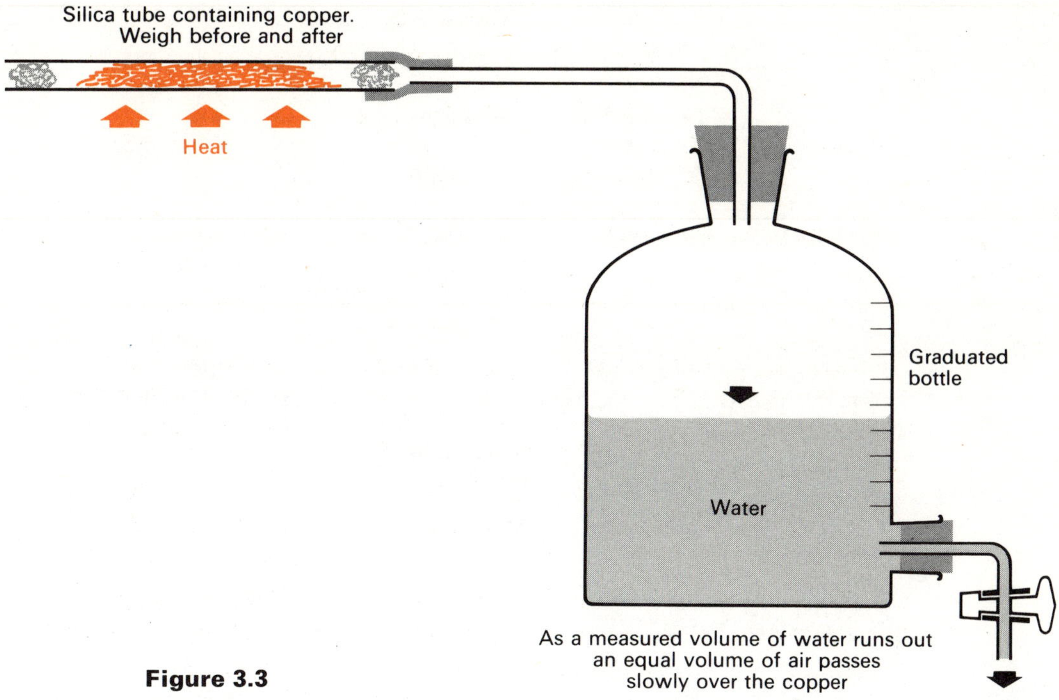

Figure 3.3

Water, acids and bases

Water is hydrogen oxide, a compound of hydrogen and oxygen. It can be made by burning hydrogen, or a hydrogen-containing substance, in air or oxygen. Table 1.2, page 2, gives some examples of this. Most of the world's water is liquid, but an important fraction is solid as ice and snow. Many mineral substances contain water of crystallization (e.g. copper sulphate) and in the atmosphere there are millions of tons of water vapour. Clouds consist of minute droplets of water or crystals of ice.

Water dissolves a very large number of substances and is the most important solvent. It does not dissolve greasy, fatty substances or most plastics. Several of its properties are listed in Table 1.4, pages 3–5. In addition to being a solvent it enters into chemical reaction with a number of elements and compounds. Its reactions with metals are particularly important: the metals corrode or **oxidize**. All the reactions are exothermic.

Table 3.2 on page 18 summarizes the action of water on some metals.

The rate at which water attacks metals is increased by:

- raising the temperature of the water or
- using metal powder, which has a larger surface area than metal lumps.

The rate of attack can also be increased by dissolving certain substances in the water. Examples of such substances are sulphur dioxide, sulphur trioxide, phosphorus pentoxide, and other oxides of non-metals. Solutions of these oxides also change the colour of litmus and other indicators. They are called **acids**.

Not all acids are solutions of non-metal oxides in water. Hydrochloric acid is made by dissolving the gas hydrogen chloride in water. Hydrobromic acid and hydriodic acid are solutions of hydrogen bromide and hydrogen iodide.

Table 3.2

Action of Water on Some Metals

Metal	*Conditions necessary for reaction*	*Description of of reaction*	*Products*
Sodium	Cold water	Vigorous, floats on water DANGEROUS and may explode	Hydrogen and sodium hydroxide solution
Potassium	Cold water	Very vigorous, floats on water DANGEROUS Always catches fire	Hydrogen and potassium hydroxide solution
Calcium	Cold water	Moderate	Hydrogen and calcium hydroxide
Magnesium (powder)	Cold water	Very slow	Hydrogen and magnesium hydroxide
Magnesium (ribbon)	Heated metal and steam	Vigorous Magnesium burns	Hydrogen and magnesium oxide
Iron	Cold water containing dissolved air	Very slow formation of rust	Hydrated iron oxide (rust)
Iron	Heated metal and steam	Vigorous	Hydrogen and tri-iron tetroxide

Acids made from the oxides of non-metals are named after the element: sulphurous and sulphuric acids from sulphur dioxide and sulphur trioxide, carbonic acid from carbon dioxide, phosphoric acid from phosphorus oxide.

In addition to turning litmus red and changing the colour of other indicators, acids have a sour taste. (It should be remembered that it is dangerous to taste acids and this test should be applied only under strict supervision.) They act on such metals as magnesium to give hydrogen gas and a solution of a magnesium salt. And they attack carbonates and bicarbonates (hydrogencarbonates) to give the gas, carbon dioxide.

The pH scale

It is useful to be able to indicate the strength of an acidic or alkaline solution. The strength is expressed as the pH of the solution.

pH numbers range from about 0 for very strong acids to about 14 for very strong alkalis. A neutral solution is of pH 7. Strongly acidic solutions are of pH 0 or 1 and weakly acidic solutions of pH 2 to 6. pH values of 8 to 12 indicate weak alkalis and 13 to 14 indicate strong alkalis.

It is important not to confuse strong and weak with concentrated and dilute. Bench sulphuric acid is dilute: it is 90 per cent water and 10 per cent acid. Its pH is about 0 and it is a strong, dilute acid. If a solution of acetic acid is made containing 90 per cent acid and 10 per cent water, its pH is about 4, although it is much more concentrated than bench sulphuric acid. Acetic acid is a weak acid.

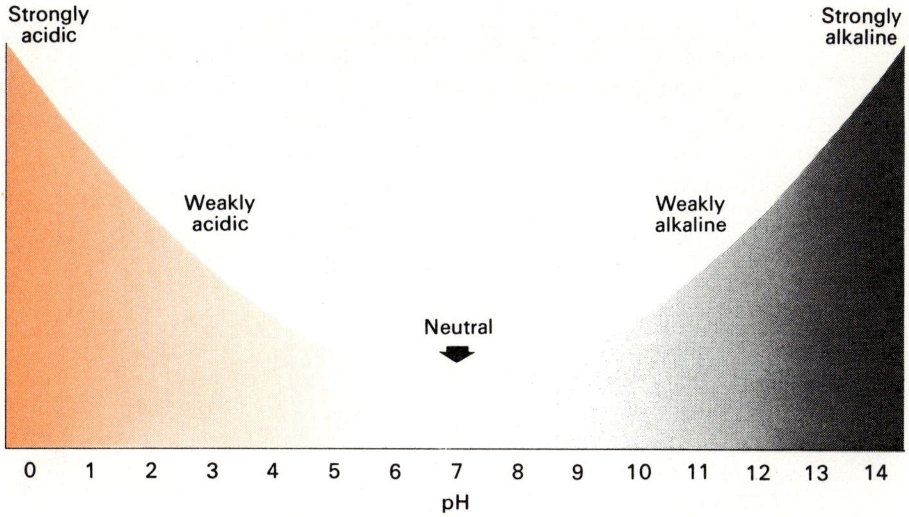

A very useful indicator is Universal Indicator. Some acids turn it bright red. These are the strong acids which act vigorously on metals, carbonates and bicarbonates. Other acids turn the indicator pink or orange. These acids are weak and their action on metals is much slower. Table 3.3 lists some acids together with colour changes produced in various indicators.

Table 3.3
Acids and Indicators*

Acid	*Strong or weak*	*Litmus*	*Phenol-phthalein*	*Methyl orange*	*Universal Indicator*
Hydrochloric	Strong	Red	Colourless	Red	Red
Sulphuric	Strong	Red	Colourless	Red	Red
Nitric	Strong	Red	Colourless	Red	Red
Sulphurous	Weak	Red	Colourless	Red	Pink
Phosphoric	Weak	Red	Colourless	Red	Pink
Acetic	Weak	Red	Colourless	Red	Pink
Carbonic	Weak	Dark red	Colourless	Orange	Orange

*It is assumed that the concentration of each acid is about the same. Strong and weak do not mean concentrated and dilute

Neutralization

Metal oxides which dissolve in water are called **alkalis**. Their solutions have a bitter or soapy taste and also change the colour of indicators. (As with acids, alkalis should not be tasted except under strict supervision.) Acids and alkalis are in a sense opposites: if an acid and an alkali are mixed a reaction takes place. Heat is given out as evidence of this and the resulting solution is found to be a solution of a **salt** in water only. Mixing an acid with an alkali so that both acidic and alkaline properties disappear is called **neutralization**.

The main alkaline substances found in a school laboratory are sodium hydroxide, potassium hydroxide, calcium hydroxide and aqueous ammonia solution. The name 'hydroxide' means hydrated oxide. Metal oxides which are soluble in water are always present in the hydrated form. Alkalis change the colour of the indicators already mentioned, as shown in Table 3.4. As a reminder the colour in strong acid is also included.

Table 3.4

Indicator	*Colour in alkali*	*Neutral colour*	*Colour in acid*
Litmus	Blue	Purple	Red
Phenolphthalein	Pink	Faint pink	Colourless
Methyl orange	Yellow	Orange	Red
Universal	Blue	Green	Red

If alkali is added carefully to acid with an indicator present, the indicator changes colour when neutralization has taken place.

Salts

Some salts formed by neutralizing acids with alkalis are shown in Table 3.5.

Table 3.5

Acid	*Alkali*	*Salt*
Sulphuric	Sodium hydroxide	Sodium sulphate
Hydrochloric	Potassium hydroxide	Potassium chloride
Nitric	Calcium hydroxide	Calcium nitrate

Using these three acids and alkalis a total of nine salts could be made by neutralization. Note that water is formed as well as a salt whenever acid and alkali react.

Most metal oxides are not soluble in water. The insoluble metal oxides also react with acids, but usually more slowly than the alkalis. If a metal oxide is stirred with an acid until all the acid is used up the resulting solution consists of water and a salt only, as in the case of acid and alkali. Metal oxides can therefore be used to make salts. Metal oxides are given the general name of **bases**. Alkalis are soluble bases. Hence

$$\text{Acid} + \text{Base} \rightarrow \text{Water} + \text{Salt}$$

The fact that water is produced as well as a salt can be shown as follows: a solution of acid containing a known mass of water is weighed. Similarly a solution of alkali is weighed. The solutions are mixed and the product put in a distillation flask, Figure 3.4, fitted with a thermometer. Heating is continued until all the water (which boils at 100 °C) is driven off. It is collected in the receiver and is found to weigh more than the water in the original solutions. If the salt which remains in the distillation flask is also weighed it will be found that

$$\text{Mass of acid solution} + \text{Mass of alkali solution} = \text{Mass of water in receiver} + \text{Mass of salt}$$

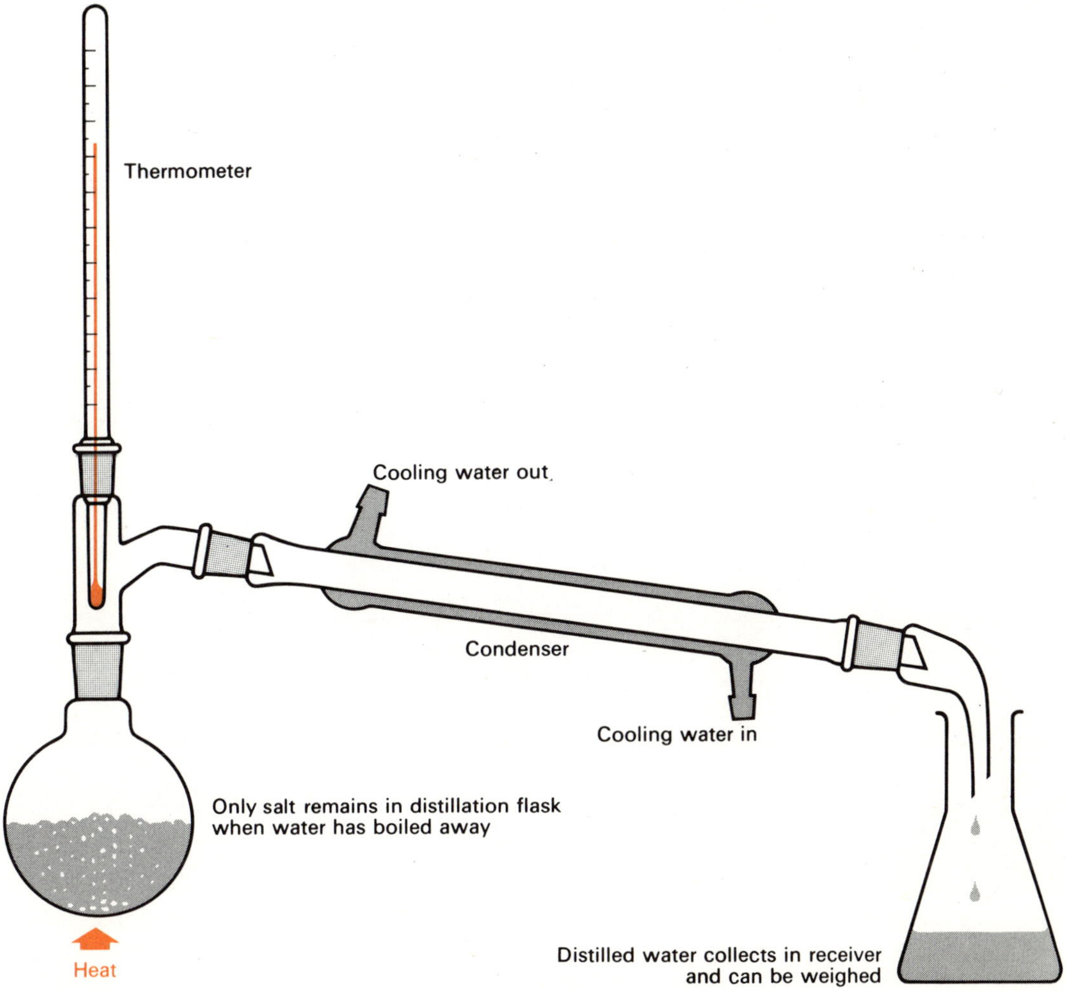

Figure 3.4 Apparatus for distillation

Summary

1. Oxygen combines with most elements to form oxides.
2. The process of combining with oxygen is an example of oxidation.
3. The slow oxidation of metals by air is called corrosion.
4. Metals may be corroded (oxidized) by water or acids.
5. Hydrogen is produced when acids oxidize metals. The other product is a salt.
6. Oxides of non-metals are acidic.
7. Oxides of metals are basic. Soluble bases are called alkalis.
8. Acids neutralize bases to give water and salts.
9. Acids attack carbonates to give carbon dioxide, water and a salt.

4 Atoms and Molecules
1: Mass and formula

The composition of oxides

One important property of an oxide is its composition by mass—this could be expressed as the mass of oxygen which is combined with 1000 mg (one gramme) of the other element. When the composition of the oxides of many elements is known, it is found that elements can be divided into two groups. Some examples of the elements in each of these groups are given in Table 4.1.

Table 4.1

First group of elements (these elements form only one oxide)	*Second group of elements* (these elements can each form more than one oxide)
Magnesium Calcium Aluminium Zinc Silicon	Carbon Lead Copper Sulphur Nitrogen
(The mass of oxygen which combines with 1000 mg of element is fixed for each element)	(At least two different masses of oxygen can combine with 1000 mg of each element)

Magnesium, for example, always combines with oxygen to form the same solid white oxide with the constant composition by mass of 667 mg of oxygen to every 1000 mg of magnesium. Lead, on the other hand, can form several oxides. These oxides of lead have different chemical properties and different, but fixed, compositions by mass. Information about the composition of a few oxides is given in Table 4.2 on page 24.

The atomic theory

Looking at the compositions of the two oxides formed between carbon and oxygen, it will be seen that there is a simple relation between the two masses of oxygen which combine with 1000 mg of carbon. The masses of oxygen are in the ratio 1:2. The two masses of oxygen which combine with 1000 mg of lead are also in the ratio 1:2. It was observations like this, made at the beginning of the last century, which gave Dalton the first experimental evidence that each element might be made up of its own identical particles called **atoms**.

Table 4.2

Oxides with only one possible composition by mass		*Oxides with more than one composition by mass*	
mass of element	*mass of oxygen*	*mass of element*	*mass of oxygen*
1000 mg of magnesium	667 mg	1000 mg of carbon	1333 mg in carbon monoxide
		1000 mg of carbon	2666 mg in carbon dioxide
1000 mg of calcium	400 mg	1000 mg of lead	39 mg in lead(II) oxide
		1000 mg of lead	78 mg in lead(IV) oxide

The atomic theory of Dalton can be summarized by the two following statements:

1. Each element is composed of its own identical atoms which are different from the atoms of all other elements.
2. A compound contains the atoms of one element combined with the atoms of one or more other elements in simple and constant whole number numerical ratio.

The constant composition by mass of magnesium oxide can be explained on this theory. Identical magnesium atoms (identical in mass is the important point) combined in a simple ratio with identical oxygen atoms will produce a compound with a constant composition by mass.

A compound's constant composition by mass is a result of its constant composition by atoms.

The relative masses of atoms

It is possible to calculate the relative masses of atoms of magnesium and oxygen if it is assumed that these elements combine in the simplest possible ratio by atoms—this is one atom of magnesium to one atom of oxygen.

By experiment:

1000 mg of magnesium combines with 667 mg of oxygen

assume:

1 atom of magnesium combines with 1 atom of oxygen

it follows that:

(1) magnesium atoms have a greater mass than oxygen atoms

(2) magnesium atoms have $\frac{1000}{667}$ or 1.5 times the mass of oxygen atoms

In a similar way inspection of Table 4.2 shows that calcium atoms have 2.5 times the mass of oxygen atoms, if it is assumed that one atom of calcium combines with one atom of oxygen.

More evidence for the atomic theory

The constant composition of compounds such as oxides supports the atomic theory, but rather more conclusive evidence comes from a study of pairs of elements which combine in more than one ratio by mass. Carbon and oxygen is one such pair of elements (*see* Table 4.2).

1000 mg of carbon can combine with either 1333 mg of oxygen in carbon monoxide

or 2666 mg of oxygen in carbon dioxide

If 1 atom of carbon combines with either 1 atom of oxygen in carbon monoxide

or 2 atoms of oxygen in carbon dioxide

it will follow that (*see* Figure 4.1):

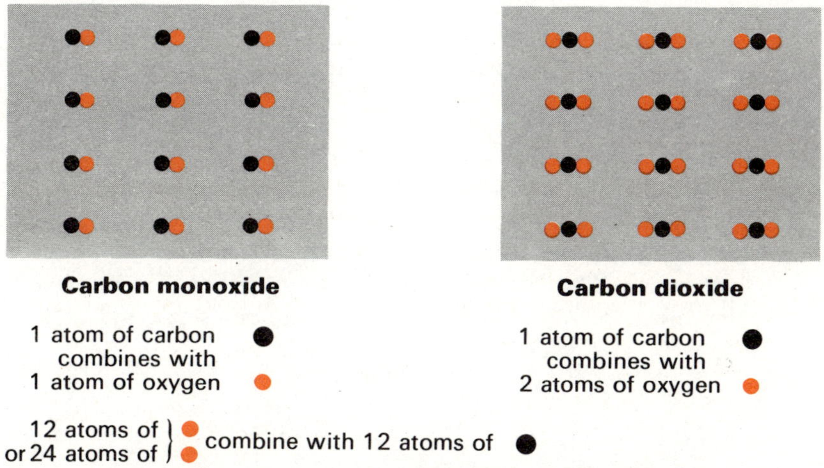

The number of atoms of oxygen (12 and 24) combining with a fixed number of atoms of carbon (12) are in the simple ratio of 1:2

Figure 4.1

1. The numbers of atoms of oxygen combining with a fixed number of atoms of carbon in the two oxides will be in the simple ratio of 1:2.

2. The masses of oxygen combining with a fixed mass of carbon in the two oxides will also be in the simple ratio of 1:2. This is in agreement with the experimental compositions by mass.

The simple ratio of the two masses of oxygen combining with a fixed mass of carbon is a result of the two simple ratios by atoms in which the elements combine.

Symbols for atoms

It is frequently convenient to have a shorthand to represent an atom of each element. Dalton introduced symbols to do this. The symbols now used to represent atoms of some of the more important elements are shown in Table 4.3. Note that if an atom has a two-letter symbol, the second letter is not a capital.

Table 4.3
Symbols of Some Important Atoms

Metallic elements			*Non-metallic elements*		
Liquid	mercury	Hg	Gases	argon	Ar
				hydrogen	H
Solids	lithium	Li		chlorine	Cl
	sodium	Na		oxygen	O
	potassium	K		nitrogen	N
	magnesium	Mg			
	calcium	Ca	Liquid	bromine	Br
	barium	Ba			
	aluminium	Al	Solids	carbon	C
	lead	Pb		silicon	Si
	iron	Fe		phosphorus	P
	copper	Cu		sulphur	S
	silver	Ag		iodine	I

The atomic mass scale

The relative masses of atoms of magnesium and oxygen, and calcium and oxygen, were determined earlier in this chapter from the experimental compositions by mass of the two oxides. It was necessary to assume that the numbers of atoms of metals and oxygen were in the simplest possible ratio of 1:1. The information in Table 4.2 can be used to determine the relative masses of other atoms. The results are shown in Table 4.4.

Table 4.4

Compound	*Assumed composition by atoms*	*Relative masses of elements* other element	oxygen
Magnesium oxide	Mg : O	1.5	1
Calcium oxide	Ca : O	2.5	1
Carbon monoxide	C : O	0.75	1
Lead(II) oxide	Pb : O	13	1

A table of atomic masses relative to oxygen = 1 can be constructed (Table 4.5). Each relative atomic mass in this table is only correct if the assumed combining ratio of 1 atom of the element to 1 atom of oxygen is also correct. This is thought to be so for the examples chosen, but early tables of relative atomic masses did contain some incorrect values due to uncertainty in the combining ratios.

Table 4.5

Element	*Mass of each atom relative to oxygen = 1*
Carbon	0.75
Oxygen	1
Magnesium	1.5
Calcium	2.5
Lead	13

Very accurate values for relative atomic masses can now be determined using a Mass Spectrometer. Measurements with this instrument do not depend on knowing the combining ratios of atoms. Atomic masses are now measured relative to the carbon atom which is given an arbitrary value of 12. Information for the twenty elements with the least massive atoms is shown in Table 4.6. The relative atomic masses of copper, silver and lead have been included in this table.

Table 4.6
Atomic Masses Relative to C = 12

Atomic number	*Element*	*Relative atomic mass*
1	Hydrogen	1.0
2	Helium	4.0
3	Lithium	6.9
4	Beryllium	9.0
5	Boron	10.8
6	Carbon	12.0
7	Nitrogon	14.0
8	Oxygen	16.0
9	Fluorine	19.0
10	Neon	20.2
11	Sodium	23.0
12	Magnesium	24.3
13	Aluminium	27.0
14	Silicon	28.4
15	Phosphorus	31.0
16	Sulphur	32.1
17	Chlorine	35.5
18	Argon	39.9
19	Potassium	39.1
20	Calcium	40.1
29	Copper	63.5
47	Silver	108
82	Lead	207

The **atomic number** in Table 4.6 is the same as the order of relative atomic masses, except for argon and potassium where the order has been reversed.

Moles of atoms

The relative atomic masses of carbon and silver are 12 and 108. A silver atom has nine times the mass of a carbon atom. If a piece of carbon of mass 12 g is considered, this will contain a certain number of atoms (the actual number is unimportant at the moment). The mass of silver which will contain an identical number of atoms of silver is 9×12 g = 108 g. This amount of silver is called **one mole of silver atoms.**

One mole of silver atoms is the amount of silver which contains the same number of silver atoms as there are carbon atoms in 12 g of carbon.

Taking another example, the relative atomic masses of carbon and oxygen are 12 and 16. This means that 16 g of oxygen contains the same number of oxygen atoms as there are carbon atoms in 12 g of carbon. One mole of oxygen atoms has a mass of 16 g. One mole of atoms is sometimes called **one gramme atom**. The number of grammes in one mole of atoms is the same as the relative mass of the element, for example:

Relative atomic mass of sodium = 23.0

1 mole of sodium atoms has a mass of 23.0 g
0.1 mole of sodium atoms has a mass of 0.1×23.0 g = 2.3 g
0.02 mole of sodium atoms has a mass of 0.02×23.0 g = 0.46 g
5 moles of sodium atoms has a mass of 5×23.0 g = 115 g

Finding the formula of a compound

Copper forms two compounds with oxygen, copper(I) oxide which is red and copper(II) oxide which is black. The relative numbers of atoms of copper and oxygen in each oxide can be found by passing hydrogen over known masses of the separate oxides. The hydrogen removes oxygen from the heated oxide leaving copper whose mass can be measured. The experiment is outlined in Figure 4.2.

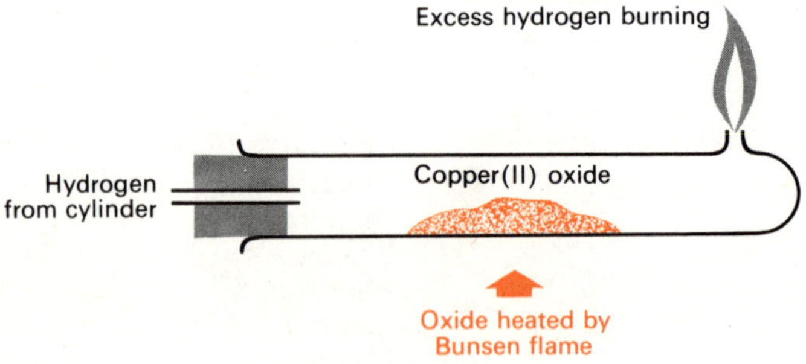

copper oxide + hydrogen → copper + hydrogen oxide (steam

Specimen masses
7.95 g of copper(II) oxide at start
6.35 g of copper remains at end
∴ 1.6 g of oxygen have been removed

Figure 4.2

By experiment, 6.35 g of copper combine with 1.6 g of oxygen

this is $\frac{6.35}{63.5}$ = 0.1 mole of copper atoms

this is $\frac{1.6}{16}$ = 0.1 mole of oxygen atoms

0.1 mole of copper atoms combines with 0.1 mole of oxygen atoms
∴ 1 mole of copper atoms combines with 1 mole of oxygen atoms
∴ 1 atom of copper combines with 1 atom of oxygen

The **formula** of copper(II) oxide would be written CuO to show that it contains 1 atom of copper : 1 atom of oxygen.

Problem: The masses recorded when the experiment was repeated with copper(I) oxide were

14.3 g of copper(I) oxide at start
12.7 g of copper at end

Use these results to determine the formula of copper(I) oxide.

The formulae of other compounds can be found in the same way if the masses of the various elements which make up a known mass of the compound have been measured experimentally. Stearic acid, for example, can be shown to contain carbon, hydrogen and oxygen with the following masses of each element in a total mass of 7.1 g of the compound.

		C	H	O
by experiment		5.4 g	0.9 g	0.8 g
number of moles of each element in 7.1 g of compound	=	$\frac{5.4}{12}$ moles	$\frac{0.9}{1.0}$ moles	$\frac{0.8}{16}$ moles
	=	0.45 mole	0.9 mole	0.05 mole
	=	45 moles	90 moles	5 moles
	=	9 moles	18 moles	1 moles
the ratio of atoms	=	9 :	18 :	1

The formula of stearic acid could be written $C_9H_{18}O$. This formula for stearic acid is called the **empirical formula**. The empirical formula represents only the **simplest ratio** of the numbers of atoms of each element in the compound.

Table 4.7
Empirical Formulae of Some Important Compounds

Oxides		*Alkalis*		*Salts*	
MgO	Magnesium oxide	$LiOH$	Lithium hydroxide	$NaCl$	Sodium chloride
CaO	Calcium oxide	$NaOH$	Sodium hydroxide	$NaBr$	Sodium bromide
Cu_2O	Copper(I) oxide	KOH	Potassium hydroxide	$NaNO_3$	Sodium nitrate
CuO	Copper(II) oxide	$Ca(OH)_2$	Calcium hydroxide	Na_2SO_4	Sodium sulphate
PbO	Lead(II) oxide			KCl	Potassium chloride
PbO_2	Lead(IV) oxide		*Acids*	$PbBr_2$	Lead bromide
Pb_3O_4	Trilead tetroxide	HCl	Hydrochloric acid	$Pb(NO_3)_2$	Lead nitrate
Fe_2O_3	Iron(III) oxide	HBr	Hydrobromic acid	$CaCO_3$	Calcium carbonate
Al_2O_3	Aluminium oxide	HNO_3	Nitric acid	$Ca(NO_3)_2$	Calcium nitrate
P_2O_5	Phosphorus pentoxide	H_2SO_3	Sulphurous acid	Ag_2SO_4	Silver sulphate
SiO_2	Silicon dioxide	H_2SO_4	Sulphuric acid	$CuSO_4$	Copper sulphate
		H_2CO_3	Carbonic acid	$KMnO_4$	Potassium permanganate
		H_3PO_4	Phosphoric acid		
				$KClO_3$	Potassium chlorate

This looks rather a frightening list of formulae to remember! It is likely, however, that you know a few of these formulae already, and you are certain to become familiar with others by meeting the compounds on numerous occasions. A simple method for working out many formulae, rather than remembering them all, is discussed in Chapter 9.

The formulae of simple molecules

Many elements and compounds are believed to be composed of small numbers of atoms held together in a regular arrangement. These groups of atoms are called **molecules**. The gas hydrogen, for example, is composed of pairs of hydrogen atoms; each pair is called a molecule and its formula is H_2. A molecule of H_2 can be separated easily from other molecules of H_2, but it is very difficult to separate the two atoms which form the molecule. Another example is the compound carbon dioxide which is composed of molecules each containing one carbon atom and two oxygen atoms: the formula of a molecule of carbon dioxide is CO_2.

Substances which are composed of small molecules can be recognized by their physical properties. They are usually gases, liquids or low melting point solids. High melting point solids do not contain simple molecules. The formulae of some important molecules are given in Table 4.8. It will be seen in this table that certain gaseous elements, called the **noble gases**, contain only one atom in their molecules.

Table 4.8
Formulae of Some substances Composed of Simple Molecules at 20 °C

Gaseous elements	*Gaseous compounds*	*Liquid compounds*
H_2 Hydrogen	HCl Hydrogen chloride	H_2O Water
He Helium	HBr Hydrogen bromide	CCl_4 Tetrachloromethane (carbon tetrachloride)
N_2 Nitrogen	NH_3 Ammonia	$SiCl_4$ Silicon tetrachloride
O_2 Oxygen	CH_4 Methane	PCl_3 Phosphorus trichloride
F_2 Fluorine	SO_2 Sulphur dioxide	C_2H_6O Ethanol
Ne Neon	CO Carbon monoxide	C_6H_{14} Hexane
Cl_2 Chlorine	CO_2 Carbon dioxide	*Solid compounds*
Ar Argon	NO_2 Nitrogen dioxide	$C_6H_{12}O_6$ Glucose
Liquid element	N_2O_4 Dinitrogen tetroxide	$C_{18}H_{36}O_2$ Stearic acid
Br_2 Bromine	NO Nitrogen monoxide	
Solid elements	C_2H_4 Ethene (ethylene)	
S_8 Sulphur	C_2H_6 Ethane	
P_4 Phosphorus		

There are certain points of interest in this table:

1. Substances composed of simple molecules contain only non-metallic elements.
2. The molecules of some gaseous elements contain only one atom. These elements are the **noble gases**.
3. There is only one element composed of simple molecules which is liquid at room temperature. (There is only one other element which is liquid at room temperature, the metallic element mercury.)
4. If a substance contains a large number of atoms in its molecule, it is either a liquid or a solid, not a gas.
5. A molecule of stearic acid contains atoms of carbon, hydrogen and oxygen in the same ratio as in the empirical formula of this compound, $C_9H_{18}O$.

5 Atoms and Molecules
2: Size and movement

Atoms and molecules under the microscope

You have probably used a microscope to watch small crystals growing in a concentrated aqueous solution. Without perhaps realizing it at the time, you will have noticed certain things about the atoms or molecules present in the solution and crystals.

Table 5.1

Atoms and molecules in solutions	*Atoms and molecules in crystals*
Aparts from specks of dust and other impurities, a solution appears to be completely uniform. There is no sign of atoms or molecules. If the solution is composed of these particles, they are too small to be seen under the microscope.	The material within each crystal is completely uniform. There is no sign of atoms or molecules. If the crystal is built up of these particles, they are too small to be seen under the microscope. The regular shapes of crystals may indicate a regular packing of particles too small to be seen.

If atoms or molecules cannot be seen it does not necessarily mean that they do not exist. It does indicate that any particles, if present, must be extremely small. A simple microscope would have detected atoms or molecules if these had been larger than about 10^{-4} m. The best optical microscopes available, which could detect atoms or molecules if they were larger than about 10^{-7} m, still show no evidence for their existence. When an electron microscope is used, which under certain circumstances could detect objects as small as about 10^{-10} m, evidence for the existence of atoms and molecules does begin to accumulate.

Figure 5.1 is an electron micrograph of a crystal of stearic acid, $C_{18}H_{36}O_2$, which shows that the crystal face has grown in a spiral. Imagine that the crystal is being illuminated from the top right-hand corner and you will see that the presence of steps in the spiral is revealed by the shadows they produce. It is possible to measure the height of these steps from the length of their shadows. The height of all the steps is found to be about 4.7×10^{-9} m (47 Å or 4.7 nm). No molecules can be seen in this electron micrograph, but the constant height

Figure 5.1 Electron micrograph of a stearic acid crystal (magnification 16 000 times). (From *Proceedings of the Royal Society*, A, **218**, 1953, pp. 225–68.)

of the steps could be explained if stearic acid is composed of molecules whose size is about 4.7×10^{-9} m (4.7 nm) in one direction. If one extra layer of stearic acid molecules is added at each complete turn of the spiral, the height of the steps would be 4.7×10^{-9} m (4.7 nm).

The oil-drop experiment

This is an experiment using simple apparatus which can be used to provide evidence that certain substances are composed of very small particles. To obtain information about stearic acid, for example, a very dilute solution of this substance in a volatile solvent is used. One drop of this solution is placed on a water surface which has been covered with a few specks of a fine powder. The powder rushes away from the place where the drop has fallen on the water, and a clean circle is formed. If the volatile solvent has evaporated, this clear circle is a thin layer of stearic acid floating on the water surface. It is possible to calculate the thickness of the stearic acid layer in the following way.

Calculation of thickness of stearic acid layer

In a certain experiment, 1000 cm^3 of solution contained 0.1 cm^3 of stearic acid. There were found to be 25 drops of this solution in 1 cm^3 (this means that the volume of a drop was 0.04 cm^3). One drop covered 200 cm^2 of the water surface with a layer of stearic acid (*see* Figure 5.2).

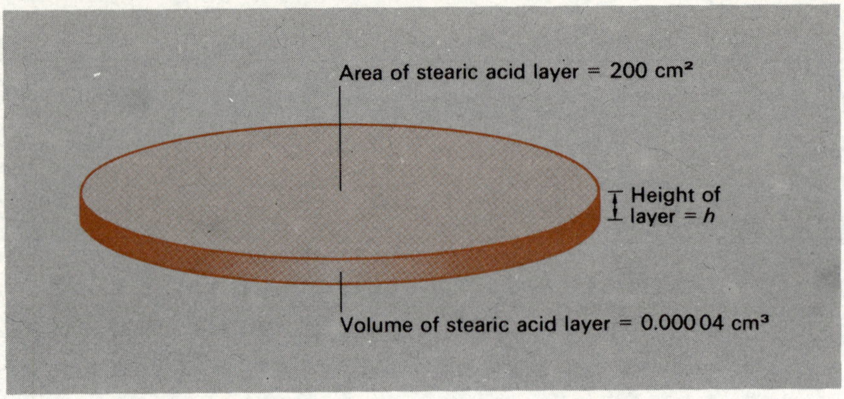

Figure 5.2 Layer of stearic acid on a water surface

1000 cm^3 of solution contains 0.1 cm^3 stearic acid
$\therefore$ 1 cm^3 of solution contains 0.0001 cm^3 stearic acid
$\therefore$ 0.04 cm^3 of solution (1 drop) contains 0.000 04 cm^3 stearic acid

$$\text{the volume of the stearic acid layer} = \text{area} \times \text{height} = 200\ \text{cm}^2 \times h$$

but it is known also that the volume of the stearic acid layer $= 0.000\,04\ \text{cm}^3$

$$200\ \text{cm}^2 \times h = 0.000\,04\ \text{cm}^3$$

$$h = \frac{0.000\,04\ \text{cm}^3}{200\ \text{cm}^2} = 2 \times 10^{-7}\ \text{cm}$$

$$\text{height of layer} = 2 \times 10^{-9}\ \text{m (2 nm)}$$

The height of the stearic acid layer gives some information about the size of the molecules of this substance. The molecules cannot be larger, in the direction of the film height, than 2×10^{-9} m (2 nm). The layer may be several molecules thick, in which case each molecule must be smaller than 2×10^{-9} m (2 nm), but they cannot be any larger. At this point it is necessary only to realize how incredibly small the molecules must be. No wonder molecules are invisible under an ordinary microscope. If one million stearic acid molecules were stacked one on top of the other, the pile would be only as high as the letter 1.

The results obtained by a group of pupils who carried out a large number of determinations of the height of the stearic acid layer are shown in the form of a histogram in Figure 5.3.

Inspection of the histogram confirms that molecules of stearic acid may have a size of about 2×10^{-9} m (2 nm) in one direction. If the results are reliable, a more exact value from the seventy determinations is about 2.4×10^{-9} m (2.4 nm).

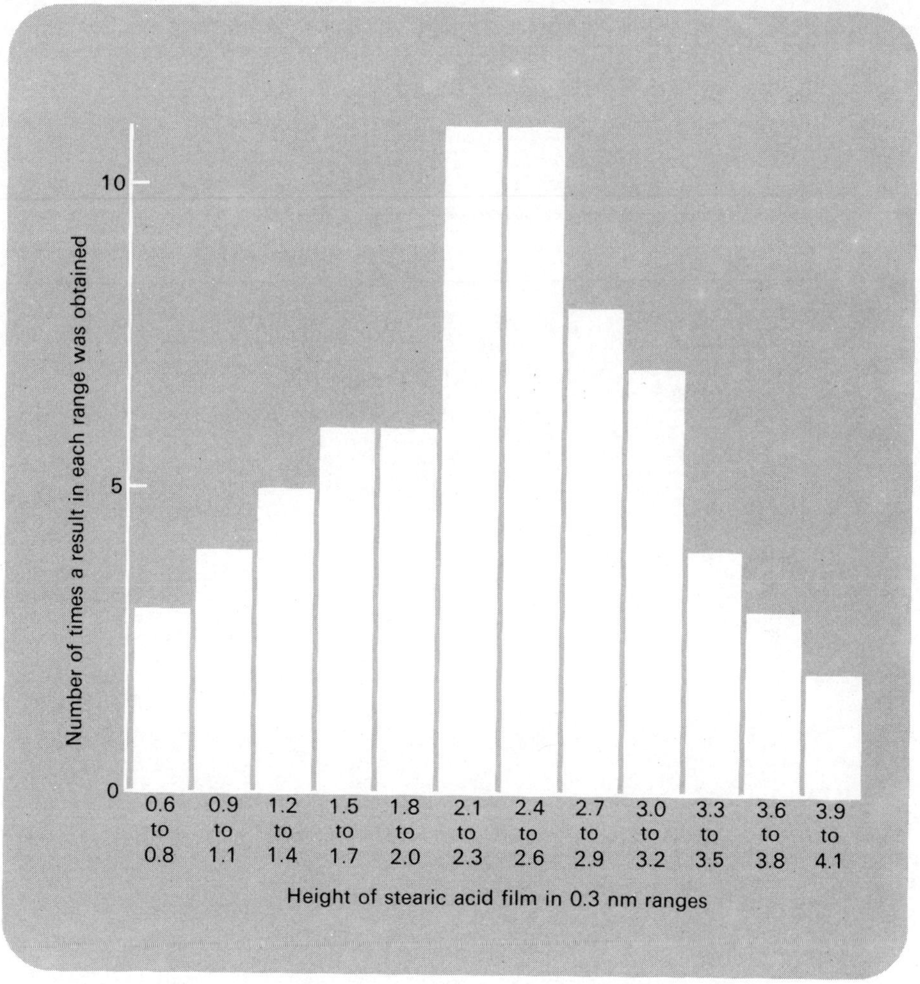

Figure 5.3 Histogram showing results of seventy determinations of the height of a stearic acid film on water

There are methods available for finding both the shape and size of a molecule. The results of these methods have been used to make the two-dimensional plan of a stearic acid molecule which is shown in Figure 5.4 on page 36.

A molecule of stearic acid really extends into three dimensions and has an infinite number of shapes, but Figure 5.4 is a reasonably good two-dimensional representation of the molecule when it is in its most extended shape. The scale in Figure 5.4 is shown beside the figure, and if you lay a ruler down the sketch you will see that the length of the molecule is just over 2×10^{-9} m (2 nm). This is very close to the value of the film height in the oil-drop experiment, and is explained on a model where one end of each stearic acid molecule (the bottom end) is held to the water surface while the other ends project up from the water surface. This is shown in Figure 5.5 on page 37.

It has been seen earlier in this chapter that investigation of stearic acid crystals with an electron microscope indicates that the size of the molecules in

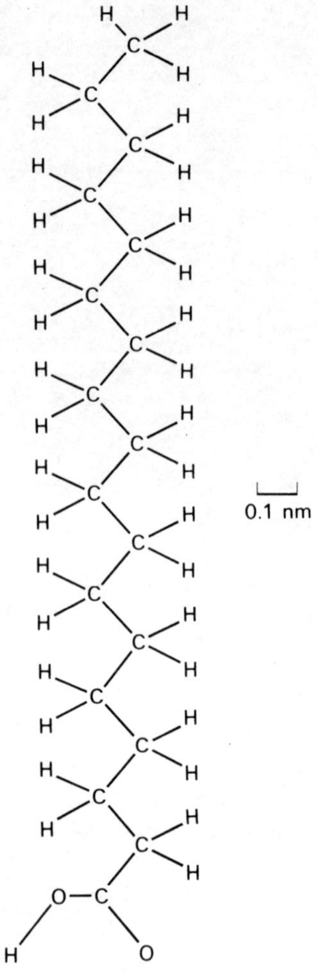

Figure 5.4 One possible arrangement for the atoms in a molecule of stearic acid, $C_{18}H_{36}O_2$

one direction is about 4.7×10^{-9} m (4.7 nm), while the oil-drop experiment gives a size of about 2.4×10^{-9} m (2.4 nm). It is no coincidence that one value is twice the other. When stearic acid molecules are on a water surface, their individual molecules have the formula $C_{18}H_{36}O_2$ and the structure given in Figure 5.4. When stearic acid molecules are packed together in a crystal, they group together in pairs to form molecules with the formula $C_{36}H_{72}O_4$—the stearic acid is said to have **dimerized**. This is an example of polymerization (*see* Chapter 25). The change is shown in Figure 5.6 on page 38.

The structure of molecules of stearic acid depends on their environment. This is an example of a situation which a chemist meets frequently. Other important examples where a substance can have more than one structure include the following.

Carbon: carbon atoms can be arranged in two completely different ways in the solid state as graphite and diamond (*see* page 102).

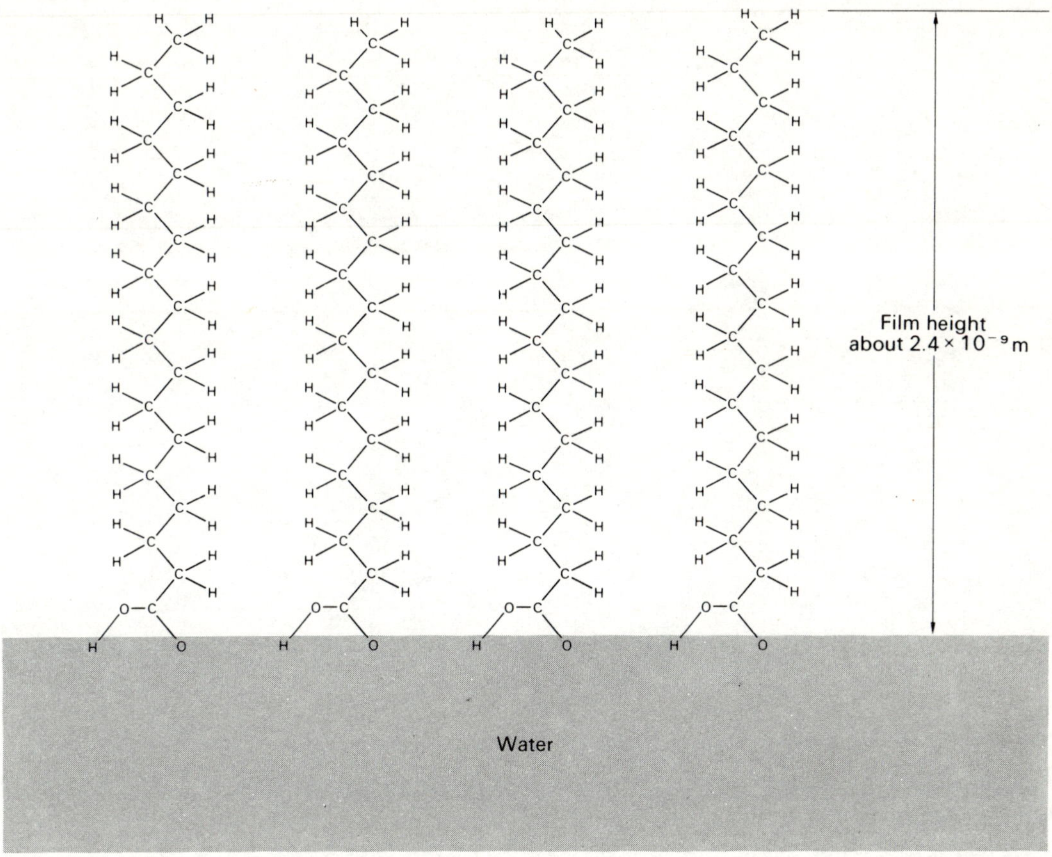

Figure 5.5 Arrangement of stearic acid molecules on a water surface

Hydrogen chloride: a solution of hydrogen chloride, HCl, in toluene contains particles which have no resemblance in structure or properties to the particles present in a solution of hydrogen chloride in water. The particles in the toluene solution do not change the colour of dry indicator paper and do not conduct electricity. The particles in the aqueous solution turn indicator paper to its acid colour and the solution conducts electricity well.

The size of atoms

It is possible to detect fairly large molecules with the electron microscope, but individual atoms are too small to show up except in the most powerful instruments. There are methods by which the sizes of atoms and their arrangement in molecules can be determined; one of these methods uses X-ray diffraction. The results of these investigations show that when atoms are in contact with other atoms in molecules or in the solid element, their radius is about 0.1×10^{-9} m (0.1 nm). Some idea of how small atoms are can be obtained by imagining one million copper atoms [radius = 0.13×10^{-9} m (0.13 nm)]. If these copper atoms were stacked one on top of the other, the pile would be as high as the full stop at the end of this sentence.

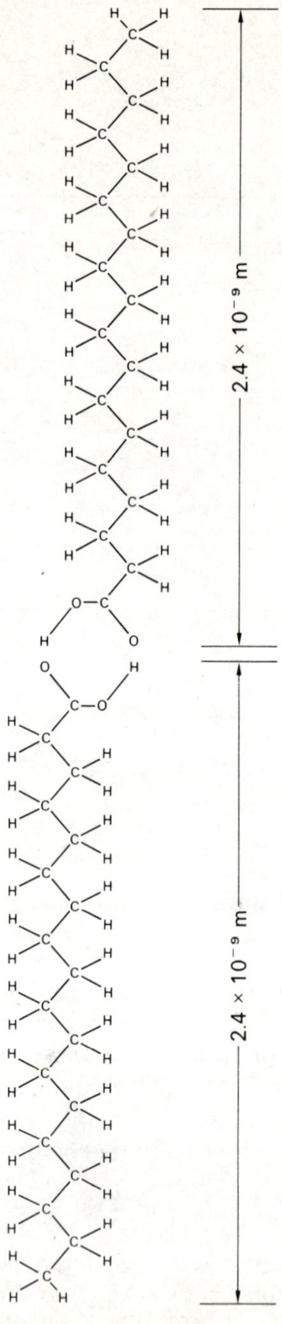

Figure 5.6 Stearic acid molecules when in contact with each other in a crystal

Atoms and molecules in motion

Some information about the motion of the atoms and molecules from which matter is built can be obtained by looking at smoke under the microscope. With sideways illumination, the smoke particles suspended in air show up as irregular bright specks. These are much bigger than atoms and molecules. A close inspection will show that the smoke particles have a motion which is sometimes called the **drunkard's walk**. They are moving first in one direction and then, suddenly, in another direction. The changes in motion are so frequent that the particles remain almost unchanged in position and the eye may take some time to realize that changes in motion are taking place. The smoke particles are not changing their motion because of collisions with each other—they are too far apart for this to happen frequently. The changes in motion occur because each smoke particle is being bombarded from all directions by the molecules of the air which are in rapid chaotic motion. The molecules of air are too small to show up in the microscope. At a certain instant, a smoke particle may be bombarded more from one direction than from any other. The particle will move away under the influence of this bombardment. A fraction of a second later, excess bombardment by air molecules may be from a completely different direction and the motion of the smoke particle will change. The molecules in a gas such as air are so numerous and fast-moving that the direction of motion of each smoke particle alters many times each second. Similar experiments observing particles suspended in a liquid show that these particles are also moving in the drunkard's walk. This can be explained if it is assumed that the molecules in a liquid, too small to be seen under the microscope, are in rapid chaotic motion.

Diffusion in gases

When a bottle containing a volatile substance with a distinctive smell is opened in a room, it is not long before the smell can be detected in all corners of the room. This happens even if draughts and all other causes of air movement have been stopped. The mixing of the molecules of 'smell' with the molecules of air is one example of a process called **diffusion**. Diffusion takes place rapidly in gases because the molecules are fast moving and are relatively far apart—the further apart molecules are, the greater is the distance each molecule can travel before being deflected from its direction of motion by collision with another molecule (*see* Figure 5.7 on page 40).

Diffusion in liquids

If a crystal of potassium permanganate is dropped into a beaker of water, a dark purple patch of solution is formed at the bottom of the beaker as the solid dissolves. The purple colour gradually spreads through the water, until after several days the solution is a uniform colour. Diffusion takes place at a much slower rate in liquids than in gases. This slow diffusion is a result of the close packing of liquid molecules—there is very little free space for movement. The molecules, like those in a gas, are moving rapidly, but continual deflection of their direction of motion by numerous collisions reduces the rate of diffusion.

Diffusion in solids

If lumps of two different metals, say copper and silver, are clamped together for a long time and are then separated, there is no visible sign that diffusion

	Gas	Liquid	Solid
Scale = 10^{-9} m (1 nm)		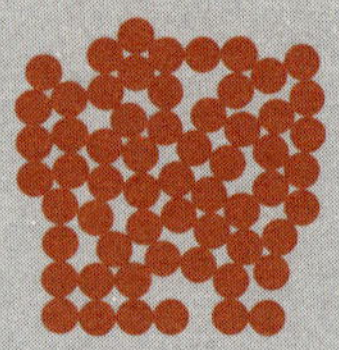	
Size	Small	Small	Small
Relative distance apart	Large	Very small	None
Arrangement	Random	Fairly regular	Regular
Movement	Rapid	Rapid	Almost none
Rate of diffusion	Rapid	Slow	Very slow

Figure 5.7 Summary of size, motion and arrangement of particles in gases, liquids and solids

has taken place. Both metal surfaces appear to be unchanged—atoms of one metal have not moved into the other metal.

The model used to account for this lack of diffusion in solids is one where the atoms or molecules of the solid are very closely packed in a regular structure which allows only a limited amount of movement to take place. Diffusion does take place in solids. This can be shown if a method for detecting the movement of a small number of atoms is used. Diffusion between two solid surfaces is normally slowed down by a layer of oxide or grease which normally covers the surface. The irregular nature of most surfaces is another factor which reduces diffusion between two solids. The solids only really touch at their high spots, most of the areas of both surfaces being out of contact. Diffusion does become important between flat clean metal surfaces in a vacuum. This diffusion can cause trouble with the moving parts of a space craft. Bearings weld to other moving parts by diffusion unless surfaces are coated with a film of non-volatile liquid. This welding of metal surfaces by diffusion is a possible way of joining together parts of a space station. The various metal parts which are to form the structure of the space station are clamped firmly together and after some time diffusion produces a strong weld.

The Avogadro constant

A mole is the amount of substance which contains the same number of particles as there are carbon atoms in 12 g of carbon. The number of particles in one mole is called the **Avogadro constant.**

It is possible to obtain an approximate value for the Avogadro constant from the oil-drop experiment with stearic acid described earlier in this chapter. The experiment indicates that the size of one molecule of stearic acid is 2.4×10^{-9} m (2.4 nm) in one direction. If a stearic acid molecule is regarded as a cube, the volume of one stearic acid molecule will be about $(2.4 \times 10^{-9})^3$ m^3, or 1.4×10^{-26} m^3. The volume of one mole of stearic acid molecules ($C_{18}H_{36}O_2$) can be found from tables of density and relative atomic masses. This volume is 3×10^{-4} m^3.

1 mole of stearic acid has a volume of $3 \times 10^{-4}\ m^3$

1 molecule of stearic acid has a volume of $1.4 \times 10^{-26}\ m^3$

$$\therefore \text{number of stearic acid molecules in 1 mole} = \frac{3 \times 10^{-4}\ m^3}{1.4 \times 10^{-26}\ m^3} \approx 2 \times 10^{22}$$

The approximate value for the Avogadro constant is 2×10^{22} particles/mole from this experiment. The true value is even larger; it is 6.0×10^{23} mole. The number of metal atoms in eighteen new pence coins, and the number of H_2O molecules in a test-tube of water, are both about 6×10^{23}. This number is so large that it is very difficult to visualize. If every person in the British Isles was asked to count the water molecules in just one test-tube of water which had been divided between them, and they each began counting at a rate of two molecules per second, they would take 1 000 000 years to finish.

The size, motion and arrangement of particles in gases, liquids and solids are summarized in Figure 5.7.

6 Periodic Classification
1: Physical properties

One of the most important ways of making sense of the world around us is to group similar things together or to classify. Scientists classify or group things in a number of ways: the biologist classifies animals into mammals, reptiles, fishes; the geographer classifies regions into desert, forest, prairie; and the astronomer classifies stars into main sequence, white dwarfs, red giants.

Substances may be grouped into solids, liquids and gases, although this grouping is not very satisfactory since a change of temperature alters the membership of each group. A more satisfactory classification might be into acids, salts, bases, fats, dyes, metals and so on. Such a classification would include elements as one class, but since all substances are made from elements it is important that the members of this class are themselves classified carefully.

A number of attempts have been made to classify elements. For example, they may be grouped into metals and non-metals. However, this is not very convenient, partly because with only two classes for over 100 elements, each class is large. Furthermore, many elements do not fit neatly into either class. Carbon, for example, has most of the properties of a non-metal, but the graphite form conducts electricity and this would be regarded as a metallic property.

By far the most important and useful classification is that suggested by Mendeléef in the late 1860s. Several experimenters had found that similar properties were possessed by a number of elements. For example, sodium and potassium share the properties of being soft, silvery, of low density, of reacting vigorously with water, and of burning with a coloured flame to give an oxide readily soluble in water and strongly alkaline. They are named the alkali metals and belong to what we shall call in this book the sodium group, Group 1. Chlorine, bromine and iodine also share similar properties, among them a readiness to combine with metals to form salts.

At the time of the Mendeléef classification, based on the Periodic Law, the number of known elements was much smaller than it is today. Mendeléef realized that some elements were as yet undiscovered and it is this in particular which made his work so important for scientific advance.

Part of the evidence for the Periodic Law is given later in this chapter in a number of histograms. Melting point, boiling point, density and atomic volume have been plotted against atomic number. Mendeléef had no knowledge of the significance of atomic number—this had to wait for discoveries about the structure of atoms. He found that there was a recurrence of similar properties; every so often, or periodically, a property reappeared if the elements are arranged in atomic number order. For example, sodium and potassium have very different atomic numbers but have melting points in the same region of temperature and these melting points differ very much from those of the elements on either side (neon and magnesium, argon and calcium).

The Periodic Law suggested by Mendeléef was that **the properties of the elements are a periodic function of their atomic masses**. When the elements known in his day were set out in what has come to be called the Periodic Table, gaps were left. Mendeléef argued that a search should be made for elements to fill the gaps and he predicted the properties which he expected those elements to have. The success of his theory was demonstrated when newly discovered elements were found to have properties very close to those he had predicted for them. Thus one important value of the periodic classification lies in its being a tool for research and discovery.

Examine the histograms on pages 44–50 carefully. On each a physical property is plotted against atomic number. In the first three only elements 1–20 (hydrogen to calcium) have been included. Later histograms show the pattern as far as element 54, xenon.

Mendeléef did not suspect the existence of helium. Why not? Hydrogen presents problems of classification but the picture for the remainder is fairly clear.

In Figures 6.1 and 6.2 on pages 44 and 45, the peaks and troughs fall in roughly the same places.

The atomic volume of an element, *see* Figure 6.3 on page 46, is the volume occupied by one mole of atoms. Since one mole of atoms of all elements contains the same number of atoms, atomic volumes give an indication of the relative sizes of atoms of different elements. However, it does not take into account the way in which the atoms are packed in the solid crystalline element.

It will be observed that there is a regular, periodic recurrence of similar properties. One period represents the fluctuation of the property through the cycle until it starts again with a similar element. Thus Li, Na and K have much larger atomic volumes than the next metals in their periods, Be, Mg and Ca.

The first period contains only hydrogen and helium.

The second period runs from lithium to neon:

Li Be B C N O F Ne

The third period runs from sodium to argon:

Na Mg Al Si P S Cl Ar

The next four histograms, Figures 6.4–6.7 on pages 47–50, give a more complete picture. All elements up to xenon, number 54, have been included. The pattern shown by the normal elements (sometimes called the main elements) is interrupted by the transition metals; scandium to zinc in the third period and yttrium to cadmium in the fourth.

Cut out an atomic volume histogram from thin card or stiff paper. Make another, on the same scale, for elements 55–92. Cut the histograms vertically so that elements 21–30, 39–48 and 57–80 can be removed. Slide the remaining pieces together to make a histogram of normal elements only. Are elements 87–92 normal?

We shall often use only the normal elements for work in this book. The table containing only these elements is called the Periodic Table of Normal (or Main) Elements to distinguish it from the Full Periodic Table which includes the transition metals.

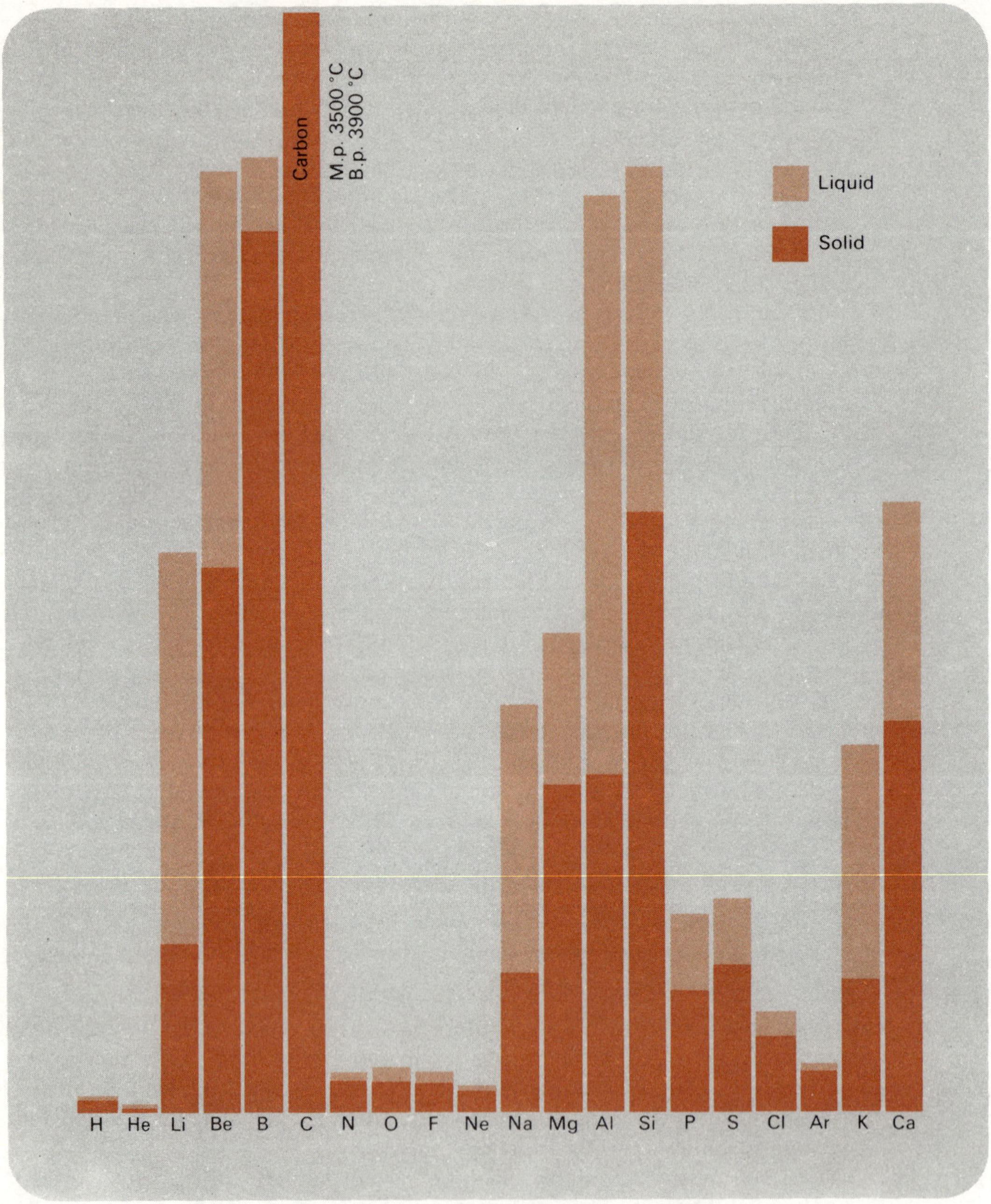

Figure 6.1 Melting and boiling points of elements 1–20

Figure 6.2 Densities of elements 1–20 in the solid state. The noble gases are tinted

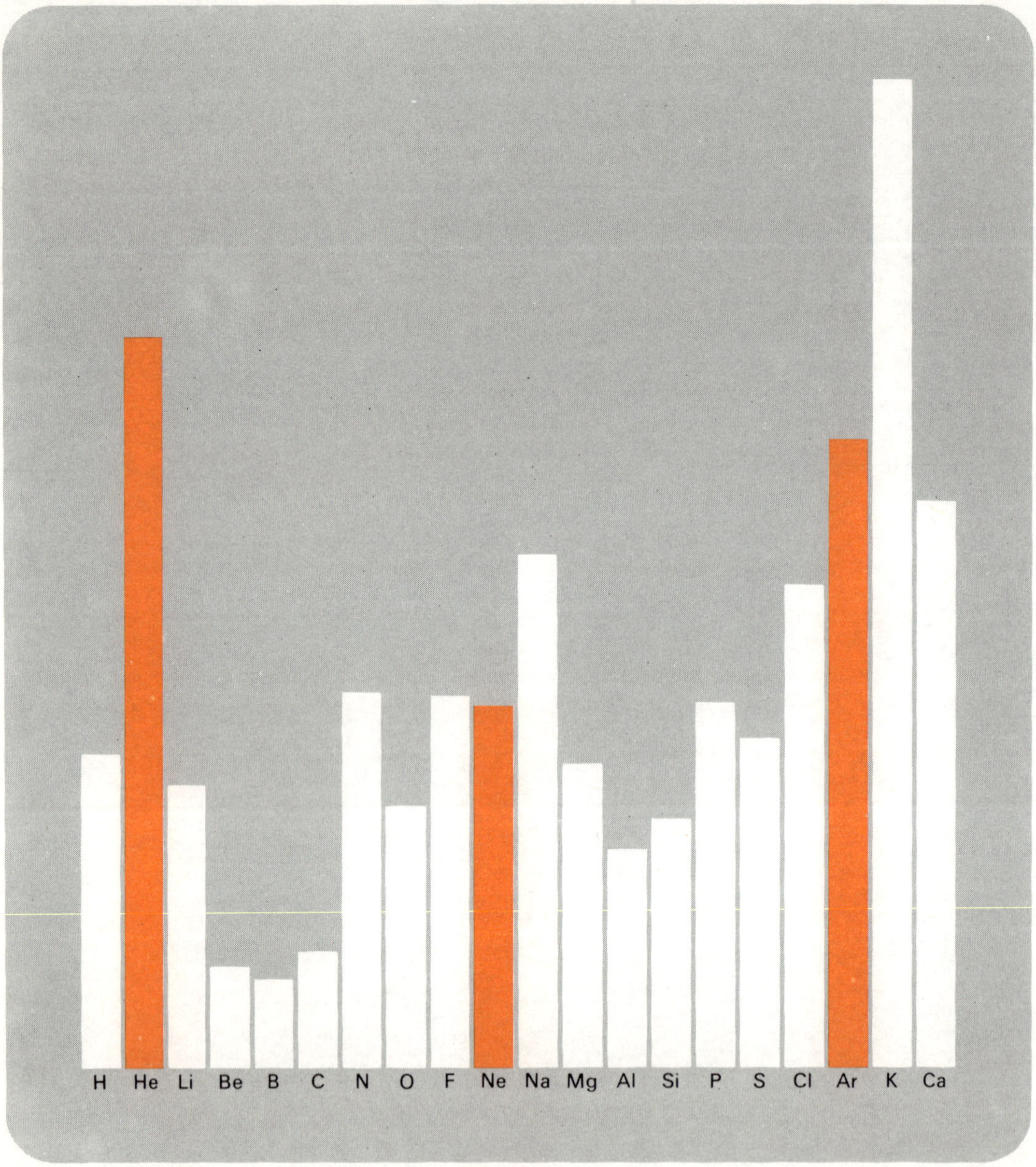

Figure 6.3 Atomic volumes of elements 1–20 in the solid state. The noble gases are tinted

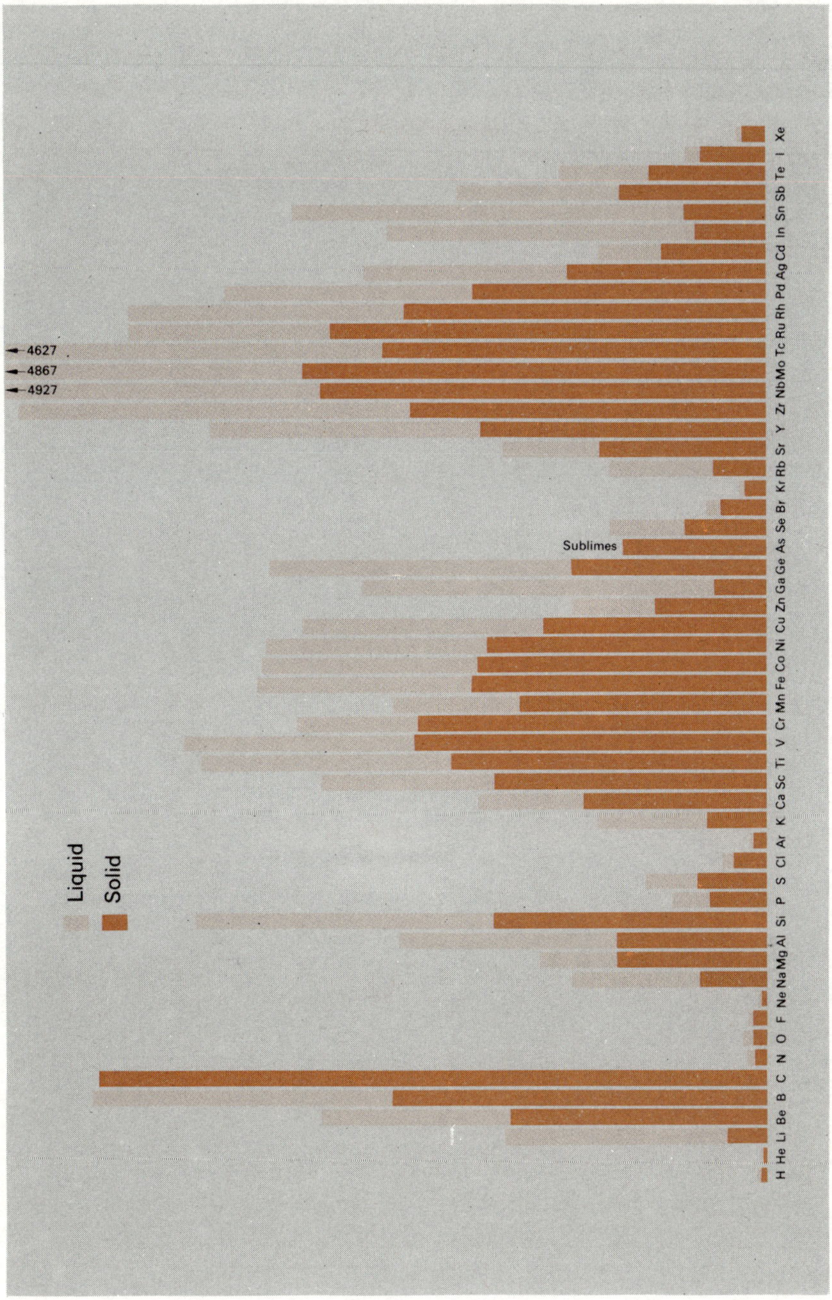

Figure 6.4 Melting and boiling points of elements 1–54. Note that the transition metals Sc to Zn and Y to Cd have been included

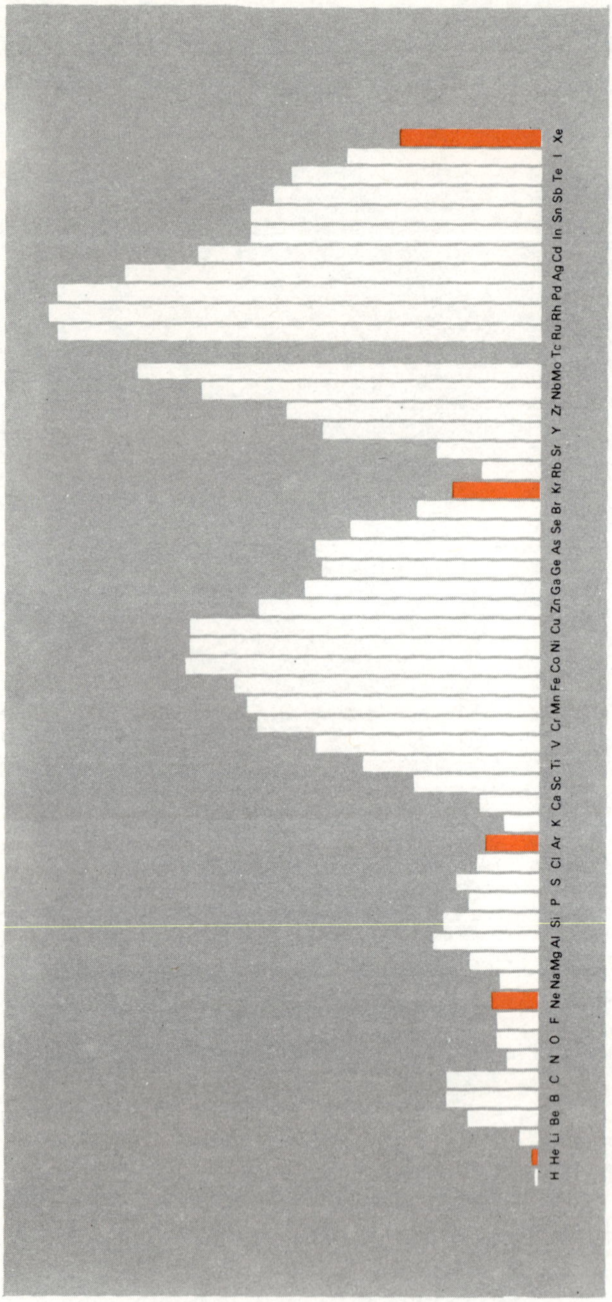

Figure 6.5 Densities of elements 1–54 in the solid state. Note that the transition metals have high densities. The noble gases are tinted

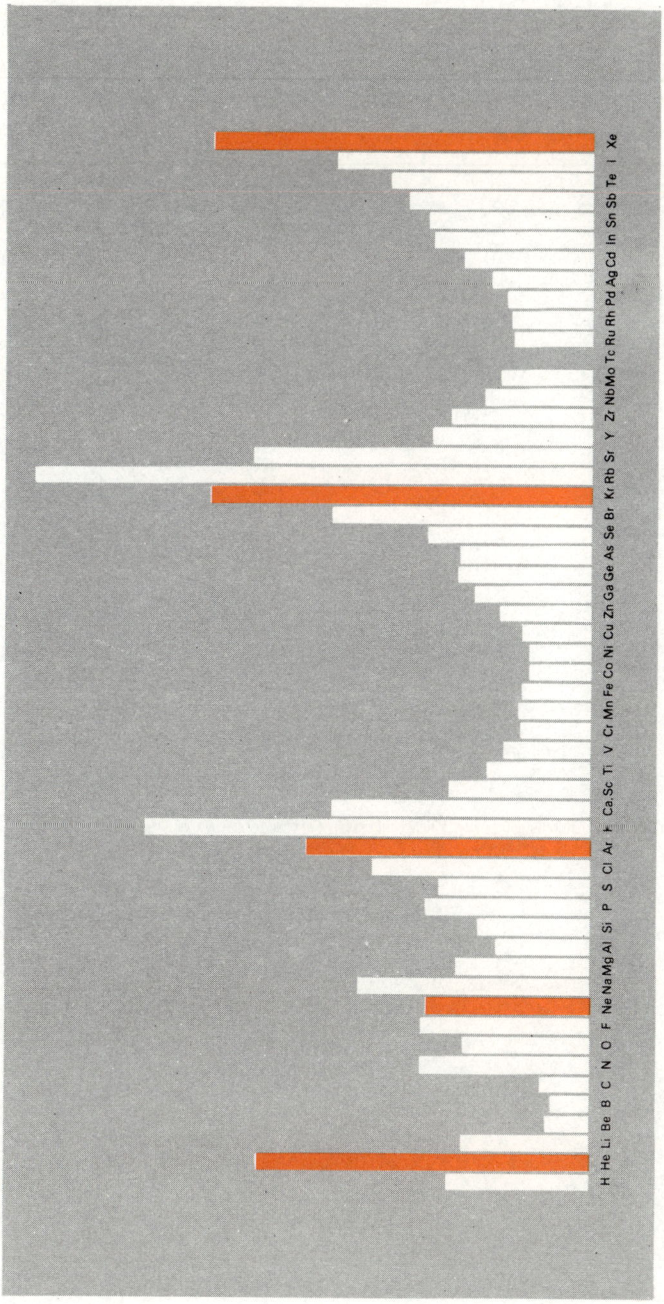

Figure 6.6 Atomic volumes of elements 1–54 in the solid state. Note that the transition metals have similar and small atomic volumes. The noble gases are tinted

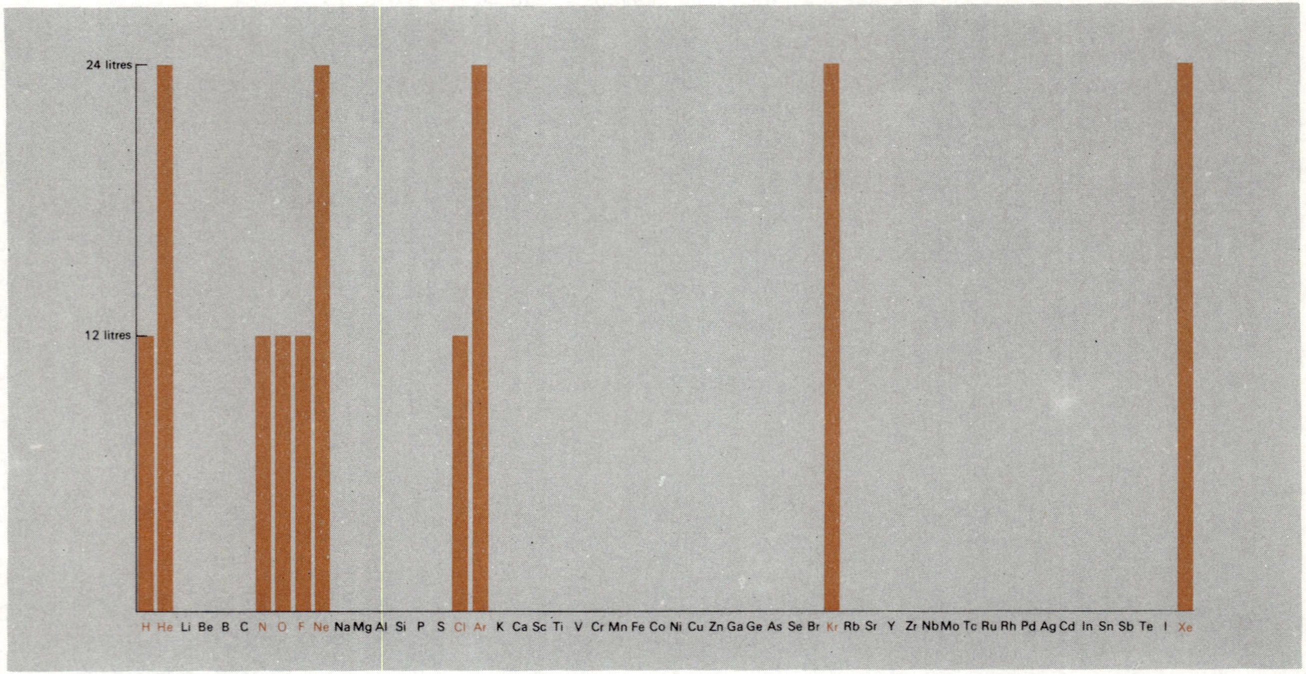

Figure 6.7 Atomic volumes of elements 1–54 at room temperature. Only the gaseous elements appear in this histogram. The atomic volumes of solid and liquid elements are so much smaller that they do not show up on the scale used

Summary

The elements can be arranged in atomic number order. When this is done, elements with similar physical properties are found to appear at periodic intervals. This is shown below for the first twenty elements by shading their symbols.

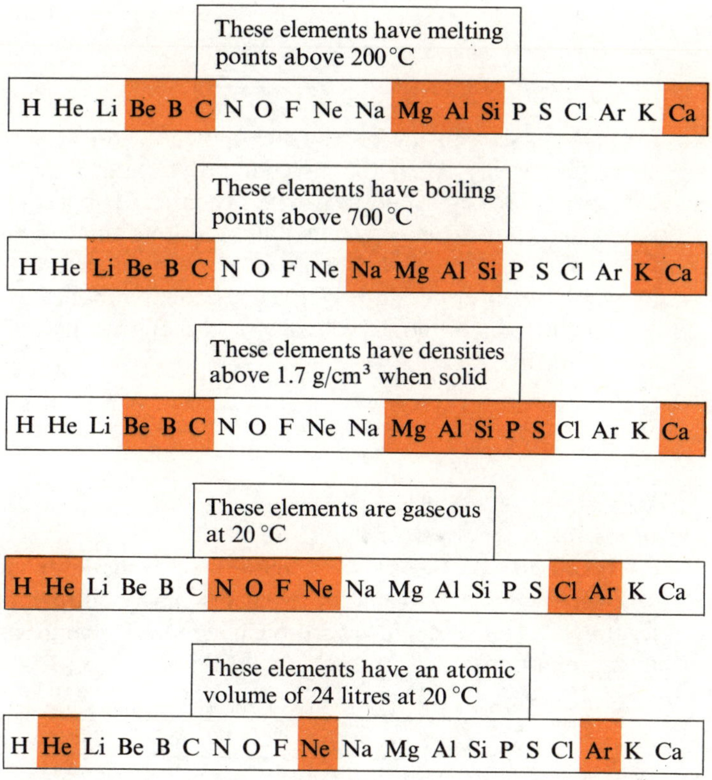

7 Periodic Classification
2: Chemical properties

It has been seen in the previous chapter that physical properties of elements show periodicity when the elements are arranged in atomic number order. This periodicity is found also when chemical properties are considered. You might like to investigate this by taking a long thin strip of paper, easily made from short strips using Sellotape or glue, and marking on the strip the symbol of each element in order of increasing atomic number. If you were to go as far as element 54, the noble gas xenon, the appearance of your strip would be something like this:

HHeLiBeBCNOFNeNaMgAlSiPSClArKCaScTiVCrMnFeCoNiCuZnGaGeAsSeBrKrRbSrYZrNbMoTcRuRhPdAgCdInSnSbTeIXe

The names of the elements in this strip will be found in the Periodic Table inside the front cover of this book.

To simplify the search for a pattern of repeating chemical properties, two sets of ten consecutive metallic elements can be cut from this strip for separate treatment. The sets, the transition metals, are scandium to zinc and yttrium to cadmium.

Sc	Ti	V	Cr	Mn	Fe	Co	Ni	Cu	Zn

Y	Zr	Nb	Mo	Tc	Ru	Rh	Pd	Ag	Cd

These twenty elements resemble each other in physical and chemical properties, and in their similar electronic structures. Their properties and electronic structures are rather different from those of the normal elements so they will be studied separately (*see* Chapter 24).

The appearance of the strip containing the remaining 34 normal elements is now:

H He Li Be B C N O F Ne Na Mg Al Si P S Cl Ar K Ca Ga Ge As Se Br Kr Rb Sr In Sn Sb Te I Xe

A number of chemical properties will now be chosen to show how the Periodic Law applies to the elements in this strip. The elements which show each chosen property will be indicated by shading their symbols. Several of the elements in the strip will be ones you have not met, and you are not expected to be familiar with their properties. What is important is that you should see the patterns which appear when the information is arranged in this way.

These elements are metals

H He **Li Be** B C N O F Ne **Na Mg Al** Si P S Cl Ar **K Ca Ga** Ge As Se Br Kr **Rb Sr In Sn** Sb Te I Xe

What properties would you look for to decide whether an element is a metal?

These elements are non-metals

H He Li Be **B C N O F Ne** Na Mg Al **Si P S Cl Ar** K Ca Ga **Ge As Se Br Kr** Rb Sr In Sn **Sb Te I Xe**

What properties would you look for to decide whether an element is a non-metal? The elements He, Ne, Ar and Kr have been included among the non-metals because of their physical properties. They have almost no chemical properties.

These elements react with cold water to produce hydrogen

H He **Li** Be B C N O F Ne **Na Mg** Al Si P S Cl Ar **K Ca** Ga Ge As Se Br Kr **Rb Sr** In Sn Sb Te I Xe

Magnesium has been marked on the strip. Is this correct? Is it possible to make aluminium react with water to produce hydrogen?

These elements have oxides and hydroxides which dissolve in water to give alkaline solutions

H He **Li** Be B C N O F Ne **Na Mg** Al Si P S Cl Ar **K Ca** Ga Ge As Se Br Kr **Rb Sr** In Sn Sb Te I Xe

Magnesium has been marked on the strip. Is this correct? You might like to investigate the solubility of magnesium hydroxide in water, and to test the water afterwards to see if an alkaline solution has been formed.

These elements can form acidic oxides

H He Li Be **B C N** O F Ne Na Mg Al **Si P S Cl** Ar K Ca Ga **Ge As Se Br** Kr Rb Sr In **Sn Sb Te I Xe**

What properties would help you to decide if an oxide is acidic?

These elements can form positive ions of charge +1, +2 or +3 in their compounds. The value of the charge carried by each ion is given above the symbol

H (+1) He **Li** (+1) **Be** (+2) B C N O F Ne **Na** (+1) **Mg** (+2) **Al** (+3) Si P S Cl Ar **K** (+1) **Ca** (+2) **Ga** (+3) Ge As Se Br Kr **Rb** (+1) **Sr** (+2) **In** (+3) **Sn** (+2) Sb Te I Xe

The value of the charge carried by ions of one of these elements does not seem to fit into the general pattern. Which element is this?

> These elements form chlorides with boiling points above 1000 °C

H He **Li** Be B C N O F Ne **Na** **Mg** Al Si P S Cl Ar **K** **Ca** Ga Ge As Se Br Kr **Rb** **Sr** In Sn Sb Te I Xe

The high boiling points arise because the chlorides of these elements are composed of ions (*see* Chapter 9).

> These elements form chlorides with boiling points above 1000 °C in which 1 mole of atoms of the element is combined with 1 mole of chlorine atoms

H He **Li** Be B C N O F Ne **Na** Mg Al Si P S Cl Ar **K** Ca Ga Ge As Se Br Kr **Rb** Sr In Sn Sb Te I Xe

The formula of lithium chloride is LiCl or (Li^+Cl^-). What will be the formulae of the chlorides of sodium, potassium and rubidium?

> These elements form chlorides with boiling points above 1000 °C in which 1 mole of atoms of the element is combined with 2 moles of chlorine atoms

H He Li Be B C N O F Ne Na **Mg** Al Si P S Cl Ar K **Ca** Ga Ge As Se Br Kr Rb **Sr** In Sn Sb Te I Xe

The formula of magnesium chloride is $MgCl_2$ or ($Mg^{2+}2Cl^-$) (*see* Chapter 9). What will be the formulae of the chlorides of calcium and strontium?

> These elements can form chlorides that are liquid or gaseous at room temperature

H He Li Be **B C N O F** Ne Na Mg Al **Si P S Cl** Ar K Ca Ga **Ge As Se Br** Kr Rb Sr In **Sn Sb** Te **I** Xe

The high boiling point chlorides are composed of ions. Very strong attractive forces between the ions makes vaporization difficult. The chlorides that are liquid or gaseous at room temperature are composed of small molecules between which only weak attractive forces operate. Their formulae would not be written ionically, e.g. (Li^+Cl^-), since they are covalent. Phosphorus trichloride, for example, is written PCl_3, not ($P^{3+}3Cl^-$) (*see* Chapter 9).

> These elements combine with hydrogen to form gaseous or liquid compounds called hydrides. Hydrogen chloride, HCl, water, H_2O, and ammonia, NH_3 are examples of hydrides

H He Li Be **B**(3) **C**(4) **N**(3) **O**(2) **F**(1) Ne Na Mg Al **Si**(4) **P**(3) **S**(2) **Cl**(1) Ar K Ca Ga **Ge**(4) **As**(3) **Se**(2) **Br**(1) Kr Rb Sr In **Sn**(4) **Sb**(3) **Te**(2) **I**(1) Xe

The number of moles of hydrogen atoms that combine with 1 mole of atoms of each element is shown above the symbols. For example, the hydride of sulphur (hydrogen sulphide) has the formula H_2S.

These elements were at one time thought to form no compounds. They are called the **noble gases**. A few compounds of krypton and a considerable number of xenon have now been made

H **He** Li Be B C N O F **Ne** Na Mg Al Si P S Cl **Ar** K Ca Ga Ge As Se Br **Kr** Rb Sr In Sn Sb Te I **Xe**

Xenon is the most reactive of the noble gases and can be classified as a non-metal on the basis of its chemical properties.

A sufficient number of chemical properties has been covered for both general and specific patterns to have appeared.

General patterns

The general patterns are found in such properties as the metallic and non-metallic natures of the elements and the positions of the elements with high boiling point chlorides. The adjacent elements in a small section of the strip possess the property, then, after a break, another small section of adjacent elements is found to share this property. The pattern repeats throughout the length of the strip.

Specific patterns

Certain properties such as the ability to form an ion El^+, with a single positive charge, are specific to one particular element in a considerable length of the strip (the symbol El stands for any element). On going a sufficient distance in either direction, however, another element is found with this property. Except at the very beginning of the strip, this repetition of specific properties is found to occur with every element which is eight elements removed from the previous element showing this property.

Every element in the strip has its own individual properties which distinguish it from all other elements. In many aspects of their chemistry, however, elements with very different atomic numbers do show considerable resemblances to each other. One way of displaying the elements so that the resemblances stand out is to divide the long strip into sections, the cuts being made at each element within a series which shows a specific property. If the strip is cut just before each metallic element which forms an El^+ ion and the sections are laid one below the other, a **Periodic Table** is produced.

Periodic Table

1	2	3	4	5	6	7	0
						H	He
Li	Be	B	C	N	O	F	Ne
Na	Mg	Al	Si	P	S	Cl	Ar
K	Ca	Ga	Ge	As	Se	Br	Kr
Rb	Sr	In	Sn	Sb	Te	I	Xe

Periods and Groups

Two terms that are frequently used in any discussion of the Periodic Tables are 'Periods' and 'Groups'.

Periods are the horizontal rows in the table. In each period there is found to be a change from metallic to non-metallic character. The change can be seen clearly in the period sodium to argon. Note that argon is excluded from the following comparison table as it has no chemical properties.

Na	Mg	Al	Si	P	S	Cl	Ar
	Metallic			*Non-metallic*			
	Oxides are basic			Oxides are acidic			
	Form solid ionic chlorides with high boiling points			Form low boiling point non-ionic chlorides			
NaCl	$MgCl_2$	$AlCl_3$	$SiCl_4$	PCl_3	SCl_2	ClCl	
	Form positive ions in most compounds			Form negative ions in compounds with metals			
(Na^+Cl^-)	$(Mg^{2+}2Cl^-)$	$(Al^{3+}3Cl^-)$	$(2Mg^{2+}Si^{4-})$	$(3Ca^{2+}2P^{3-})$	$(2Na^+S^{2-})$	(Na^+Cl^-)	

The division between metallic and non-metallic character is not as sharp as the table might appear to suggest. Some of the chemical properties of aluminium put this element on the borderline between metallic and non-metallic.

Look at the table and note any patterns that you can see in:

the formulae of the chlorides
the sign of the charge carried by ions of the elements
the number of charges carried by ions of the elements

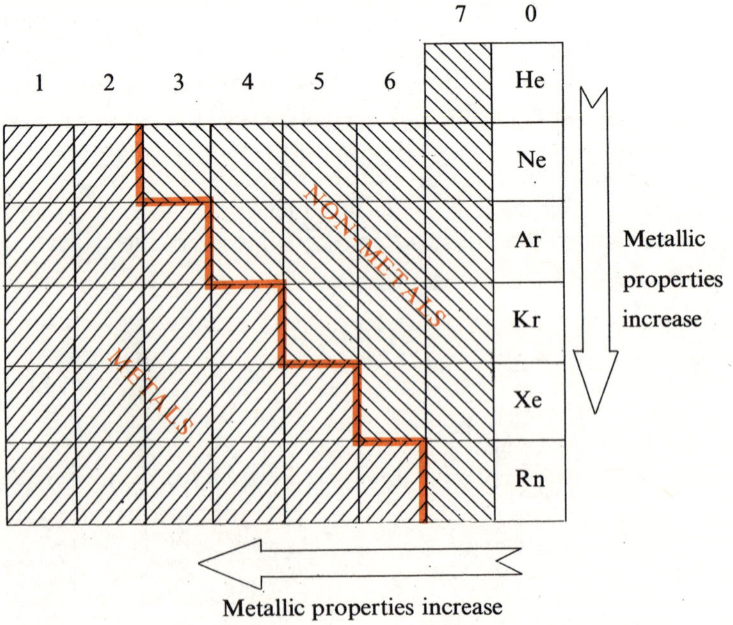

Groups are the vertical numbered columns in the Periodic Table, the division in the present table being confined to the normal elements. Similarities and trends are found within each group. One clear trend is the increase in metallic character on going down a group.

Metals and non-metals

The red **diagonal** or **stair-case** which has been drawn through the Periodic Table is an attempt to separate elements which show metallic properties from those which show non-metallic properties. The properties of elements to the left and to the right of the diagonal are summarized in Table 7.1.

Table 7.1*

Elements to the LEFT of the diagonal	*Elements to the RIGHT of the diagonal*
Metallic	Non-metallic
Solids	Solids when near the diagonal, liquids or gases when far from diagonal
Relatively high boiling points	Relatively low boiling points (is this true for B, C and Si?)
Good conductors of heat and electricity	Usually poor conductors of heat and electricity, but some elements near the diagonal have two forms (polymorphs or allotropes), one of which can show metallic properties
Oxides and hydroxides are frequently soluble in water to give alkaline solutions	Oxides frequently dissolve in water to give acidic solutions
All form an oxide that can act as a base. Oxides combine with acids to give a salt where the metal is present as a positive ion, e.g. MgO forms ($Mg^{2+}SO_4^{2-}$)	Oxides combine with bases to form salts where the element and oxygen are in the negative ion, e.g. SO_3 forms ($Mg^{2+}SO_4^{2-}$)
Compounds formed with hydrogen are solids	Compounds formed with hydrogen are gases or low boiling point liquids, e.g. HCl, H_2S, H_2O, CH_4
Compounds with hydrogen frequently contain ions, e.g. (Na^+H^-)	Compounds formed with hydrogen do not contain ions. Ions are formed if the compound with hydrogen can dissolve in water, e.g. aqueous solutions of NH_3, HCl, HBr, and HI do contain ions

*Table 7.1 continues on page 58.

Table 7.1—*continued*

Elements to the LEFT of the diagonal	*Elements to the RIGHT of the diagonal*
Do not combine with other elements to the left of the diagonal	Combine with other elements to the right of the diagonal. These compounds frequently have low boiling points, e.g. HCl, SO_2, CO_2. When these compounds are in the liquid state, they are poor conductors of electricity as they do not contain ions. If the compound dissolves in water, there is reaction with the formation of new substances which are ionic, e.g. SO_3 forms H^+ and SO_4^{2-} ions in aqueous solution
Elements to the left of the diagonal combine with elements to the right of the diagonal to form compounds with high boiling points, e.g. NaCl, MgO, Na_2SO_4. The molten compounds frequently conduct electricity, as do the aqueous solutions if the compound is soluble. These compounds are ionic (see Chapter 9), the metallic part being the the positive ion and the non-metallic part the negative ion	

Summary

Some of the periodic features found in chemical properties are summarized in the tables which follow.

						H	He
Li	**Be**	B	C	N	O	F	Ne
Na	**Mg**	**Al**	Si	P	S	Cl	Ar
K	**Ca**	**Ga**	Ge	As	Se	Br	Kr
Rb	**Sr**	**In**	**Sn**	Sb	Te	I	Xe

Metallic elements
Combine with non-metallic elements only.
Usually form positive ions in compounds.
Oxides are basic.
Chlorides are solids, usually composed of ions.

						H	He
Li	Be	B	C	N	O	F	Ne
Na	**Mg**	Al	Si	P	S	Cl	Ar
K	**Ca**	Ga	Ge	As	Se	Br	Kr
Rb	**Sr**	In	Sn	Sb	Te	I	Xe

Elements react with cold water to produce hydrogen.
Hydroxides form alkaline aqueous solutions.
Chlorides have boiling points above 1000 °C and contain ions.

						H	He
Li	Be	**B**	**C**	**N**	**O**	**F**	Ne
Na	Mg	Al	**Si**	**P**	**S**	**Cl**	Ar
K	Ca	Ga	Ge	**As**	**Se**	**Br**	Kr
Rb	Sr	In	Sn	Sb	**Te**	**I**	Xe

Non-metallic elements.
Combine with both metals and other non-metals.
Oxides are frequently acidic.
Chlorides are usually liquids or gases.

8 Atomic Structure

One of the most striking patterns in chemistry is the periodic repetition of physical and chemical properties which is seen when the elements are arranged in atomic number order. Every element retains its individuality while at the same time being closely linked in the Periodic Table to its vertical and horizontal neighbours. The pattern of the Periodic Table stands out too clearly to be simply the result of chance. The reason for this pattern is based on the structure of the atoms of which each element is composed. Atoms possess a structure, changing from element to element, in which a repeating feature is recurring at periodic intervals as the atomic number increases. This repeating feature involves the electrons which are found within the atoms of all elements.

Electrons

Gases are very poor conductors of electricity when at atmospheric pressure. If the pressure is reduced to far below atmospheric, gases begin to conduct electricity more readily. This conductivity of gases at low pressure is made use of in various forms of lighting—fluorescent tubes, mercury and sodium lamps and neon signs. The electrical conductivity of gases is due in part to the rapid movement through the gas of **negatively charged** units called **electrons**. The same negatively charged units are involved whatever the nature of the gas used. All gaseous elements and compounds can be broken down during the process of conduction into identical electrons. **This shows that the atoms of all elements contain electrons.** The pattern of the arrangement of these electrons within atoms of different elements produces the pattern of the Periodic Table.

Radioactivity

Early chemists spent much of their time attempting to turn readily available elements and compounds into precious metals such as gold. All these attempts were unsuccessful and it became accepted that the conversion of one element into another was impossible. At the end of the nineteenth century, however, it was found that atoms of certain elements were changing spontaneously into atoms of other elements in the phenomenon called **radioactive decay**. An example of this type of change is:

Thorium atom → Radon atom + High velocity helium atom with double positive charge (an α-particle)

This change of one element into another is possible because atoms of all elements have very similar structures. The similarity in structure is much closer than the possession of one common unit, the electron.

The nucleus

Atoms are uncharged particles so they must contain positively charged units within their structure to balance the effect of the negatively charged units (electrons). Information about the arrangement of the positive and negative units can be obtained by probing inside atoms with the high velocity α-particles produced during radioactive decay. A thin sheet of a solid element placed in the path of a beam of high velocity α-particles is found to have almost no effect on the beam. Most of the α-particles pass straight through the sheet. Closer investigation shows, however, that a few α-particles are deflected from their

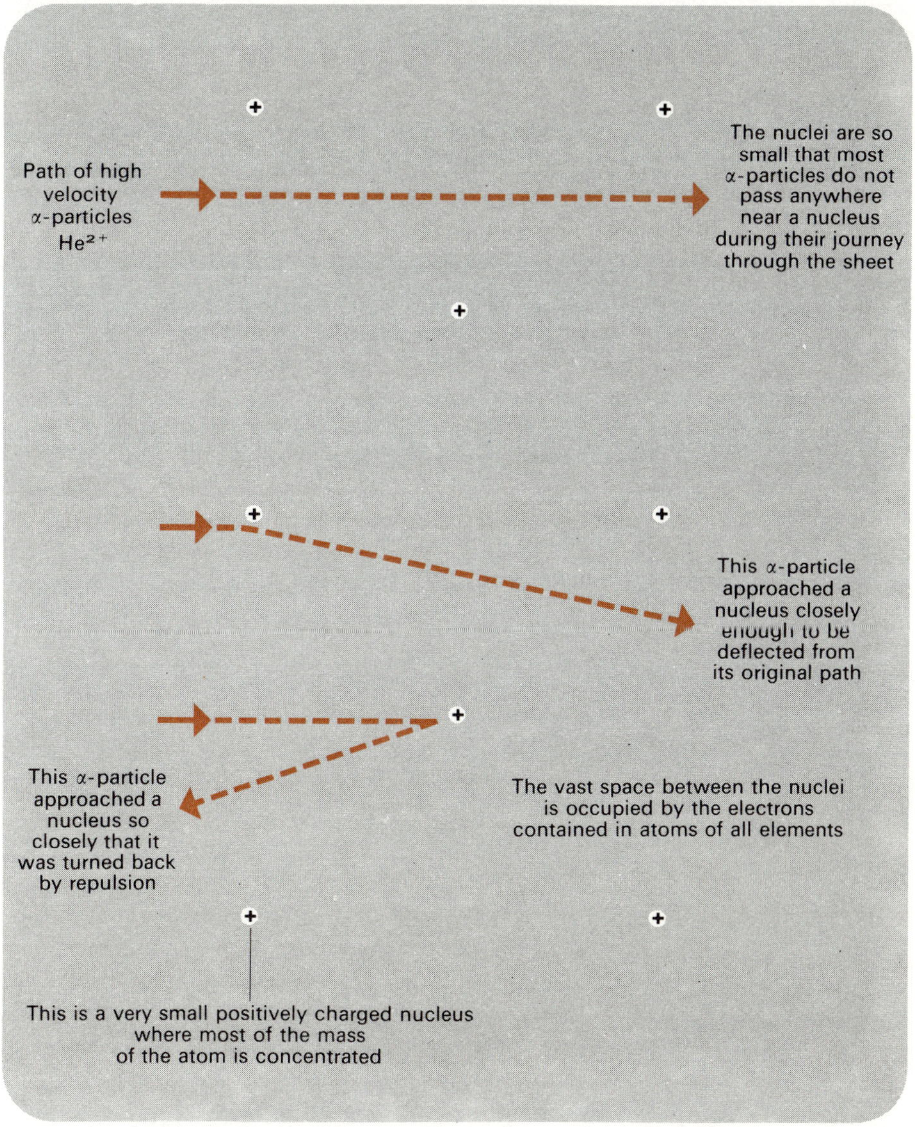

Figure 8.1 Electrons and nuclei in atoms composing a thin sheet of element. The diameter of the nucleus is about 10^{-15} m. The distance between nuclei in a solid is about 10^{-10} m. If a portion of a solid could be magnified so that the nuclei became the size of footballs, the distance between the nuclei would be about 20 miles

straight course as they pass through the sheet, while a very small fraction return in almost the opposite direction rather than going through the sheet. The unexpected rebounding of a small fraction of α-particles made Lord Rutherford, an early investigator, say, 'It is as though a 15-inch shell has been fired at a piece of tissue paper and has come straight back.'

The atoms composing the sheet behave as though their structure is almost completely 'tissue paper', offering negligible resistance to the beam of α-particles. Embedded in this 'tissue paper' are minute portions which exert a massive influence on those α-particles which on rare occasions approach too closely. The atomic model which accounts for these observations is shown in Figure 8.1, page 61.

Atomic number

It is possible to find the value of the positive charge held by the nuclei of atoms—one method for doing this involves measuring the extent to which α-particles are deflected by nuclei. If a scale of charge is used in which the electron has charge −1 unit, **the magnitude of the positive charge on the nucleus is called the atomic number of the element.** The positive charge on the nuclei of atoms of the normal elements is shown in the Periodic Table below.

1	2	3	4	5	6	7	0
						1 H	2 He
3 Li	4 Be	5 B	6 C	7 N	8 O	9 F	10 Ne
11 Na	12 Mg	13 Al	14 Si	15 P	16 S	17 Cl	18 Ar
19 K	20 Ca	31 Ga	32 Ge	33 As	34 Se	35 Br	36 Kr
37 Rb	38 Sr	49 In	50 Sn	51 Sb	52 Te	53 I	54 Xe

Atoms are electrically neutral, so **the atomic number is also the number of electrons which fill the relatively vast space around the nucelus of each atom.** It is somewhere within this vast space that an atom can come into contact with other atoms. This suggests that the physical and chemical properties of an element will be determined in some way by the number of electrons contained in its atoms. The numbers of electrons for atoms of the alkali metals (sodium group) and the halogens (chlorine group) are shown below.

Alkali metals	Li	3	Na	11	K	19	Rb	37
Halogens	F	9	Cl	17	Br	35	I	53

Why should very similar properties be shown by elements containing 3, 11, 19 and 37 electrons in their atoms? These numbers of electrons are all odd, but beyond this there appears to be very little connection between them. There also seems to be very little connection between the numbers of electrons which give the halogens their similar properties. Looking at the rest of the elements in the Periodic Table, no obvious connection can be found between the total numbers of electrons and the properties of elements. This is because the properties of elements are decided not by the total number of electrons in their atoms, but by the small number of **outer electrons** (*see* page 66).

Electrons, protons and neutrons

The nucleus is believed to have a complex structure in which a large number of structural units is held together by various forces of interaction. A simple model of the nucleus which is sufficient for chemical discussion pictures the nucleus as built up only from protons and neutrons. The relative charges and masses of the three structural units in which a chemist is interested are given in Table 8.1.

Table 8.1
Relative Charges and Masses of Electrons, Protons and Neutrons

		Charge on scale, electron charge = −1 unit	*Relative mass on scale, C = 12*
Electron	e	−1	very small
Proton	p	+1	1
Neutron	n	0	1

The numbers of electrons, protons and neutrons in an atom of an element can be calculated if the atomic number and relative atomic mass of the element are known.

Number of electrons	=	Number of protons	=	Atomic number of element
Number of protons	+	Number of neutrons	=	Relative atomic mass of element

The numbers of electrons, protons and neutrons in atoms of hydrogen, fluorine and aluminium are shown in Table 8.2 on page 64.

Isotopes

The atomic number of chlorine is 17, so the nucleus contains 17 protons while 17 electrons surround the nucleus. The chemical properties of chlorine are a result of each atom having 17 electrons in the space around the nucleus—it is in this space that chlorine atoms come into contact with further chlorine atoms or the atoms of other elements. No other element has properties identical to chlorine because the atoms of no other element contain 17 electrons in the space around the nucleus.

Table 8.2

Hydrogen atom *relative atomic mass*=1	*Fluorine atom* *relative atomic mass*= 19	*Aluminium atom* *relative atomic mass*=27
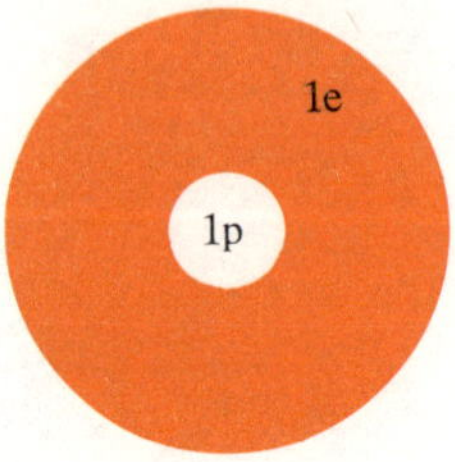	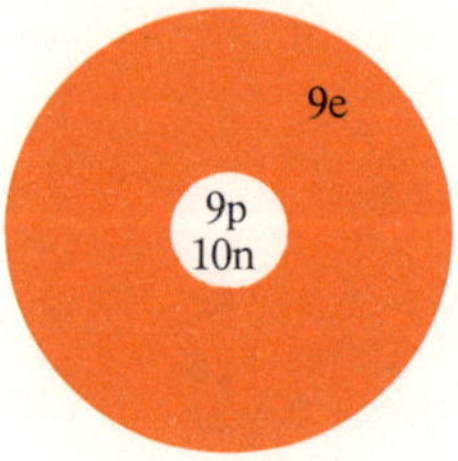	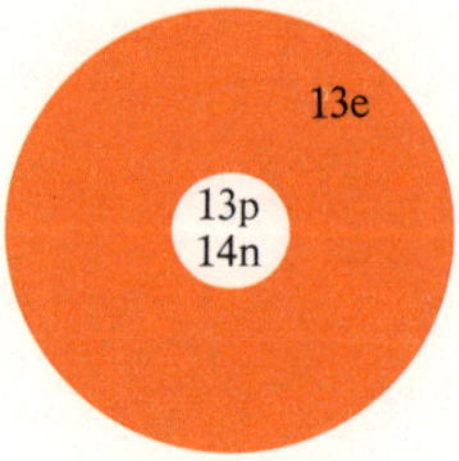
The atomic number is 1 so the atom contains 1 e and 1 p. No neutrons are present as the relative atomic mass is 1	The atomic number is 9 so the atom contains 9 e and 9 p. The nucleus contains 10 n to make the relative atomic mass 19	The atomic number is 13 so the atom contains 13 e and 13 p. The nucleus contains 14 n to make the relative atomic mass 27
The diagrams show only the numbers of each structural unit and whether they are in the nucleus or outside the nucleus. The diagrams do not show how the protons and neutrons are arranged within the nucleus, or how the electrons occupy the vast volume around the nucleus		

The number of neutrons in the nucleus will not influence the properties of chlorine. There are found to be two different types of chlorine atoms, called **isotopes**, one with 18 neutrons in the nucleus and the other with 20 neutrons in the nucleus.

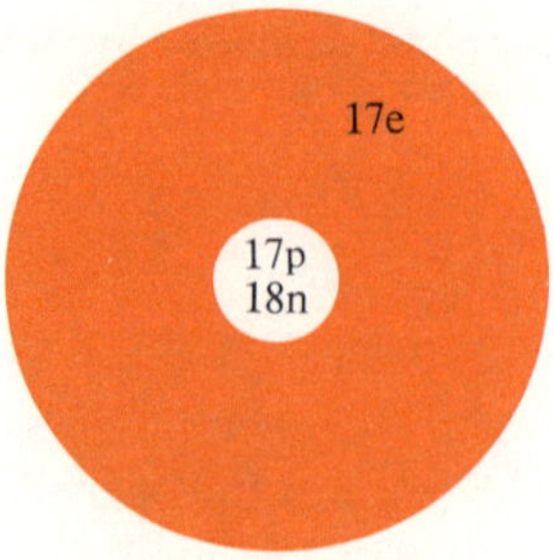

The relative atomic mass of this isotope is 35. This isotope is ^{35}Cl.

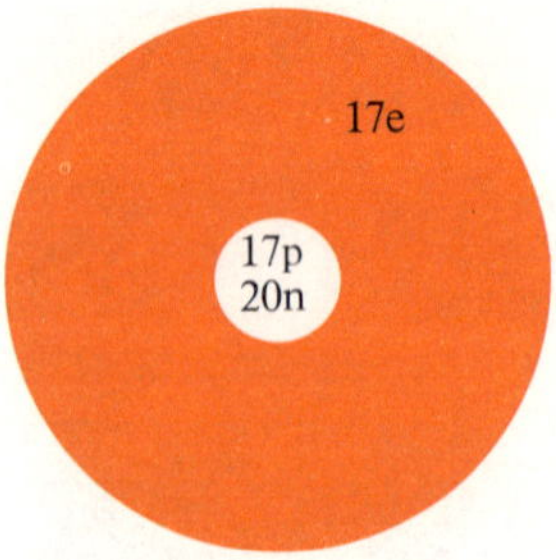

The relative atomic mass of this isotope is 37. This isotope is ^{37}Cl.

About 75 per cent of all chlorine atoms are the isotope ^{35}Cl, while the remaining 25 per cent are ^{37}Cl. The relative atomic mass of chlorine is given as 35.5 in Table 4.6 (*see* page 27) because this is the average relative atomic mass of chlorine atoms.

Many elements are composed of isotopes. Tin is an example of an element with a large number of isotopes. The atomic number of tin is 50, so the nucleus contains 50 protons while 50 electrons surround the nucleus. There are ten isotopes of tin in which the numbers of neutrons in the nucleus are 62, 64, 65, 66, 67, 68, 69, 70, 72, and 74. The chemical properties of these isotopes are identical because they all contain 50 electrons.

The arrangement of electrons in atoms

It has been seen in this chapter that an atom is composed of a very small nucleus with a vast surrounding volume containing electrons. The negatively charged electrons are held strongly within the volume of the atom by the attraction of the positively charged nucleus—considerable energy is required to pull electrons away from atoms against the attraction of the nucleus.

A potassium atom contains 19 electrons. The relative values of the energies needed to remove these electrons, one by one, from gaseous potassium atoms are shown in Figure 8.2 (these energies are called **ionization energies**). The area of each coloured square in Figure 8.2 is proportional to one of the ionization energies—the energy needed to remove the first electron is represented by the very small square at the far right.

$K(g)$	$\rightarrow$	$K^+(g)$	+	e	First ionization energy = 420 kJ/mole
Gaseous potassium atom		Gaseous potassium ion		First electron removed	

Figure 8.2 Ionization energies of potassium. The value of each ionization energy is proportional to the area of the tinted square

It appears from Figure 8.2 that the 19 electrons in a potassium atom can be divided into four groups, the electrons within each group requiring comparable energies for their removal. These groups of electrons are:

2	8	8	1
electrons require vast energies for their removal	electrons require high and comparable energies for their removal	electrons require comparable energies of a rather lower value for removal	electron requires low energy for removal

It is for this reason that the **electronic configuration** of potassium is written 2. 8. 8. 1.

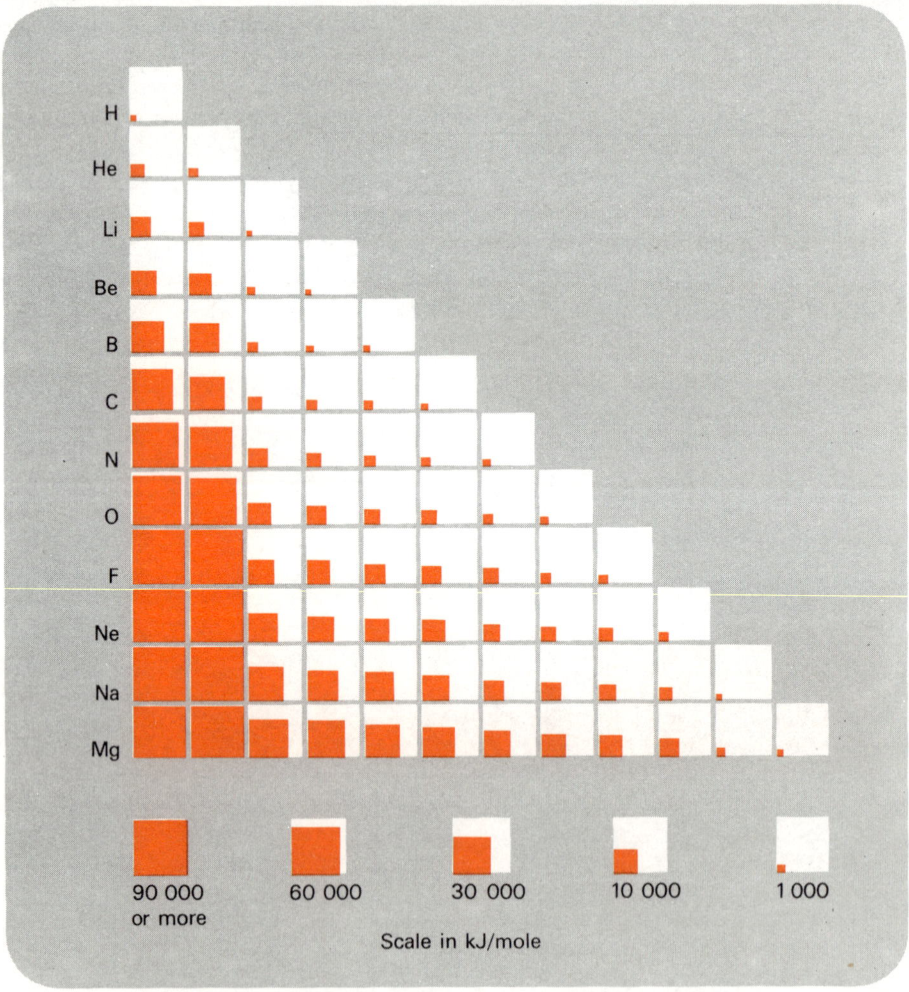

Figure 8.3 Ionization energies of elements with atomic numbers 1–12. The value of each ionization energy is proportional to the area of the tinted square

Table 8.3
Electronic Configuration of Elements with Atomic Number 1–20

Atomic number	*Element*	*Electronic configuration*
1	H	**1**
2	He	**2**
3	Li	2. **1**
4	Be	2. **2**
5	B	2. **3**
6	C	2. **4**
7	N	2. **5**
8	O	2. **6**
9	F	2. **7**
10	Ne	2. **8**
11	Na	2. 8. **1**
12	Mg	2. 8. **2**
13	Al	2. 8. **3**
14	Si	2. 8. **4**
15	P	2. 8. **5**
16	S	2. 8. **6**
17	Cl	2. 8. **7**
18	Ar	2. 8. **8**
19	K	2. 8. 8. **1**
20	Ca	2. 8. 8. **2**

The energies needed to remove all the electrons, one by one, from atoms of the elements with atomic number 1–12 are shown in Figure 8.3. The area of each coloured square is proportional to the ionization energy.

The ionization energies of each element in Figure 8.3 show abrupt changes at certain points similar to those seen earlier for potassium. The positions of these abrupt changes can be used to decide the electronic configurations of atoms of these elements. The electronic configurations are given in Table 8.3 which includes also elements of atomic number 13–20.

The electrons which can be removed relatively easily from atoms are shown in bold type in Table 8.3. These electrons are called the **outer electrons**. The term outer electrons is used to indicate that certain electrons can be removed from atoms far more easily than any others. The number of outer electrons in atoms of the normal elements is shown in the Periodic Table on the next page.

There is a simple relation between the number of outer electrons and the position of the element in the Periodic Table.

The number of outer electrons in the atoms of a normal element is the same as

1	2	3	4	5	6	7	0
						1 H	2 He
1 Li	2 Be	3 B	4 C	5 N	6 O	7 F	8 Ne
1 Na	2 Mg	3 Al	4 Si	5 P	6 S	7 Cl	8 Ar
1 K	2 Ca	3 Ga	4 Ge	5 As	6 Se	7 Br	8 Kr
1 Rb	2 Sr	3 In	4 Sn	5 Sb	6 Te	7 I	8 Xe

the element's group number in the periodic table. (This relation does not apply to hydrogen and the noble gases.)

The pattern of the Periodic Table is a result of the pattern in the number of outer electrons. The elements of the sodium group (the alkali metals Li, Na, K and Rb) have similar properties because their atoms all have one outer electron. The similar properties found for all the members of the chlorine group (the halogens) arise from the identical electronic configuration of seven outer electrons. The noble gases, another group of elements with very similar properties, all have the highest possible number of outer electrons—two in the case of helium and eight for the other noble gases.

Summary

1. Atoms contain the following structural units.

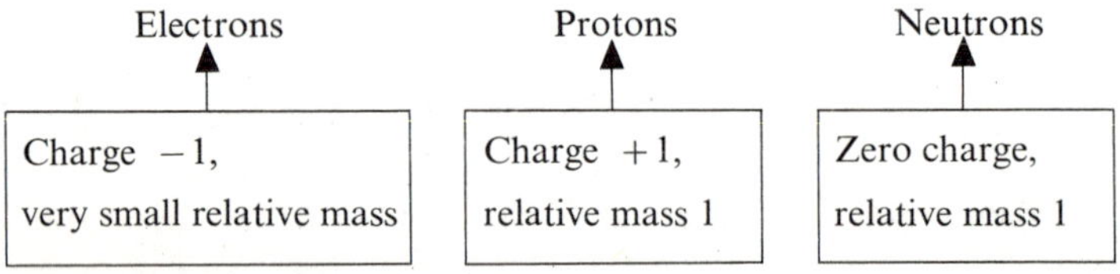

2. The positions and numbers of these structural units in an atom is shown below.

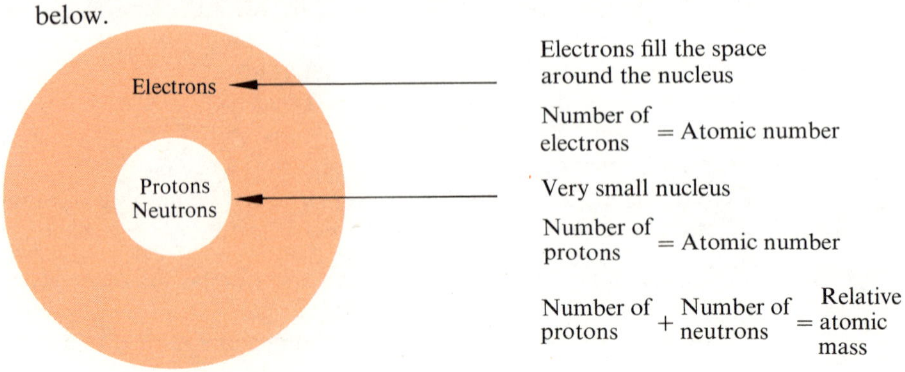

3. Isotopes of an element have identical chemical properties but different relative atomic masses. Their atoms contain the same numbers of electrons and protons, but different numbers of neutrons.

4. The pattern of the Periodic Table is the result of the pattern in the numbers of outer electrons in atoms of different elements. The elements of Group 1 (the sodium group) have similar properties because their atoms all have one outer electron.

Li	2. **1.**
Na	2. 8. **1.**
K	2. 8. 8. **1.**

The elements of Group 7 (the halogens) have similar properties because their atoms all have seven outer electrons.

F	2. **7.**
Cl	2. 8. **7.**

9 Structure and Bonding
1: Ionic compounds

The formulae of some chlorides

Chlorine is a very reactive element and combines with most other elements to form compounds called chlorides. Chlorine is a non-metallic element, and like other non-metals it combines both with metals and with non-metals (*see* Table 7.1, page 57). The formulae of the compounds formed between chlorine and the elements of the period sodium to argon are summarized in Figure 9.1. Note that argon does not appear in Figure 9.1 as no chloride of this element has yet been made.

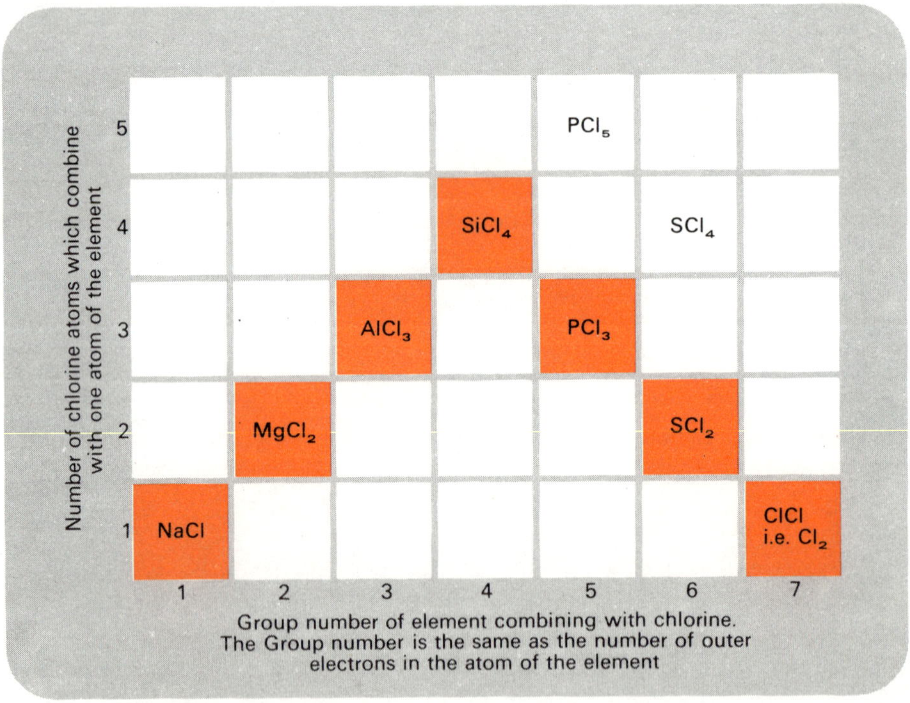

Figure 9.1 Formulae of the chlorides formed by the elements in the period sodium to argon

A definite pattern is apparent in the formulae of these chlorides. The number of atoms of chlorine combining with one atom of the element is the same as the atom's number of outer electrons for Na, Mg, Al, and Si. The pattern continues with P which can form a chloride PCl_5. Phosphorus forms a second chloride PCl_3 in which the number of atoms of chlorine combining with one atom of phosphorus is 3. This is the same as the number of electrons which an atom of phosphorus would need to have eight electrons in its outer shell (the electronic configuration of the noble gas argon). This second pattern continues with SCl_2 and ClCl (i.e. Cl_2).

The physical properties of some chlorides

The chlorides whose structure and bonding will be discussed first are those marked with a coloured background in Figure 9.1. The properties of these chlorides are summarized in Figure 9.2.

Number of chlorine atoms combining with one atom of element equals number of outer electrons in atom of the element (brace spanning the first four boxes)	Crystalline white **solid** High melting and boiling points	
	$MgCl_2$ Crystalline white **solid** High melting and boiling points	
	$AlCl_3$ White **solid** Vaporizes readily on heating	
	$SiCl_4$ Colourless **liquid** Low boiling point	
	PCl_3 Colourless **liquid** Low boiling point	Number of chlorine atoms combining with one atom of element equals 8 minus number of outer electrons in atom of the element (brace spanning the last four boxes)
	SCl_2 Red **liquid** Low boiling point	
	ClCl *(Cl_2)* Green **gas**	

Figure 9.2 Properties of some chlorides formed by the elements in the period sodium to argon

There is a great difference in physical properties between the chlorides of the first two elements (NaCl, $MgCl_2$) and the chlorides of the final four elements ($SiCl_4$, PCl_3, SCl_2, Cl_2). The chloride of aluminium, $AlCl_3$, has properties between the two extremes. Certain physical properties of a compound give information about structure and bonding in the compound. Among the most important of these properties are:

- electrical conductivity of molten compound
- melting point and boiling point of compound
- energy needed to vaporize one mole of molten compound (called the molar heat of vaporization)

The relative values of each of these properties for the elements in the period sodium to argon are summarized in Figures 9.3–9.6. Note how in each case the properties of NaCl and $MgCl_2$ are very different from the properties of the chlorides of the later elements in the period.

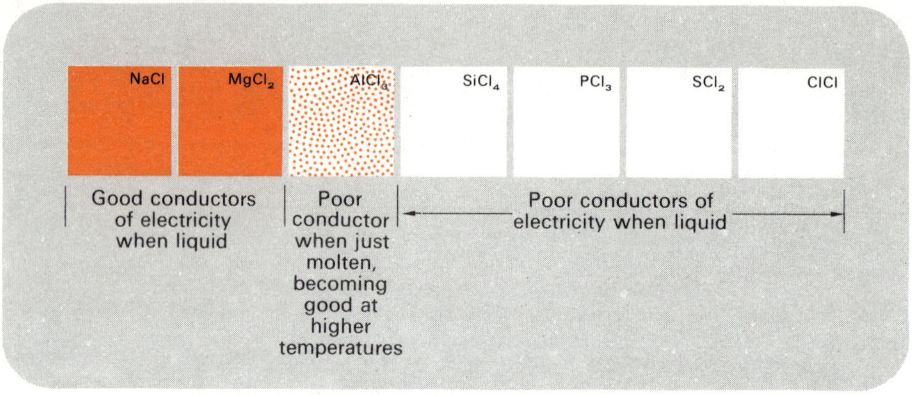

Figure 9.3 Electrical conductivity of molten chlorides

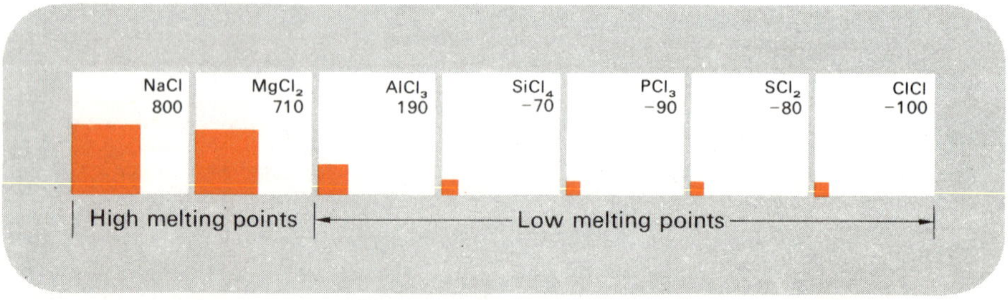

Figure 9.4 Melting points of chlorides in °C

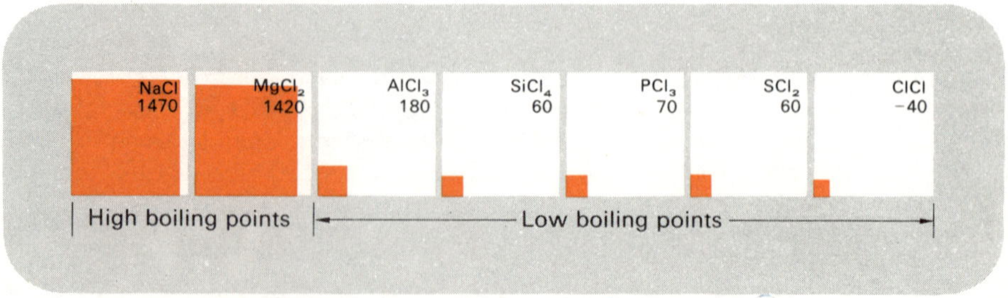

Figure 9.5 Boiling points of chlorides in °C

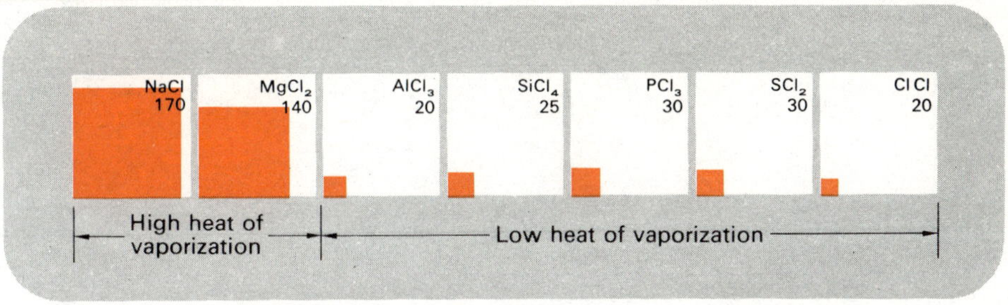

Figure 9.6 Molar heat of vaporization in kJ/mole

The chlorides of the elements in the period sodium to argon can be divided into two types, the division being decided by the physical properties of the compounds. These two types of chloride will be called **ionic** and **non-ionic**.

Table 9.1

Ionic chlorides	*Non-ionic chlorides*
Crystalline solids at room temperature, never liquid or gas	Frequently liquid or gas at room temperature
Conduct electricity when molten	Do not conduct electricity when molten
High melting point	Low melting point
High boiling point	Low boiling point
High molar heat of vaporization	Low molar heat of vaporization
$NaCl$ and $MgCl_2$ have all these properties	$SiCl_4$, PCl_3, SCl_2, and Cl_2 have all these properties

The ionic chlorides NaCl and $MgCl_2$

Sodium chloride when solid or molten is composed of ions Na^+ and Cl^-. The relation between the electronic configurations of these ions and neutral atoms is shown in Figure 9.7 on page 74.

A sodium ion has the electronic configuration of the noble gas neon; a chloride ion has the electronic configuration of the noble gas argon (*see* page 67). The formation of (Na^+Cl^-) from atoms of sodium and chlorine could be represented in the following way.

$Na\bullet$	$+$	$:\ddot{\underset{\cdot\cdot}{Cl}}\bullet$	$\rightarrow$	Na^+	$+$	$:\ddot{\underset{\cdot\cdot}{Cl}}:^-$
Sodium atom with one electron in excess of the electronic configuration of the noble gas neon		Chlorine atom with one electron less than the electronic configuration of the noble gas argon		Sodium ion with the electronic configuration of neon		Chloride ion with the electronic configuration of argon

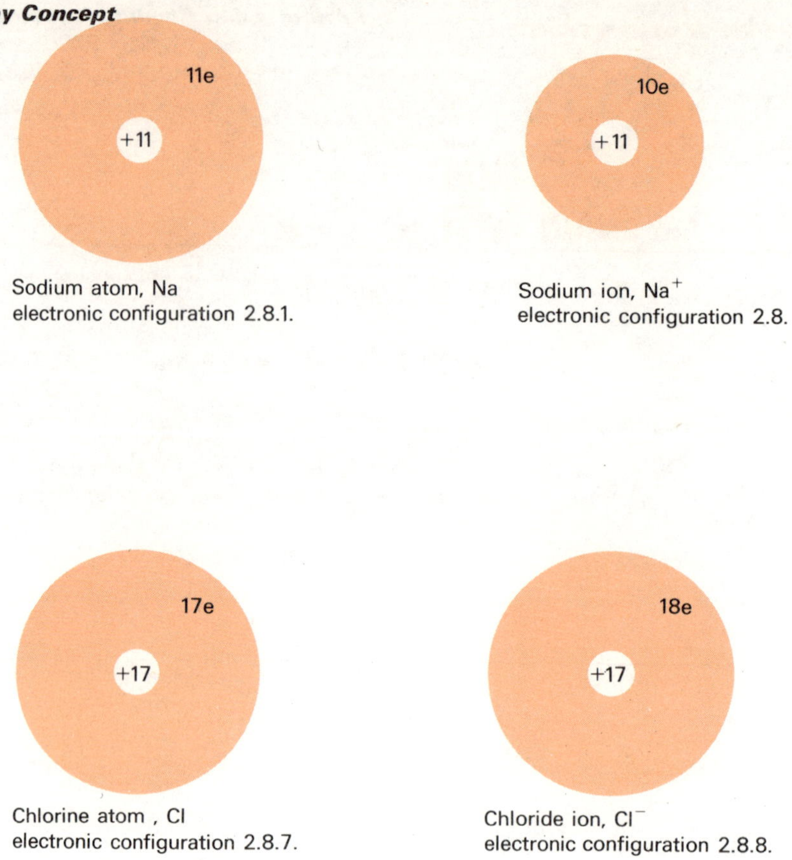

Figure 9.7

This electronic representation of the formation of sodium chloride shows only the **numbers** of outer electrons in the atoms and ions—it is not used to show where the electrons are to be found within the atoms or ions, or how the ions are arranged within a portion of the substance. It should be noted that the electrons have been gathered into pairs in the diagram. The significance of grouping electrons in pairs will be seen when the bonding and shape of non-ionic chlorides is discussed in Chapter 10.

Magnesium chloride when solid or molten is composed of ions Mg^{2+} and Cl^-. The formation of $(Mg^{2+}2Cl^-)$ can be represented in a similar way to that used to represent the formation of (Na^+Cl^-).

$$\mathrm{Mg}{:} \quad + \quad 2\,{:}\ddot{\mathrm{Cl}}{\cdot} \quad \rightarrow \quad \mathrm{Mg}^{2+} \quad + \quad 2\,{:}\ddot{\mathrm{Cl}}{:}^{-}$$

Magnesium atom with two electrons in excess of the electronic configuration of the noble gas neon

Magnesium ion with the electronic configuration of neon

Giant structures of ions

In solid (Na^+Cl^-) and ($Mg^{2+}2Cl^-$) the ions are arranged in a closely packed regular three-dimensional structure called a **giant structure of ions**. The structure of solid sodium chloride is described in Figures 9.8–9.10.

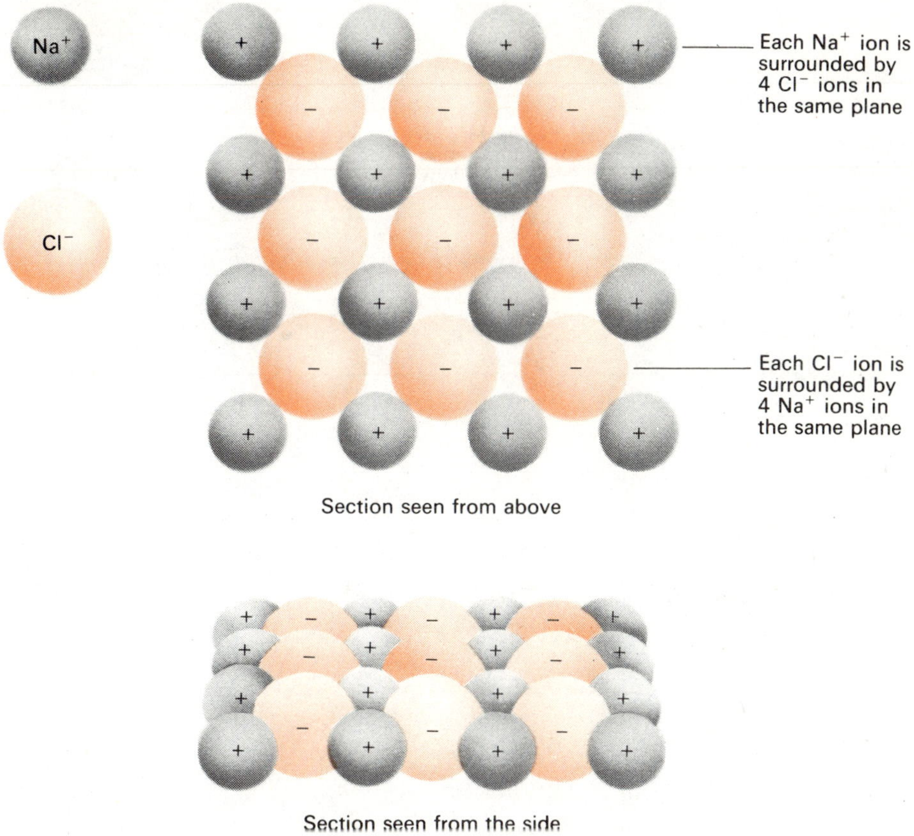

Figure 9.8 Two views of sections through regularly packed ions in solid (Na^+Cl^-)

All ionic compounds have a regular giant structure of ions when solid. Many of the halides formed by the elements of the sodium group (Group 1) have the structure of sodium chloride. The halides of the magnesium group (Group 2), for example ($Mg^{2+}2Cl^-$), have structures more complex than that of (Na^+Cl^-) and no attempt will be made to show them here.

Attraction and repulsion between ions in a giant structure

Owing to the regularity of the arrangement of the ions within a giant structure of ions they are crystalline substances. **Electrostatic attraction** between ions of opposite charge gives rise to:

- high melting point
- high boiling point
- high molar heat of vaporization

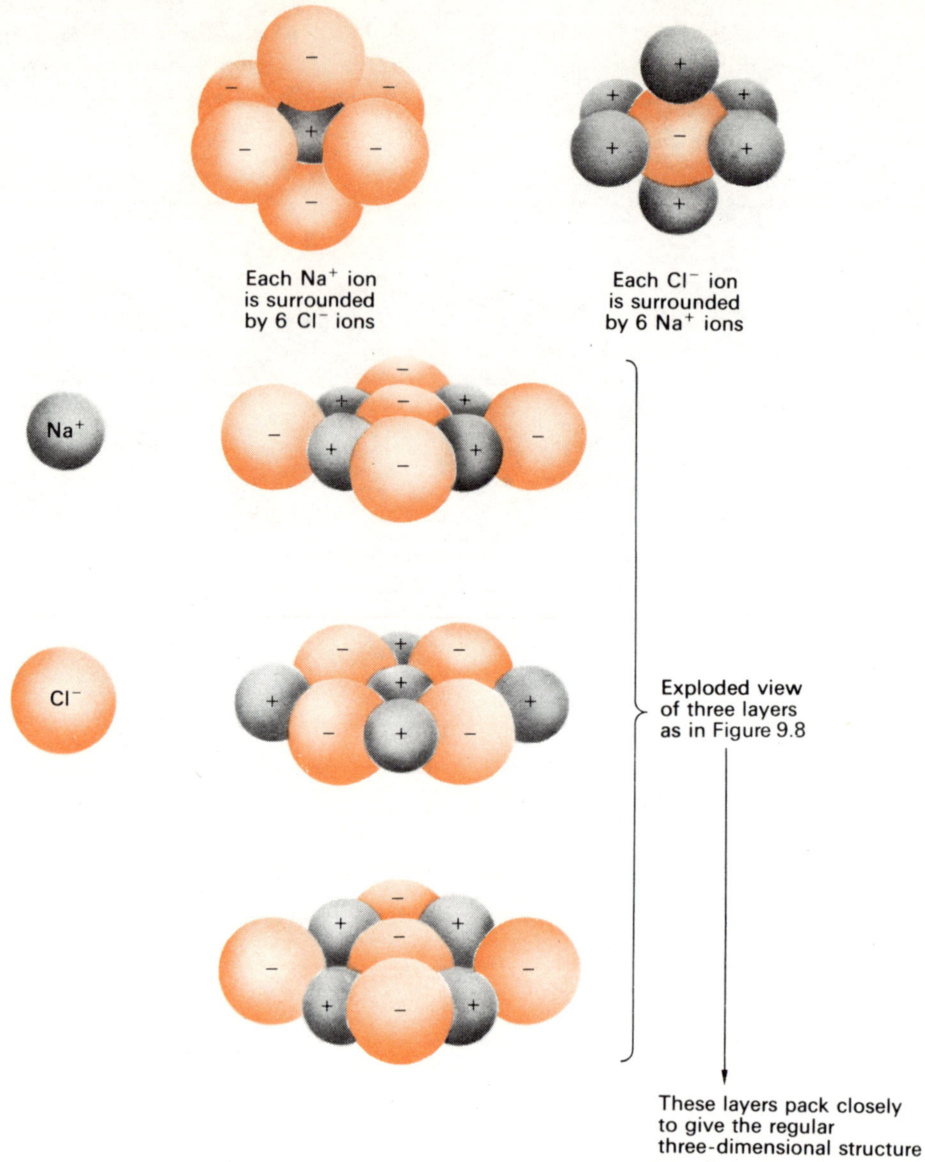

Figure 9.9 The regular three-dimensional packing of ions in solid (Na^+Cl^-)

In sodium chloride (*see* Figure 9.10) the nearest ions to each Na^+ ion are the Cl^- ions with opposite charge; the nearest ions to each Cl^- ion are the Na^+ ions with opposite charge. Attraction between ions of opposite charge holds the structure together.

Each Na^+ ion is surrounded most closely by Cl^- ions, but slightly further away are more Na^+ ions. In a similar way, each Cl^- ion is surrounded most closely by Na^+ ions but slightly further away are more Cl^- ions. There is strong electrostatic repulsion between ions of the same charge within a giant structure. Attraction is greater than repulsion because ions of opposite charge are closer to each other than ions of the same charge.

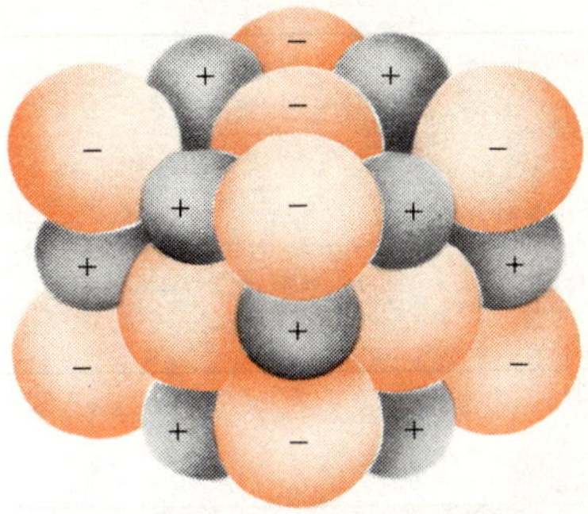

Na⁺ and Cl⁻ ions regularly and closely packed

Each Na^+ ion is surrounded by 6 Cl^- ions

Each Cl^- ion is surrounded by 6 Na^+ ions

Figure 9.10 In solid (Na^+Cl^-), each ion is surrounded most closely by ions of opposite charge

The physical properties of ionic chlorides listed in Table 9.1 on page 73 can be accounted for in the following way (Table 9.2).

Table 9.2

Properties of ionic chlorides	*Reasons for ionic chlorides having these properties*
Crystalline	Due to regular arrangement of ions
Solids	Due to strong electrostatic attraction between ions of opposite charge
Conduct electricity when molten	The giant structure of ions is partially broken down on melting, and each ion can escape sufficiently from the attraction of ions of opposite charge to move towards the electrodes during electrolysis
High melting points, high boiling points, high molar heats of vaporization	The high values for all these quantities indicate how difficult it is to separate ions of opposite charge in a giant structure of ions

Ionic chlorides and the Periodic Table

It is possible to decide whether a chloride has an ionic structure if certain important properties of the chloride are known. These properties of the chloride are:

- electrical conductivity when molten
- melting point
- boiling point
- energy needed to vaporize one mole of molten compound

Information on the relative values of each of these properties is set out in Figures 9.11–9.14 for some of the chlorides formed by the normal elements.

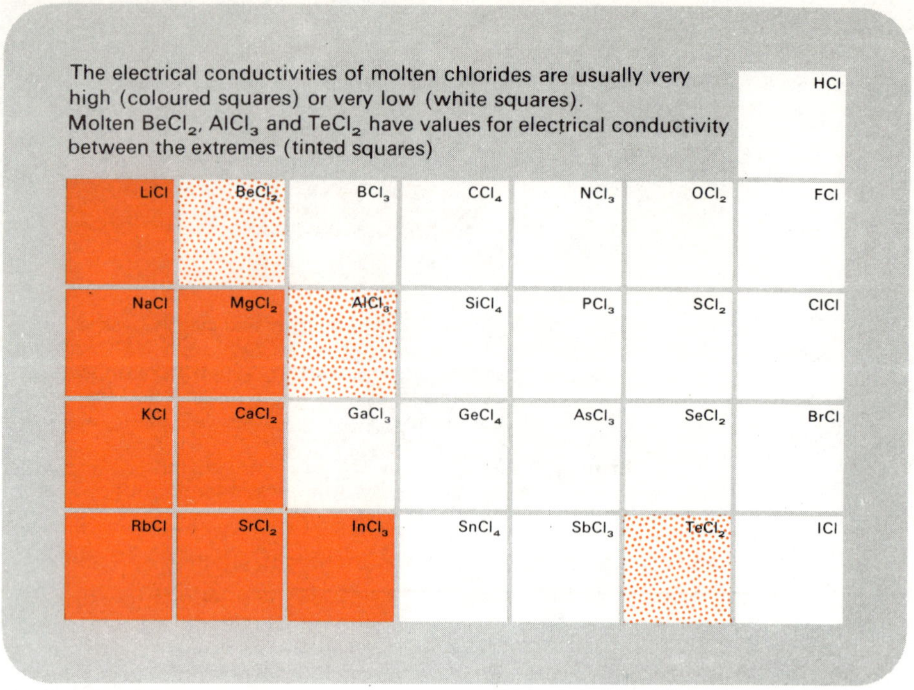

Figure 9.11 Electrical conductivities of molten chlorides

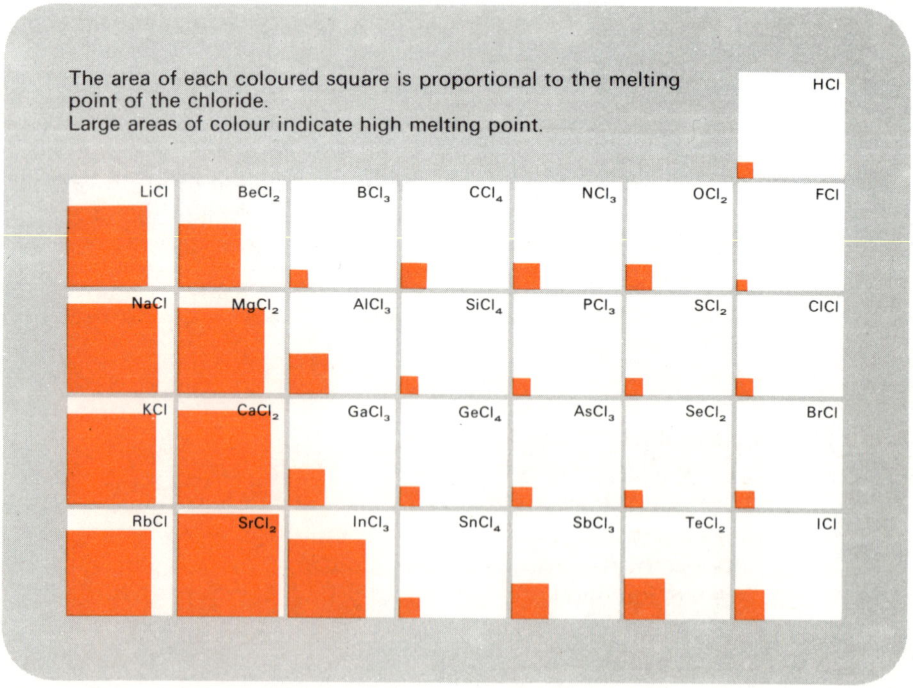

Figure 9.12 The melting points of chlorides.

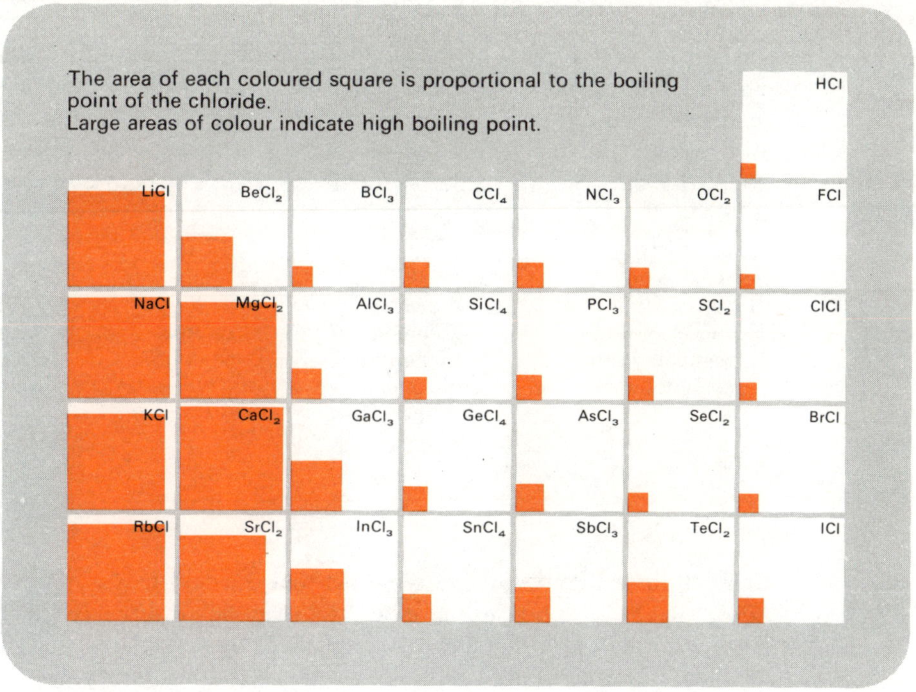

Figure 9.13 The boiling points of chlorides.

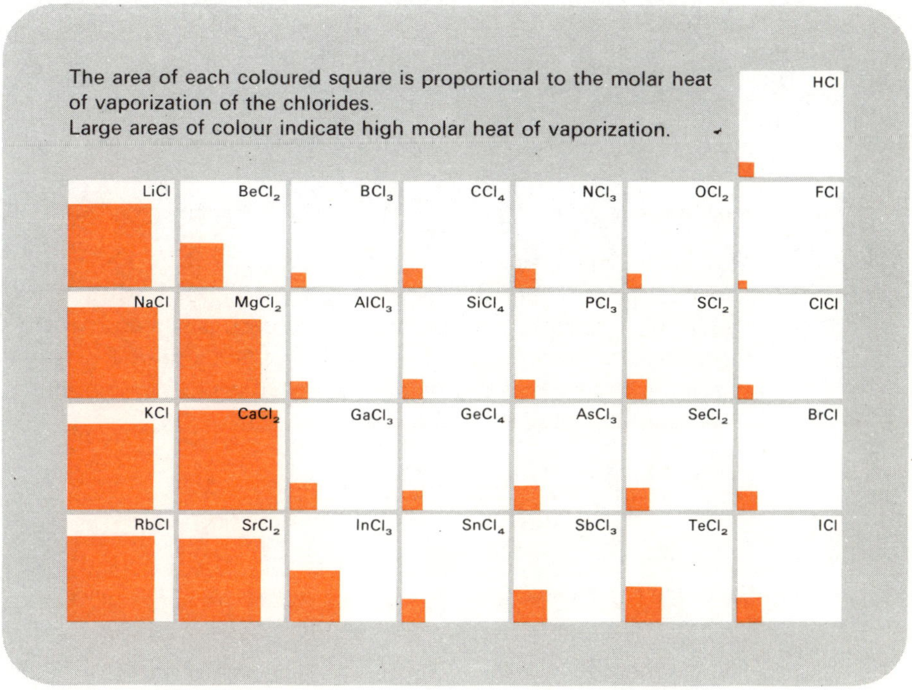

Figure 9.14 The molar heat of vaporization of chlorides.

Chloride of Group 1

Inspection of Figures 9.11–9.14 shows that the chlorides of these elements are all ionic; the general formula is (El^+Cl^-), where El is used as the general symbol for an element. The elements in Group 1 all have one electron in the outer shell, and the formation of their chlorides can be represented in the following way. *See* also page 73 for the electronic structure of (Na^+Cl^-).

El•	+	:C̈l̤•	→	El^+	+	:C̈l̤:$^-$
One electron in excess of the electronic configuration of a noble gas		One electron less than the electronic configuration of the noble gas argon		El^+ ion with the electronic configuration of a noble gas		Chloride ion with the electronic configuration of argon

Chlorides of Group 2

Inspection of Figures 9.11–9.14 shows that the chlorides of these elements are ionic ($BeCl_2$ is a borderline case); the general formula is ($El^{2+}2Cl^-$). The elements in Group 2 all have two electrons in the outer shell, and the formation of their chlorides can be represented in the following way. *See* also page 74 for the electronic structure of ($Mg^{2+}2Cl^-$).

El:	+	2:C̈l̤•	→	El^{2+}	+	2:C̈l̤:$^-$
Two electrons in excess of the electronic configuration of a noble gas				El^{2+} ion with the electronic configuration of a noble gas		

Chlorides of Group 3

All the elements in this group form a chloride $ElCl_3$. The chloride of indium is definitely ionic, ($In^{3+}3Cl^-$), *see* Figures 9.11–9.14. Molten $AlCl_3$ is a poor conductor of electricity at the melting point, but its conductivity increases to a high value as the temperature of the liquid is raised. If $AlCl_3$ is also regarded as an ionic chloride, the formation of the chlorides of aluminium and indium can be represented in the following way.

Ä̇l•	+	3:C̈l̤•	→	Al^{3+}	+	3:C̈l̤:$^-$
Ïn•	+	3:C̈l̤•	→	In^{3+}	+	3:C̈l̤:$^-$
Aluminium and indium atoms each with three electrons in excess of the electronic configuration of a noble gas				Aluminium and indium ions each with the electronic configuration of a noble gas		

Indium forms another chloride, InCl, which is also ionic (In^+Cl^-). This chloride contains the In^+ ion in which the indium atom has lost only one of its three outer electrons. The In^+ ion does not have the electronic configuration of a noble gas.

$$\ddot{\mathrm{In}}\cdot \quad + \quad :\ddot{\underset{..}{\mathrm{Cl}}}\cdot \quad \rightarrow \quad \ddot{\mathrm{In}}^{+} \quad + \quad :\ddot{\underset{..}{\mathrm{Cl}}}:^{-}$$

Chlorides of Group 4

All the elements in this group form a chloride $ElCl_4$. These chlorides are non-ionic (*see* their physical properties in Figures 9.11–9.14). The considerable energy required to remove all four outer electrons from atoms of the Group 4 elements prevents the formation of El^{4+} ions. The energy required to remove outer electrons from an atom decreases on going down a group in the Periodic Table. The Group 4 elements which lose outer electrons most easily are those at the bottom of the group, Sn and Pb. Even with atoms of Sn and Pb, loss of all four outer electrons is not observed in compound formation. Loss of two outer electrons does, however, occur in the formation of the ionic chlorides ($Sn^{2+}2Cl^-$), tin(II) chloride, and ($Pb^{2+}2Cl^-$), lead(II) chloride.

$$\ddot{\mathrm{Sn}}: \quad + \quad 2:\ddot{\underset{..}{\mathrm{Cl}}}\cdot \quad \rightarrow \quad \ddot{\mathrm{Sn}}^{2+} \quad + \quad 2:\ddot{\underset{..}{\mathrm{Cl}}}:^{-}$$

$$\ddot{\mathrm{Pb}}: \quad + \quad 2:\ddot{\underset{..}{\mathrm{Cl}}}\cdot \quad \rightarrow \quad \ddot{\mathrm{Pb}}^{2+} \quad + \quad 2:\ddot{\underset{..}{\mathrm{Cl}}}:^{-}$$

The formation of Sn^{2+} and Pb^{2+} resembles the formation of In^+ in Group 3. In each case the ion has two electrons in excess of the electronic configuration of a noble gas.

Chlorides of Groups 5–7

The physical properties of the chlorides of these elements given in Figures 9.11–9.14 indicate that the compounds are non-ionic. Loss of outer electrons from atoms of these elements is more difficult than from atoms of elements in earlier groups. Atoms of elements in Groups 5–7 would have to lose a large number of outer electrons if the electronic configuration of a noble gas was going to be achieved by ionic compound formation. The difficulty of removing electrons, and the large number of electrons to be removed, both contribute to the non-ionic nature of the chlorides of Groups 5–7.

Bismuth, the element below antimony in Group 5, does form an ionic chloride ($Bi^{3+}3Cl^-$).

$$\ddot{\underset{.}{\mathrm{Bi}}}: \quad + \quad 3:\ddot{\underset{..}{\mathrm{Cl}}}\cdot \quad \rightarrow \quad \ddot{\mathrm{Bi}}^{3+} \quad + \quad 3:\ddot{\underset{..}{\mathrm{Cl}}}:^{-}$$

Note that the Bi^{3+} ion contains two electrons in excess of the electronic configuration of a noble gas. The Bi^{3+} ion resembles the ions Tl^+, Sn^{2+} and Pb^{2+} in electronic configuration.

Ionic fluorides

The elements which form ionic chlorides also form ionic fluorides. Information about the fluorides of the period sodium to argon is given below (Table 9.3).

Table 9.3
Properties of Fluorides

(Na^+F^-)	$(Mg^{2+}2F^-)$	$(Al^{3+}3F^-)$	SiF_4	PF_3	SF_2 SF_4 SF_6	ClF ClF_3
Ionic			**Non-ionic**			
High melting point, high boiling point, conduct electricity when molten			Low melting point, low boiling point, do not conduct electricity when molten			
Solids consist of giant structures of ions						

The formation of the ionic fluorides can be represented by electronic diagrams.

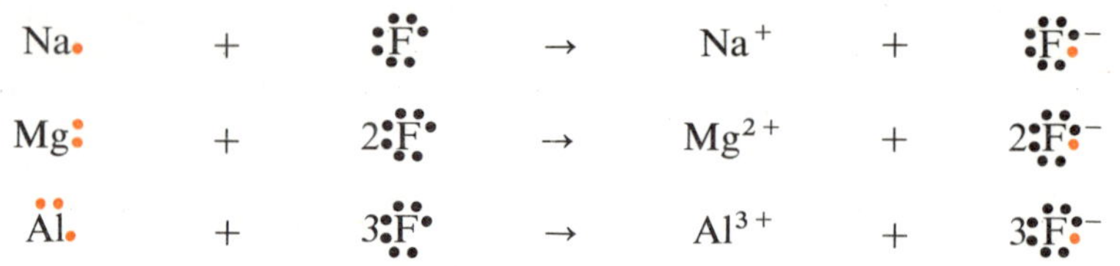

It has been seen earlier in this chapter that $AlCl_3$ is on the borderline between ionic and non-ionic. Fluorides tend to be more ionic than other halides, and this shows in the ionic nature of $(Al^{3+}3F^-)$.

Ionic bromides and iodides

The elements which form ionic chlorides usually form ionic bromides and iodides. Bromides and iodides have slightly less tendency to be ionic, and this shows in the non-ionic nature of $AlBr_3$ and AlI_3. Information about some of these compounds is given in Table 9.4.

Table 9.4
Properties of Bromides and Iodides

(Na^+Br^-) (Na^+I^-)	$(Mg^{2+}2Br^-)$ $(Mg^{2+}2I^-)$	$AlBr_3$ AlI_3	$SiBr_4$ SiI_4	PBr_3 PI_3	SBr_2 SI_2	$BrCl$ $BrCl_3$ ICl ICl_3
Ionic		**Non-ionic**				
High melting point, high boiling point, conduct electricity when molten		Low melting point, low boiling point, do not conduct electricity when molten				
Solids consist of giant structures of ions						

The formation of the ionic bromides and iodides can be represented by the following electronic diagrams.

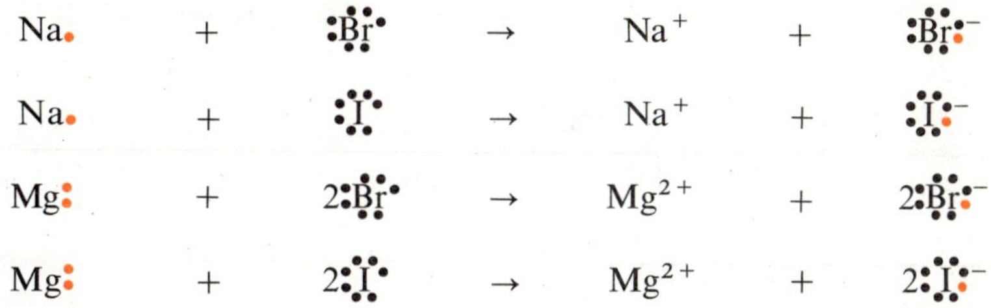

Ionic oxides

Ionic oxides contain the O^{2-} ion which is formed when an oxygen atom accepts two electrons from a metallic element, usually an element of Group 1 or 2.

The physical properties of the oxides of the elements in the period sodium to argon are summarized in Table 9.5.

Table 9.5
Properties of Oxides

$(2Na^+O^{2-})$	$(Mg^{2+}O^{2-})$	Al_2O_3	SiO_2	P_4O_{10} P_4O_6	SO_2 gas	Cl_2O gas
Solids with high melting and boiling points		Solids with high melting and boiling points		Solids with low melting and boiling points	SO_3 Solid with low melting and boiling point	ClO_2 gas; Cl_2O_7 liquid
Ionic Conduct electricity when molten		**Non-ionic** Do not conduct electricity when molten				

Note that Al_2O_3 and SiO_2 have some of the properties of an ionic compound. They do not, however, conduct electricity when molten.

The formation of the ionic oxides $(2Na^+O^{2-})$ and $(Mg^{2+}O^{2-})$ can be represented by the following electronic diagrams.

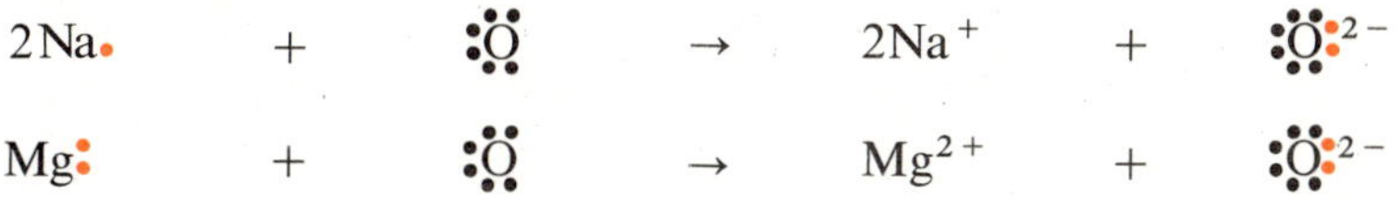

The formula of ionic compounds

Ionic compounds are formed when a metallic element (one to the left of the diagonal in the periodic table of normal elements) combines with

either a single non-metallic element	$(2Na^+S^{2-})$	Sodium sulphide
	$(Ca^{2+}2Br^-)$	Calcium bromide
or with several non-metallic elements	$(2Na^+SO_4^{2-})$	Sodium sulphate
	$(Ca^{2+}2NO_3^-)$	Calcium nitrate
	(K^+OH^-)	Potassium hydroxide

The positive ion is also called the **cation**, and the negative ion is also called the **anion**. It is useful to know the number of charges carried by cations and anions as these can be used to work out (rather than remember) the formulae of compounds. The charges are shown for important cations and anions in Table 9.6. This table contains some transition metals.

Table 9.6

Charges Carried by Important Cations and Anions

+1	+2	+3
Metals of Group 1, hydrogen and ammonium. Li^+, Na^+, K^+, H^+, NH_4^+ and Ag^+ silver (I) Cu^+ copper(I)	Metals of Group 2 Mg^{2+}, Ca^{2+}, Ba^{2+} and Pb^{2+} lead(II) Cu^{2+} copper(II) Fe^{2+} iron(II)	Al^{3+} Fe^{3+} iron(III)
−1	**−2**	**−3**
Halide ions (Group 7), F^-, Cl^-, Br^-, I^- and OH^- hydroxide NO_3^- nitrate CH_3COO^- acetate HCO_3^- bicarbonate (hydrogen carbonate) MnO_4^- permanganate	SO_4^{2-} sulphate SO_3^{2-} sulphite CO_3^{2-} carbonate O^{2-} oxide S^{2-} sulphide $Cr_2O_7^{2-}$ dichromate	PO_4^{3-} phosphate

The ratio of the numbers of cations and anions in a compound can be decided from the values of the ionic charges and the knowledge that the compound is electrically neutral.

Taking sodium sulphate as an example:

the ions present in sodium sulphate are Na^+ and SO_4^{2-}

for electrical neutrality, relative numbers must be $2Na^+$ and SO_4^{2-}

The formula would be written either as $(2Na^+SO_4^{2-})$ or as Na_2SO_4.

Taking iron(III) sulphate as a second example:

the ions present in iron(III) sulphate are Fe^{3+} and SO_4^{2-}

for electrical neutrality, relative numbers must be $2Fe^{3+}$ and $3SO_4^{2-}$

The formula would be written either as $(2Fe^{3+}3SO_4^{2-})$ or as $Fe_2(SO_4)_3$.

Summary

1. Ionic compounds are formed when an element on the left of the Periodic Table (a metallic element) combines with an element or group of elements from the right of the Periodic Table (non-metallic elements).
2. Atoms of metallic elements usually lose all the electrons in their outer shell when they form positive ions. This produces a positive ion with the electronic configuration of a noble gas. When this happens, the charge on the positive ion is the same magnitude as the group number of the element.
   Group 1 elements form an ion El^{+} (e.g. Na^{+})
   Group 2 elements form an ion El^{2+} (e.g. Mg^{2+})
   Group 3 elements can sometimes form an ion El^{3+} (e.g. Al^{3+})
3. Elements in the halogen group can form an ion El^{-} (e.g. Cl^{-}) by each atom gaining one electron from a metallic element. This gives the halide ion is the same magnitude as the group number of the element. configuration of argon).

   Elements in Group 6 can form an ion El^{2-} (e.g. O^{2-}) by each atom gaining two electrons from a metallic element. This gives the El^{2-} ion the electronic configuration of a noble gas (e.g. O^{2-} has the electronic configuration of neon).
4. Ionic compounds can be recognized by their physical properties of high melting point, high boiling point, high molar heat of vaporization and by their high electrical conductivity when molten.
5. The physical properties of solid ionic compounds arise from strong electrostatic forces of attraction between ions of opposite charge. This gives the solid a close and regularly packed structure called a giant structure of ions.
6. If the values of the charges carried by the ions in a compound are known, it is possible to work out the formula of the compound. It is only necessary to know the values of a small number of ionic charges to be able to work out, rather than remember, the formulae of a vast number of ionic compounds.

10 Structure and Bonding
2: Non-ionic bonding

The energy needed to break up an ionic lattice

Strong forces of electrostatic attraction hold together the ions in a giant structure of ions. Considerable energy is needed to pull the ions apart. Some values for the energies involved are given in Table 10.1. This table shows also the energy needed to melt one mole of ice and to raise the temperature of one mole of water from the melting point to the boiling point. Comparison of the values will give some idea of the strength of the electrostatic attraction within a giant structure of ions.

Table 10.1

Change			*Energy needed to bring about the change in kJ/mole*
(Na^+Cl^-) ions in 1 mole of solid sodium chloride	⟶	$Na^+ + Cl^-$ separate gaseous ions, 1 mole of Na^+ and 1 mole of Cl^-	765
$(Mg^{2+}2Cl^-)$ ions in 1 mole of solid magnesium chloride	⟶	$Mg^{2+} + 2Cl^-$ separate gaseous ions, 1 mole of Mg^{2+} and 2 moles of Cl^-	2500
$H_2O(s)$ 1 mole of ice at 0 °C	⟶	$H_2O(l)$ 1 mole of liquid water at 0 °C	6
$H_2O(l)$ 1 mole of liquid water at 0 °C	⟶	$H_2O(l)$ 1 mole of liquid water at 100 °C	8

The energy needed to break up a non-ionic substance

Non-ionic substances are composed usually of small molecules such as $SiCl_4$, Cl_2, I_2 and H_2O. If energy is supplied to these non-ionic substances, they vaporize readily as widely separated molecules. Some values for the energy changes are shown in Table 10.2.

Comparison of Tables 10.1 and 10.2 shows that the particles in a non-ionic substance can be separated relatively easily. This is because **the attractive forces**

Table 10.2

1 Mole of closely packed molecules in the liquid or solid state	⟶	*1 Mole of widely separated gaseous molecules*	*Energy needed to bring about the change in kJ/mole*
$SiCl_4(l)$	⟶	$SiCl_4(g)$	26
$PCl_3(l)$	⟶	$PCl_3(g)$	30
$SCl_2(l)$	⟶	$SCl_2(g)$	31
$Cl_2(l)$	⟶	$Cl_2(g)$	23
$H_2(l)$	⟶	$H_2(g)$	1
$I_2(s)$	⟶	$I_2(g)$	62
$H_2O(l)$	⟶	$H_2O(g)$	40

between the molecules of a non-ionic substance are weak. The energy changes of Table 10.2 give no information, however, about the attractive forces **within** the molecules of a non-ionic substance. To obtain information about the attractive forces within molecules it is necessary to look at the energy needed to break up gaseous molecules into separate gaseous atoms. Some values for this energy change are given in Table 10.3.

Table 10.3

1 Mole of separate gaseous molecules	⟶	*Separate gaseous atoms*	*Energy needed to break up molecules into separate gaseous atoms in kJ/mole*
$SiCl_4(g)$	⟶	$Si(g) + 4Cl(g)$	1540
$PCl_3(g)$	⟶	$P(g) + 3Cl(g)$	970
$SCl_2(g)$	⟶	$S(g) + 2Cl(g)$	480
$Cl_2(g)$	⟶	$2Cl(g)$	240
$H_2(g)$	⟶	$2H(g)$	435
$I_2(g)$	⟶	$2I(g)$	150
$H_2O(g)$	⟶	$2H(g) + O(g)$	925

Compare the energies involved in the changes of Tables 10.1 and 10.3. It will be seen that the energies required to separate the atoms in a molecular structure are comparable with the energies required to separate the ions in an ionic structure. The bonding forces within non-ionic molecules are of the same order of magnitude as the bonding forces in an ionic structure.

Bonding in non-ionic substances: covalent bonding

The bonding in ionic compounds is electrostatic in nature. Ionic bonding arises from the attraction between positive and negative ions. The bonding in non-ionic substances also is electrostatic. The atoms within each molecule are held together by attraction between positive nuclei and shared electrons. One of the simplest non-ionic molecules is hydrogen, H_2. The bonding in the hydrogen molecule is shown in Figure 10.1.

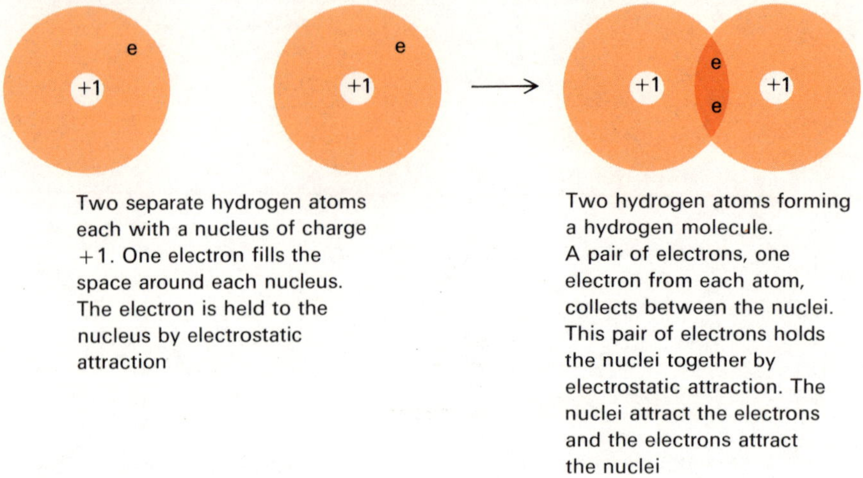

Other ways of representing the bonding in the hydrogen molecule include those shown below:

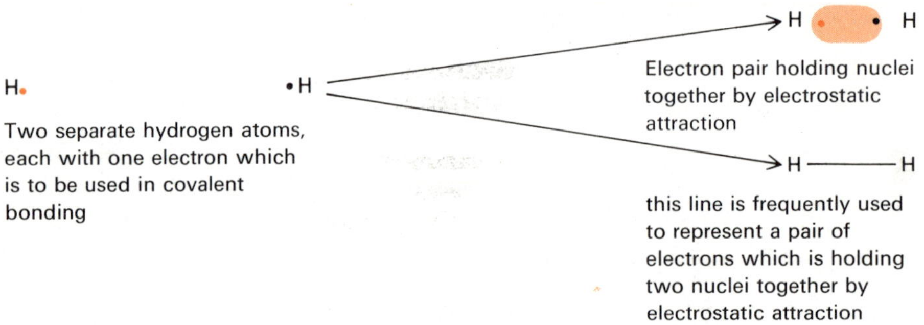

Figure 10.1 Covalent bonding in the hydrogen molecule

The bonding in non-ionic molecules is called **covalent bonding**. A covalent bond holds two atoms together. The atoms are linked by the attraction between their nuclei and a shared pair of electrons. The electrons which are shared come from the outer shells of the two atoms. Covalent bonding in the chlorides $SiCl_4$, PCl_3, SCl_2 and Cl_2 is illustrated in Figure 10.2. In all cases the atoms are held together by the electrostatic attraction between nuclei and shared electron pairs.

The shape of covalent molecules

The shape taken up by a covalent molecule is the one in which there is minimum repulsion between pairs of electrons. The pairs of electrons which decide the shape are the pairs in covalent bonds and also any remaining lone pairs. A lone pair is a pair of electrons in the outer shell of an element which do not combine with electrons from another element when forming a compound.

This general rule can be used with reasonable certainty for purposes of

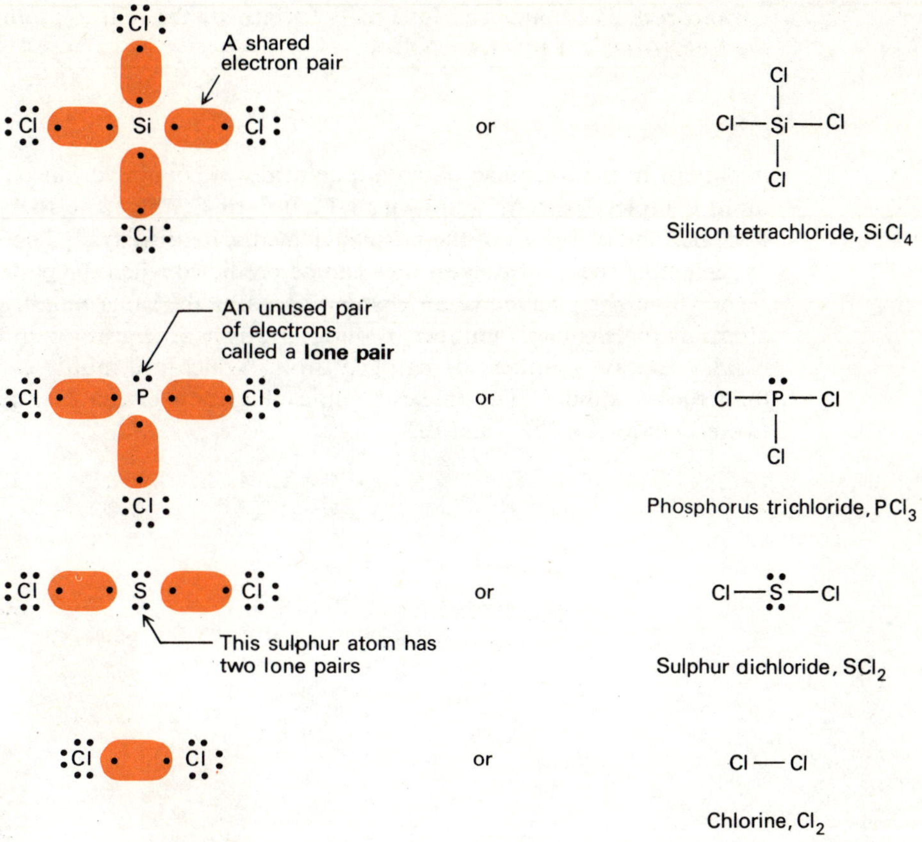

Figure 10.2 Covalent bonding in certain chlorides

prediction. The chlorides of Si, P and S are used as examples in Figure 10.3. These chlorides all have a **tetrahedral** shape, this being the shape in which minimum repulsion between electron pairs is achieved. The tetrahedral shape is the one in which the central atom can be imagined to be at the middle of a

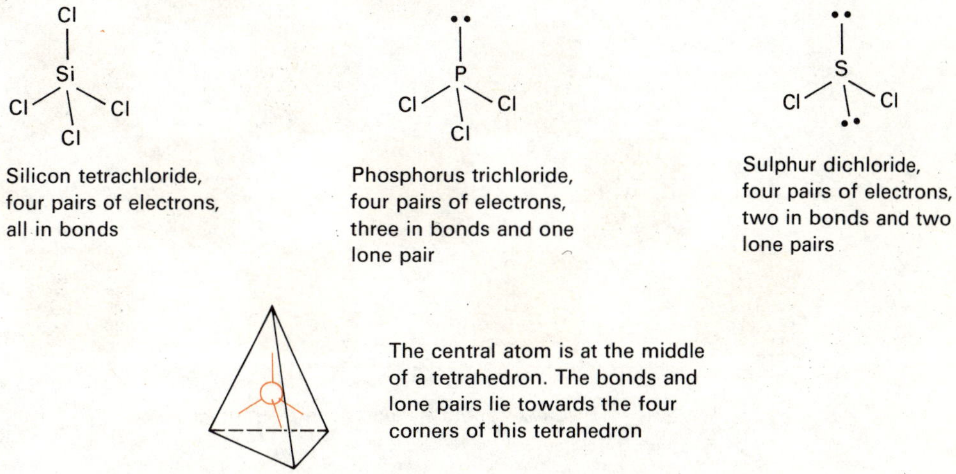

Figure 10.3 The shapes of certain covalent molecules

tetrahedron. The bonds and lone pairs formed by the central atom lie towards the four corners of this tetrahedron.

Patterns in the formulae of halides

A pattern in the formulae of certain chlorides was pointed out on page 70. A more complete pattern is shown in Figure 10.4, where the formulae of the fluorides and chlorides of the normal elements are displayed. There is no need to remember these formulae—they can be predicted when the pattern has been seen. The highest halide of an element contains the same number of halogen atoms as the element's number of outer electrons (the group number). Lower halides contain numbers of halogen atoms which are multiples of two less than the maximum. The highest halide of Group 6, for example, is ElX_6. Lower halides are ElX_4 and ElX_2.

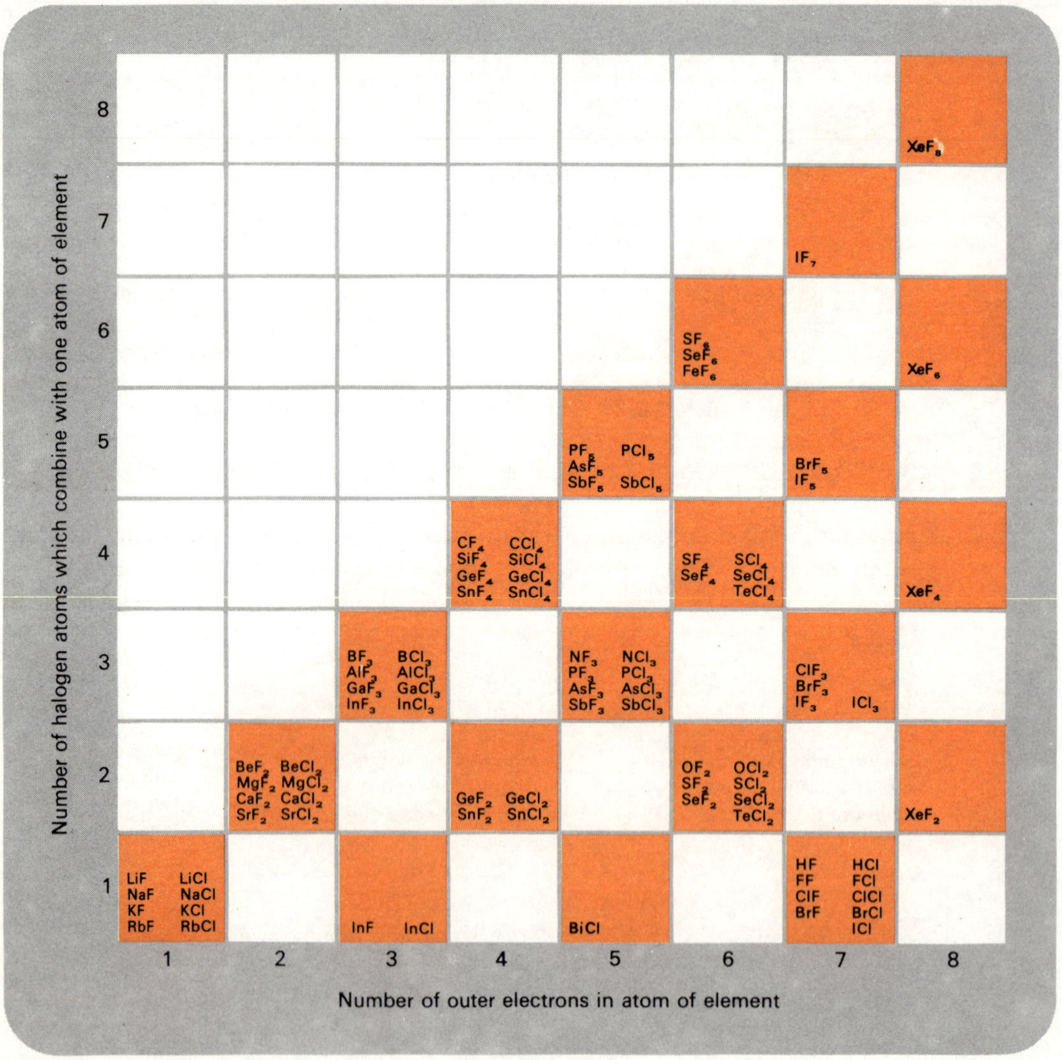

Figure 10.4

Halides of Groups 1 and 2

These are ionic, with the exception of the halides of beryllium (*see* page 80). The vapour of $BeCl_2$ is composed of linear molecules. The molecules are linear because this is the shape in which repulsion between electron pairs is a minimum.

Compare $BeCl_2$ with $(Mg^{2+}2Cl^-)$. Magnesium is lower in Group 2 and loses its outer electrons more readily than does beryllium.

Ionic bonding

Halides of Group 3

These are frequently ionic (*see* Figures 9.11–9.14, page 78). All compounds formed by boron are, however, covalent. The halides of boron are composed of flat symmetrical molecules of formula BX_3. The molecules have this shape because this is the one in which repulsion between electron pairs is a minimum.

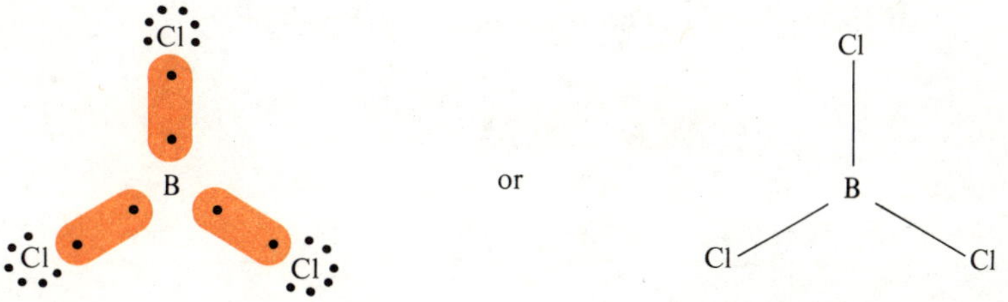

Boron trichloride, BCl_3, a flat symmetrical molecule

Halides of Group 4

Loss of two of the four outer electrons is relatively easy with the elements of higher atomic number. Germanium, tin and lead form ionic halides $(El^{2+}2X^-)$, *see* page 81. Loss of all four outer electrons requires considerable energy and is not observed in compound formation. Group 4 elements can share all four outer electrons in covalent bonds. The halides of Group 4 with formula ElX_4 all have a tetrahedral shape because this is the one in which repulsion between electron pairs is a minimum.

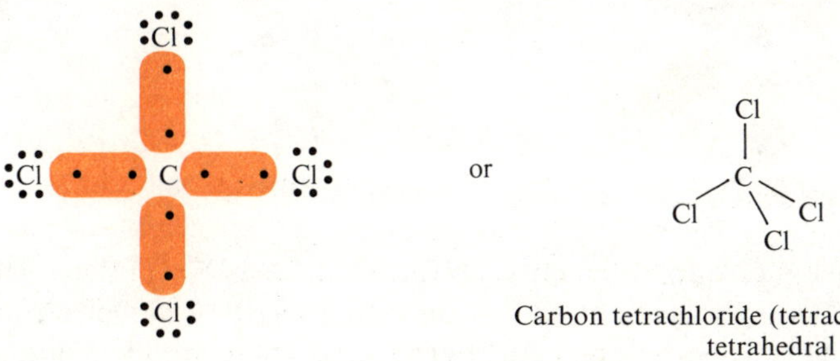

Carbon tetrachloride (tetrachloromethane), CCl_4, tetrahedral shape

Halides of Group 5

The halides of Groups 5, 6, 7 and 0 are usually covalent. Atoms of Group 5 have five outer electrons. If all these electrons are used in sharing, five covalent bonds can be formed. The shape of the molecules ElX_5 is decided by repulsion between electron pairs in the five bonds. In PCl_5 vapour, for example, the shape is that of a trigonal bipyramid.

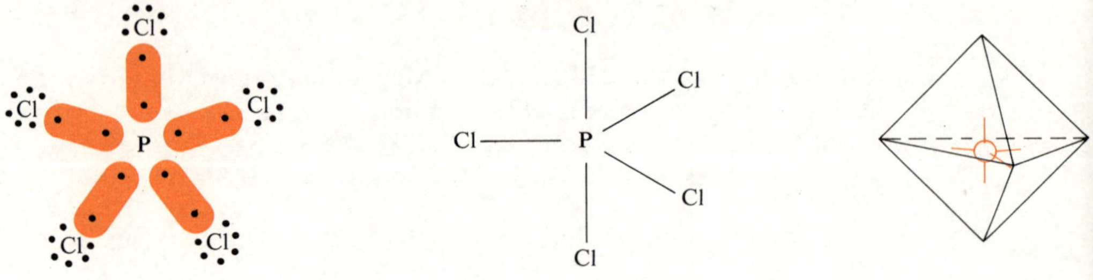

Phosphorus pentachloride, PCl_5

Group 5 atoms frequently use only three of their outer electrons in covalent bond formation. The remaining two electrons form a lone pair. Repulsion between pairs of electrons—lone pairs and pairs in bonds—decides the shape of the molecule. The shape is tetrahedral.

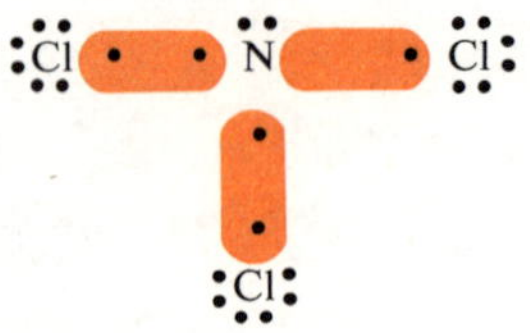

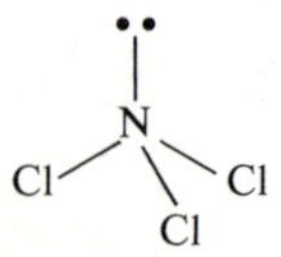

Nitrogen trichloride, NCl_3

Halides of Group 6

Atoms of elements in Group 6 have six outer electrons which can be used in covalent bond formation. Symmetrical arrangement of six bonds is achieved in an **octahedral** shape—the bonds from the central atom are directed towards the six corners of an octahedron.

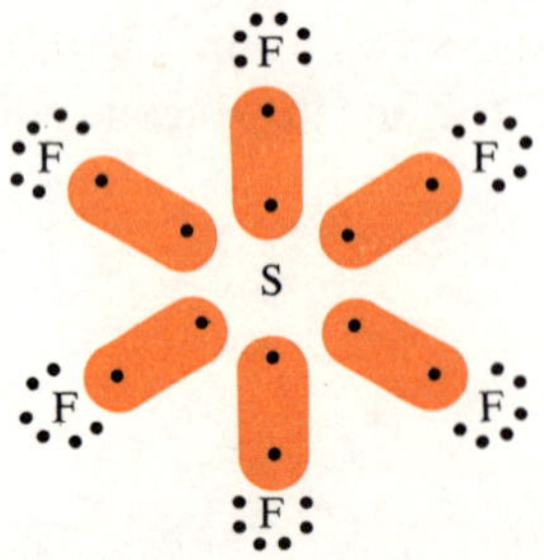

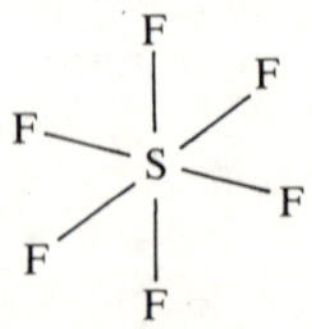

Sulphur hexafluoride, SF_6

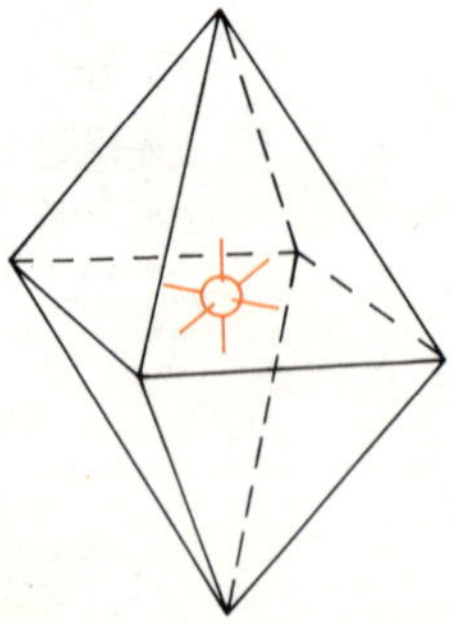

Atoms of Group 6 elements can form halides ElX_4. In these halides the atom of the Group 6 element shares only four of its outer electrons in covalent bonds. This leaves a lone pair of electrons. There are five pairs of electrons to arrange symmetrically around the central atom. The shape of the molecule is similar to that of PCl_5 vapour.

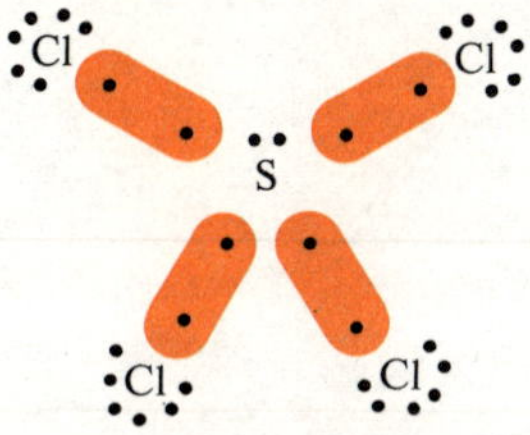
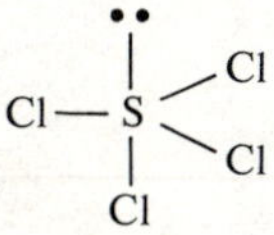

Sulphur tetrachoride, SCl_4

Atoms of Group 6 elements can use only two of their outer electrons in covalent bond formation. This occurs in the halides ElX_2. Two lone pairs of electrons remain. In such compounds there are four pairs of electrons to arrange symmetrically around the central atom. The shape of such molecules is tetrahedral.

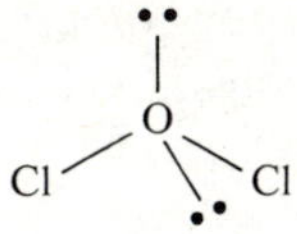

Oxygen dichloride, OCl_2

Halides of Group 7

Atoms of elements in this group have seven outer electrons which can be used in covalent bond formation. The formulae of compounds which can be formed between halogen atoms are ElX_7, ElX_5, ElX_3 and ElX.

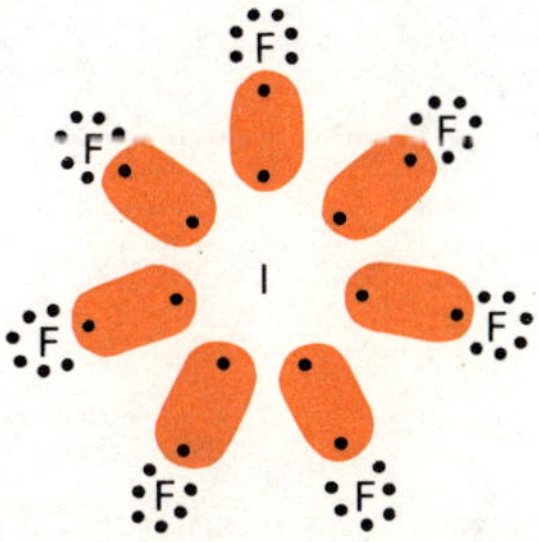
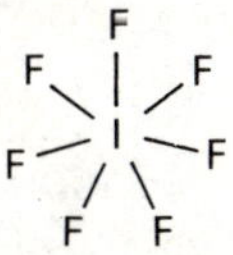

Iodine heptafluoride, IF_7

In IF_7 the iodine atom shares all seven outer electrons in covalent bond formation with fluorine. It is not possible to obtain a completely symmetrical shape when seven pairs of electrons are arranged around a central atom.

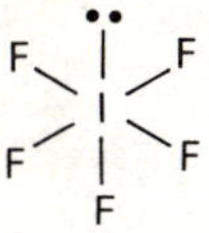

Iodine pentafluoride, IF_5

In IF_5 the iodine atom shares five outer electrons in covalent bond formation with fluorine. This leaves one lone pair. The shape of the molecule is similar to that of SF_6.

Iodine trichloride, ICl_3

In ICl_3 the iodine atom shares three outer electrons in covalent bond formation with chlorine. This leaves two lone pairs. The shape of the molecule is similar to that of PCl_5.

In many compounds formed by the halogens only one of the seven outer electrons is used in covalent bond formation. This leaves three lone pairs as in chlorine, Cl_2.

Halides of Group O

Atoms of elements in this group have eight outer electrons. Most of the noble gases only form compounds with difficulty. Xenon does form compounds with reactive elements such as fluorine.

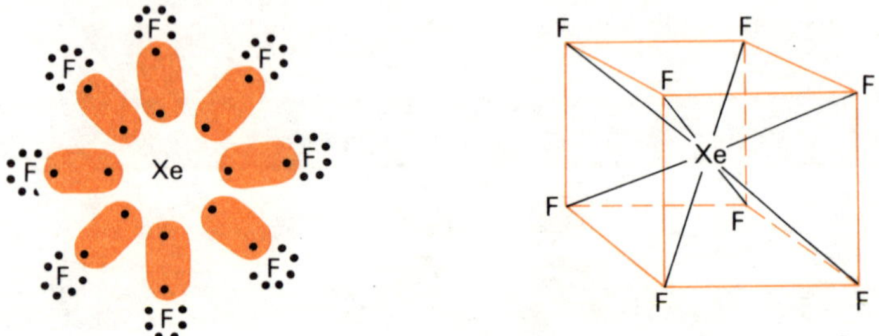

Xenon octafluoride, XeF_8

In XeF_8 the xenon atom shares all eight outer electrons in covalent bond formation with fluorine. The shape of the molecule is one in which the xenon atom can be imagined at the centre of a cube. The eight fluorine atoms are at the corners of this cube.

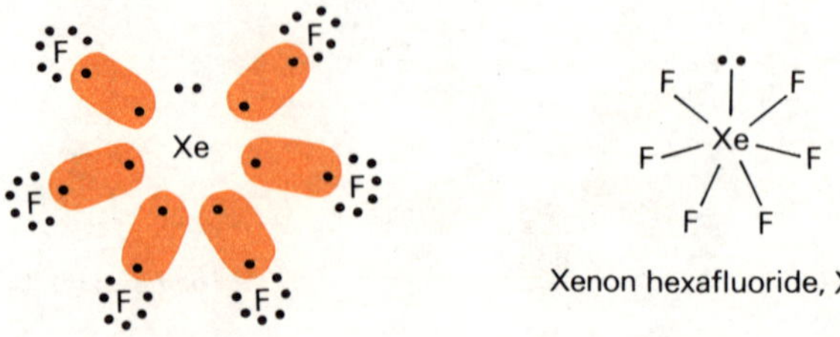

Xenon hexafluoride, XeF_6

In XeF_6 the xenon atom shares six outer electrons in covalent bond formation with fluorine. This leaves one lone pair. The shape of the molecule is similar to that of IF_7.

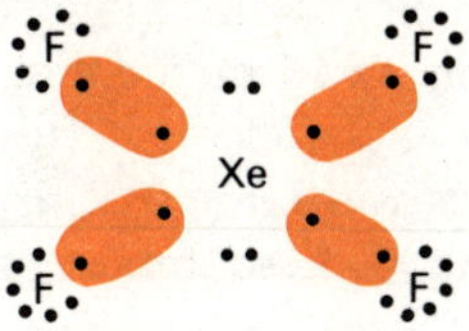

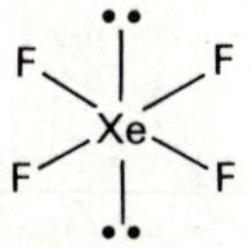

Xenon tetrafluoride, XeF_4

In XeF_4 the xenon atom shares four outer electrons in covalent bond formation with fluorine. This leaves two lone pairs. The shape of the molecule is similar to that of SF_6.

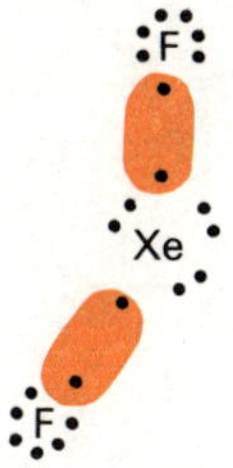

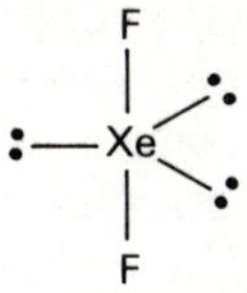

Xenon difluoride, XeF_2

In XeF_2 the xenon atom shares two outer electrons in covalent bond formation with fluorine. This leaves three lone pairs. The shape of the molecule is similar to that of PCl_5.

Covalent molecules containing double and triple bonds

The atoms in some molecules are linked by more than one pair of shared electrons. These molecules are said to contain double or triple bonds. Three examples of such molecules are given in Figure 10.5. Further examples of molecules containing double and triple bonds will be found later in this chapter.

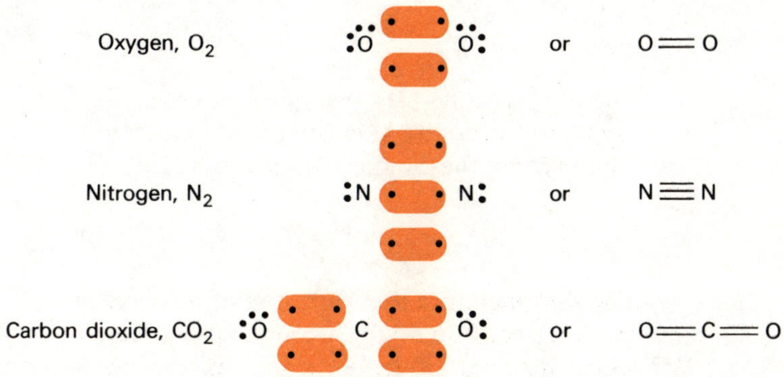

Figure 10.5 Molecules containing double and triple bonds

Patterns in the number of covalent bonds formed by normal elements

The formulae of the halides of the normal elements have been used already in this chapter to give information about the number of covalent bonds each atom is able to form. The patterns in formulae and number of bonds are not restricted to the halides. Atoms of the elements to the right of the diagonal (non-metals) form covalent bonds with atoms of hydrogen, oxygen and other non-metals. The bonding in some of these important compounds will now be discussed.

Covalent bonding in the period Li–Ne

The number of covalent bonds formed by atoms of boron and carbon is the same as their number of outer electrons. The number of bonds for nitrogen, oxygen, and fluorine is eight minus the number of outer electrons. Compounds which illustrate this pattern are shown below.

The boron atom uses all three outer electrons to form flat symmetrical molecules.

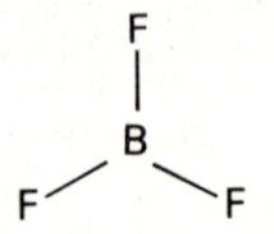

Boron trifluoride, BF_3

The carbon atom uses all four outer electrons to form tetrahedral molecules.

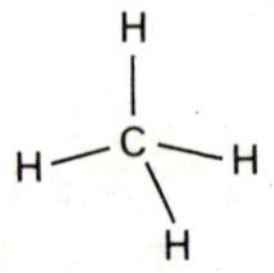

Methane, CH_4

The nitrogen atom uses three of its outer electrons for bonding, leaving one lone pair. Molecules of this type are tetrahedral.

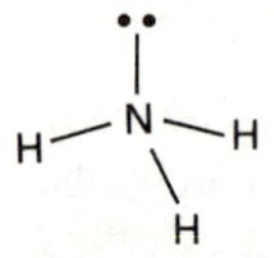

Ammonia, NH_3

The oxygen atom uses two of its outer electrons for bonding, leaving two lone pairs. The shape of these molecules is tetrahedral.

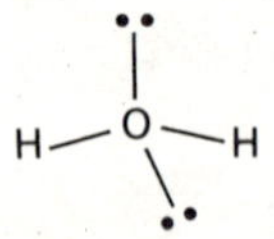

Water, H_2O

The fluorine atom uses only one of its outer electrons for bonding. Three lone pairs remain which have no influence on the shape of the molecule.

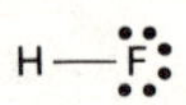

Hydrogen fluoride, HF

Carbon atoms form very strong covalent bonds with other carbon atoms, and also with atoms of hydrogen, oxygen, nitrogen, and the halogens. This strong bonding gives a vast number of stable compounds called carbon compounds or organic compounds. Examples of important carbon compounds are shown below.

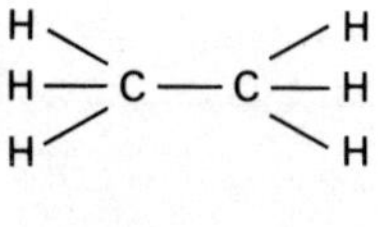

Ethane, C_2H_6

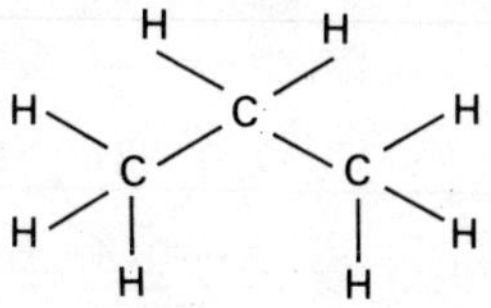

Propane, C_3H_8

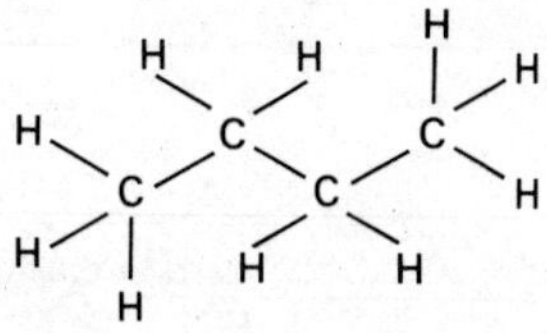

Butane, C_4H_{10} (m.p. − 135 °C, b.p. + 1 °C)

Carbon compounds containing several atoms can frequently exist in two or more different forms called **isomers**. Butane, C_4H_{10}, has another isomeric form shown below. Isomers may have very different physical and chemical properties.

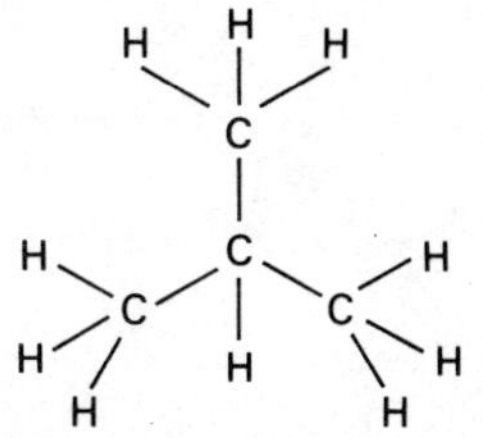

Methyl propane, an isomer of butane, C_4H_{10} (m.p. −160 °C, b.p. −11 °C)

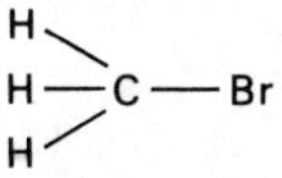

Bromomethane, CH_3Br

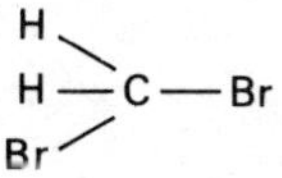

Dibromomethane, CH_2Br_2

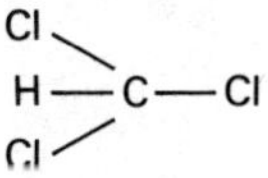

Trichloromethane, $CHCl_3$

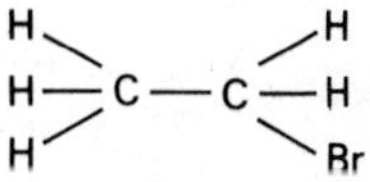

Bromoethane, C_2H_5Br

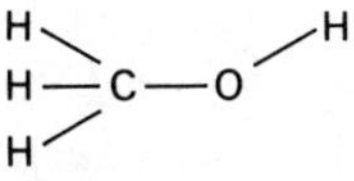

Methanol, CH_3OH

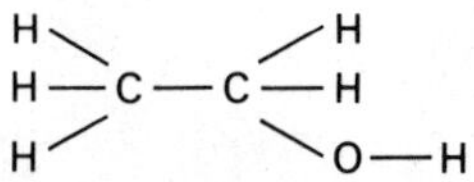

Ethanol, C_2H_5OH
(m.p. −115 °C, b.p. +79 °C)

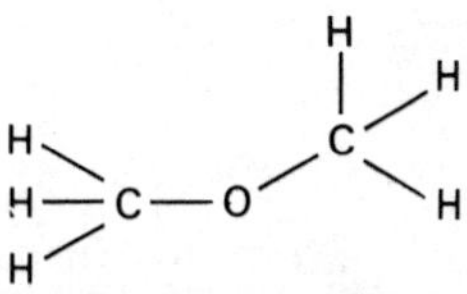

Dimethyl ether, CH_3OCH_3
an isomer of ethanol
(m.p. −142 °C, b.p. −25 °C)

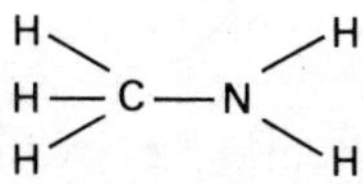

Methylamine, CH_3NH_2

Carbon atoms are able to form strong double or triple bonds with other carbon atoms, and with atoms of oxygen and nitrogen. This extends even further the range of possible carbon compounds. Some important examples are shown in Figure 10.6.

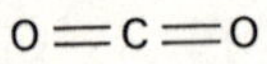

Carbon dioxide, CO_2

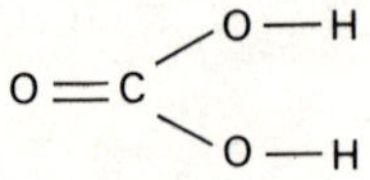

Carbonic acid, H_2CO_3

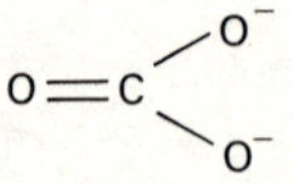

The carbonate ion, CO_3^{2-}

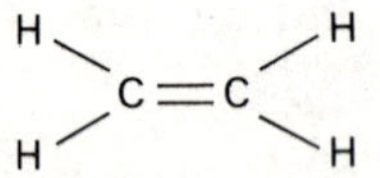

Ethylene (ethene), C_2H_2, a flat symmetrical molecule

Acetylene (ethyne), C_2H_2

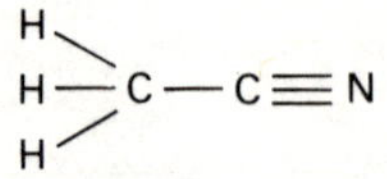

Methyl cyanide (ethanonitrile), CH_3CN

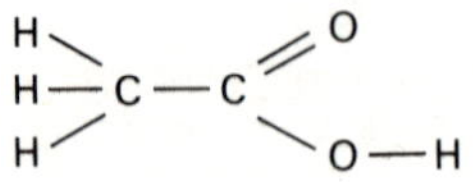

Acetic acid (ethanoic acid), CH_3COOH

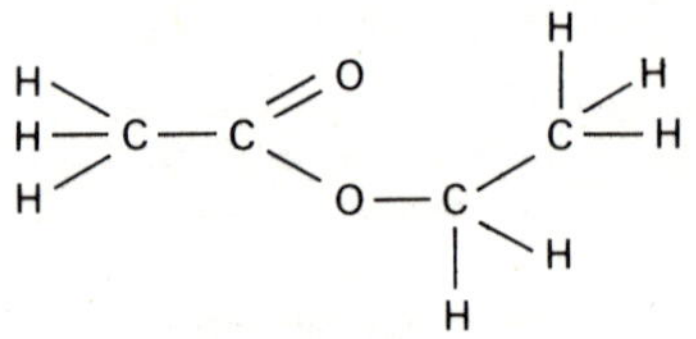

Ethyl acetate (ethyl ethanoate), $CH_3COOC_2H_5$, an example of an ester

Figure 10.6

Covalent bonding in the period Na–Ar and later periods

The maximum number of covalent bonds formed by atoms of these elements is the same as their number of outer electrons. All the outer electrons are then being used to form covalent bonds. Examples of compounds where all outer electrons are being used are given in Figure 10.7.

The atoms of elements in later periods form many compounds where some of the outer electrons are not used in covalent bond formation. These unused electrons remain as lone pairs which influence the shape of the molecule.

Group 3

The elements of higher atomic number in this group can form compounds

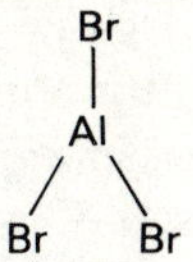

Aluminium tribromide, $AlBr_3$

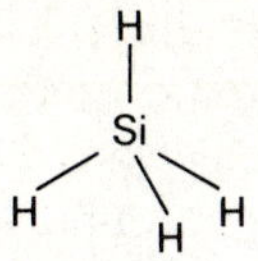

Silane, SiH_4

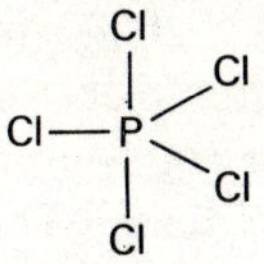

Phosphorous pentachloride, PCl_5

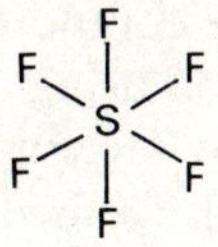

Sulphur hexafluoride, SF_6

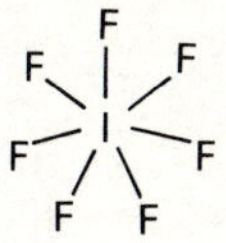

Iodine heptafluoride, IF_7

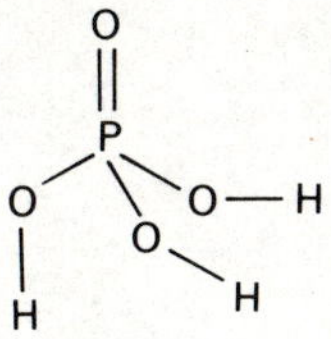

Phosphoric acid, H_3PO_4

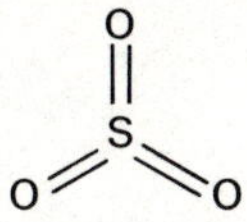

Sulphur trioxide, SO_3

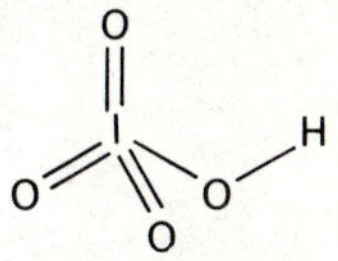

Periodic acid, HIO_4

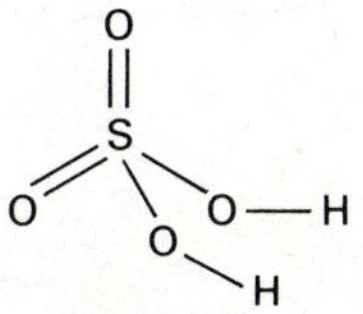

Sulphuric acid, H_2SO_4

Figure 10.7

where only one of the three outer electrons is being used in bond formation. These compounds are, however, ionic (*see* page 81).

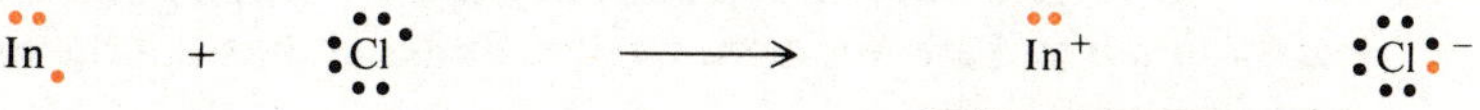

Indium ion with one lone pair

The compound (In^+Cl^-) is called indium(I) chloride to show that the indium atom is using one electron in ionic bond formation. The name distinguishes the compound from indium(III) chloride ($In^{3+}3Cl^-$) where the indium atom is using all three outer electrons in ionic bond formation.

Group 4

The elements of higher atomic number in this group can form compounds where only two of the four outer electrons are being used in bond formation. These compounds are, however, ionic (*see* page 81).

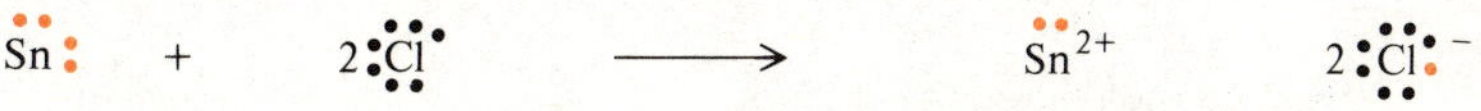

Tin(II) ion with one lone pair

The compound ($Sn^{2+}2Cl^-$) is called tin(II) chloride to show that the tin atom is using two outer electrons in ionic bond formation. When atoms of Group 4 elements use all four outer electrons in bond formation, the compounds formed are covalent. When a compound is composed of small **covalent** molecules, the number of atoms of each element in the molecule is shown in the name of the compound.

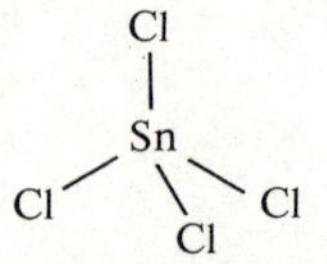

Tin tetrachloride is a tetrahedral covalent molecule

Group 5

Phosphorus and elements of higher atomic number in this group frequently use only three outer electrons in covalent bond formation. One lone pair remains and these molecules have a tetrahedral shape. This has been seen already with phosphorus trichloride, PCl_3 (*see* page 89). Phosphine, PH_3, is another example.

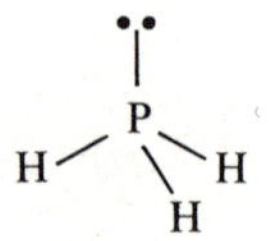

Tetrahedral PH_3 molecule

Group 6

Sulphur and elements of higher atomic number in this group frequently use only four or two outer electrons in covalent bond formation. One or two lone pairs remain and are important in deciding the shape of the molecules. Examples are given in Figure 10.8.

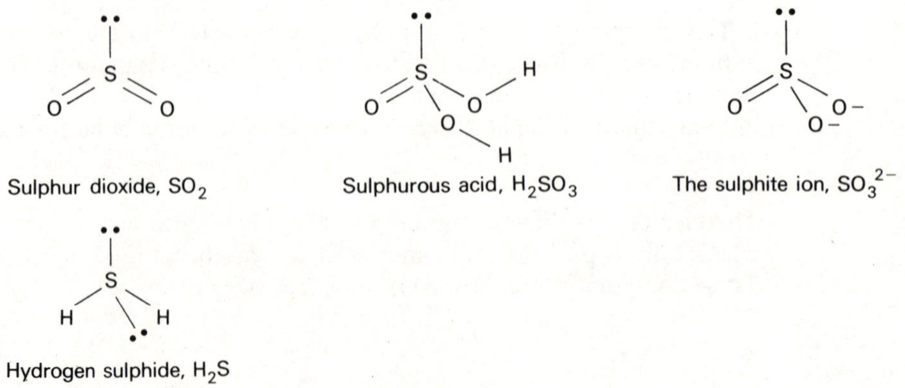

Figure 10.8

Group 7

Chlorine and elements of higher atomic number in this group frequently use only five, three or one electron in covalent bond formation. One, two or three lone pairs of electrons remain and are important in deciding the shape of the molecule. Examples are given in Figure 10.9.

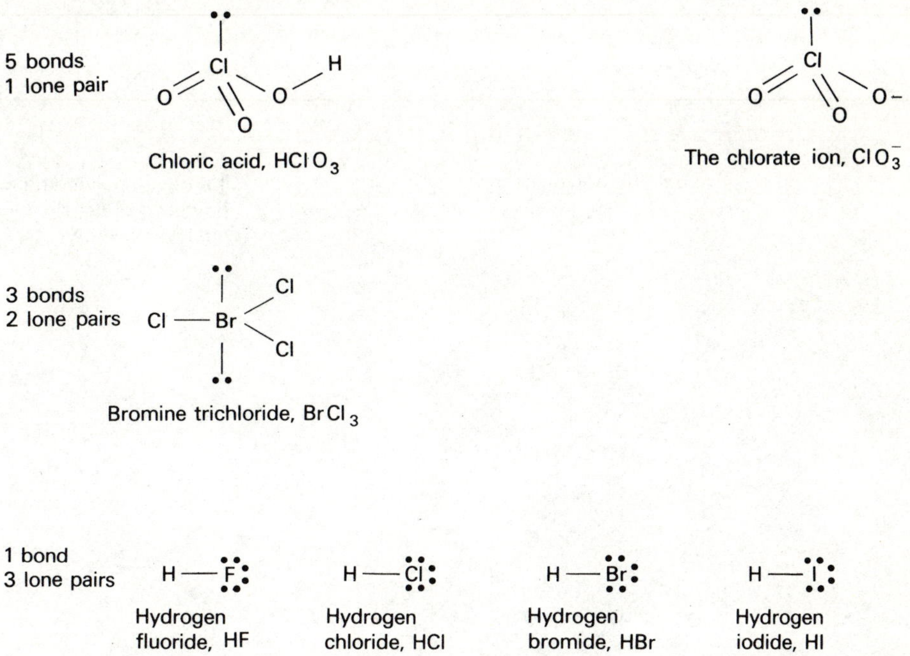

Figure 10.9

Giant molecular structures

Covalent substances frequently are composed of fairly small molecules. The bonding between atoms in these molecules is strong—it is difficult to break up the molecule into its separate atoms. Attraction between molecules is weak—it is relatively easy to separate the molecules from each other. Covalent substances usually have low melting points and low boiling points because this attraction between molecules is weak. The high melting points and high boiling points of ionic compounds arise from the strong electrostatic forces which attract ions of opposite charge in giant structures of ions.

Some non-ionic substances have high melting and boiling points. The oxides of aluminium and silicon, Al_2O_3 and SiO_3, are examples (*see* Table 9.5, page 83). These substances do not conduct electricity when molten. This shows that they are not composed of ions. In these non-ionic substances, covalent bonding is used to link atoms in an infinite network called either a **giant structure of atoms** or a **giant molecule**. The atoms in a giant structure are linked by covalent bonds in chains or sheets, or the bonding can extend into a three-dimensional network. Some examples of giant structures are given in Figure 10.10 on page 102.

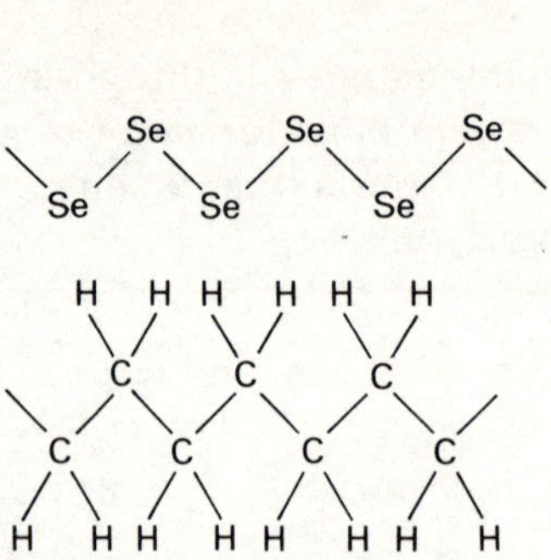

Selenium and polythene are composed of long chains of atoms.

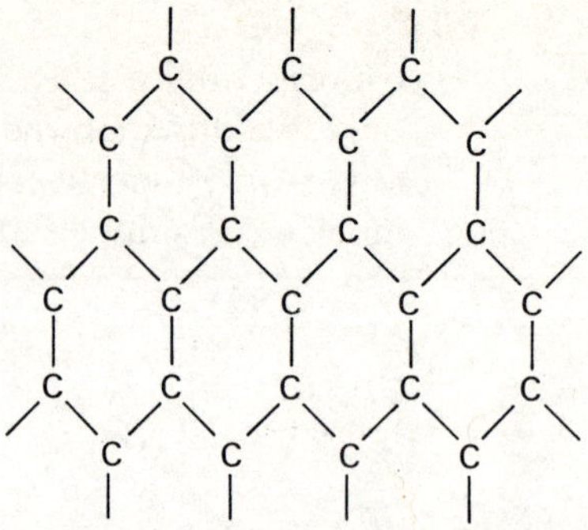

Graphite, a polymorph of carbon, is composed of flat sheets of atoms. One layer is shown here.

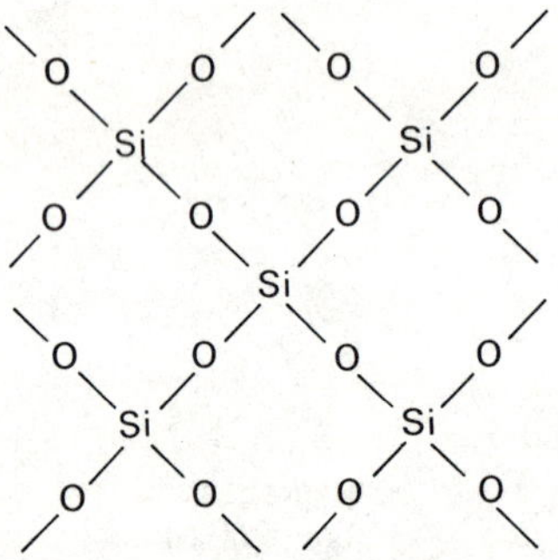

In silicon dioxide, SiO_2, each silicon atom is surrounded by four oxygen atoms and each oxygen atom by two silicon atoms. A three-dimensional network is formed. No attempt is made in this diagram to show the three-dimensional nature of the structure.

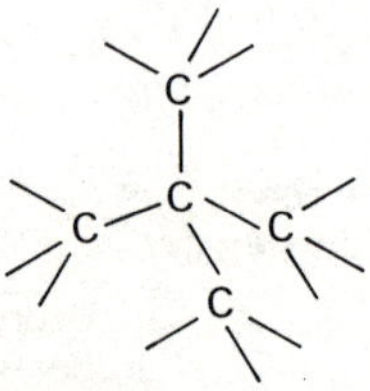

Diamond, another polymorph of carbon, is composed of a three-dimensional network of atoms. Each carbon atom is surrounded tetrahedrally by four other carbon atoms. No attempt is made in this diagram to show the three-dimensional nature of the structure.

Figure 10.10

Bonding in normal elements

Metallic bonding

The metallic elements are shown in the Periodic Table on page 103. Transitional elements are all metallic and have not been included. Metallic elements are recognized by certain properties. Among the most important of these properties are high electrical and thermal conductivity, and high boiling point. Many metals possess the property of malleability—these metals can be readily altered in shape by, for example, hammering or the application of pressure.

Li	Be		C				
Na	Mg	Al		P			
K	Ca	Ga	Ge	As			
Rb	Sr	In	Sn	Sb			

The atoms of metallic elements are closely packed in regular giant structures. The outer electrons of each atom are used to hold the structure together. These electrons act as an 'electron cement' which is able to move through the whole of the metallic structure. This mobile electron cement is responsible for the high electrical and thermal conductivity of metals. The strength of the electron cement is responsible for the high boiling points of metals.

Some elements exist in more than one crystalline form. These forms are called **polymorphs** or **allotropes**. Carbon, tin, phosphorus, arsenic and antimony are examples of such elements. One polymorph of each of these elements has metallic properties, the other polymorph has non-metallic properties. This is the reason for indicating these elements with fainter symbols in the Periodic Table above. Some electrons in the metallic polymorph are mobile.

Carbon in the form of graphite is composed of layers of atoms (*see* Figure 10.14). The bonding within each layer is covalent, but the layers are held together by mobile electron cement. The 'electron cement' can move along the plane of the layers, making graphite a conductor of heat and electricity in this direction only. In most metals the attraction between atoms is due to the electron cement. No definite bonds have to be broken when the shape of a piece of metal is altered—the metal is malleable. In graphite, and in certain other elements with some metallic properties, covalent bonds are present also. Altering the shape of a piece of these metals involves the breaking of covalent bonds. These metals are not malleable, they are brittle.

Tin can exist as a non-metallic polymorph, but the usual polymorph is a soft metal. The usual polymorphs of **phosphorus** are red or yellow non-metallic solids. A metallic polymorph of phosphorus is known in which phosphorus atoms are arranged in buckled sheets of infinite extension. The more common polymorphs of **arsenic** and **antimony** are metallic with the same structure of buckled sheets as in phosphorus. Covalent bonding is present in the sheets and these metals are brittle.

Small molecules

Elements to the right of the diagonal frequently exist as small molecules. The covalent bonding within these molecules is strong. Attractive forces between molecules are weak and these elements have low melting and boiling points. Many of them are gaseous at room temperature. The bonding in these elements is summarized in Figure 10.11 on page 104.

Atoms of these elements each supply one outer electron to form **one** covalent bond. The single bonds link pairs of atoms.

H—H F—F Cl—Cl Br—Br I—I

Attractive forces between the molecules are weak. H_2, F_2 and Cl_2 are gaseous at room temperature while Br_2 is a liquid. I_2 molecules pack in a regular solid structure called a **molecular crystal** which has a low melting and boiling point.

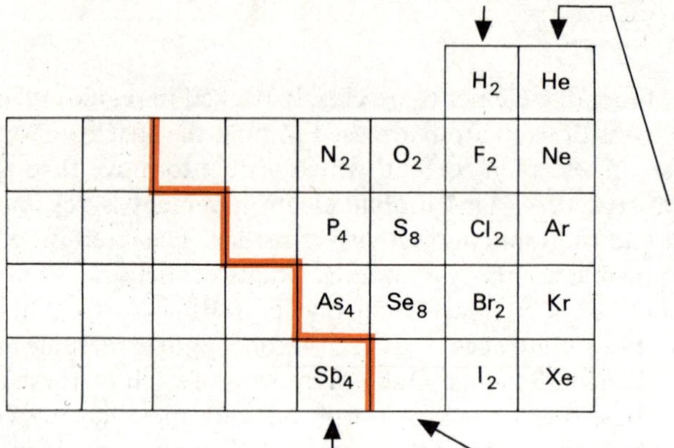

Atoms of these elements do not combine. They exist as widely separated gaseous atoms.

Atoms of these elements each supply three electrons to form **three** covalent bonds. These bonds are used in nitrogen to form a diatomic molecule.

N≡N

In yellow phosphorus, and in the non-metallic polymorphs of arsenic and antimony, the three bonds are used as single bonds to form El_4 molecules. These molecules form molecular crystals.

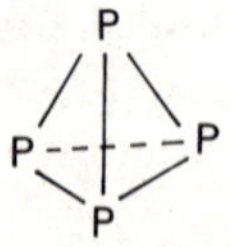

Atoms of these elements each supply two outer electrons to form **two** covalent bonds. These bonds are used in oxygen to form a diatomic molecule.

O=O

In sulphur, and in one polymorph of selenium, these two bonds are used to form a ring of eight atoms. The S_8 and Se_8 molecules form molecular crystals.

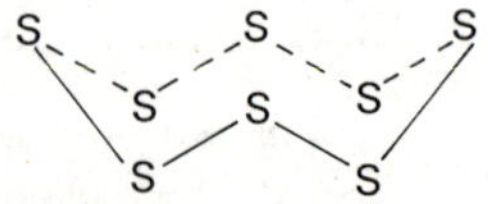

In the other polymorph of selenium, the two bonds are used to form infinite chains.

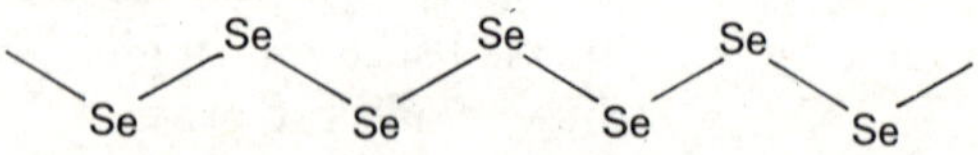

Figure 10.11 Elements composed of small molecules

Giant networks

Giant networks can be formed by non-metallic elements close to the diagonal. Atoms of these elements are linked by single covalent bonds to two or more other atoms. The bonding in these elements is summarized in Figure 10.12.

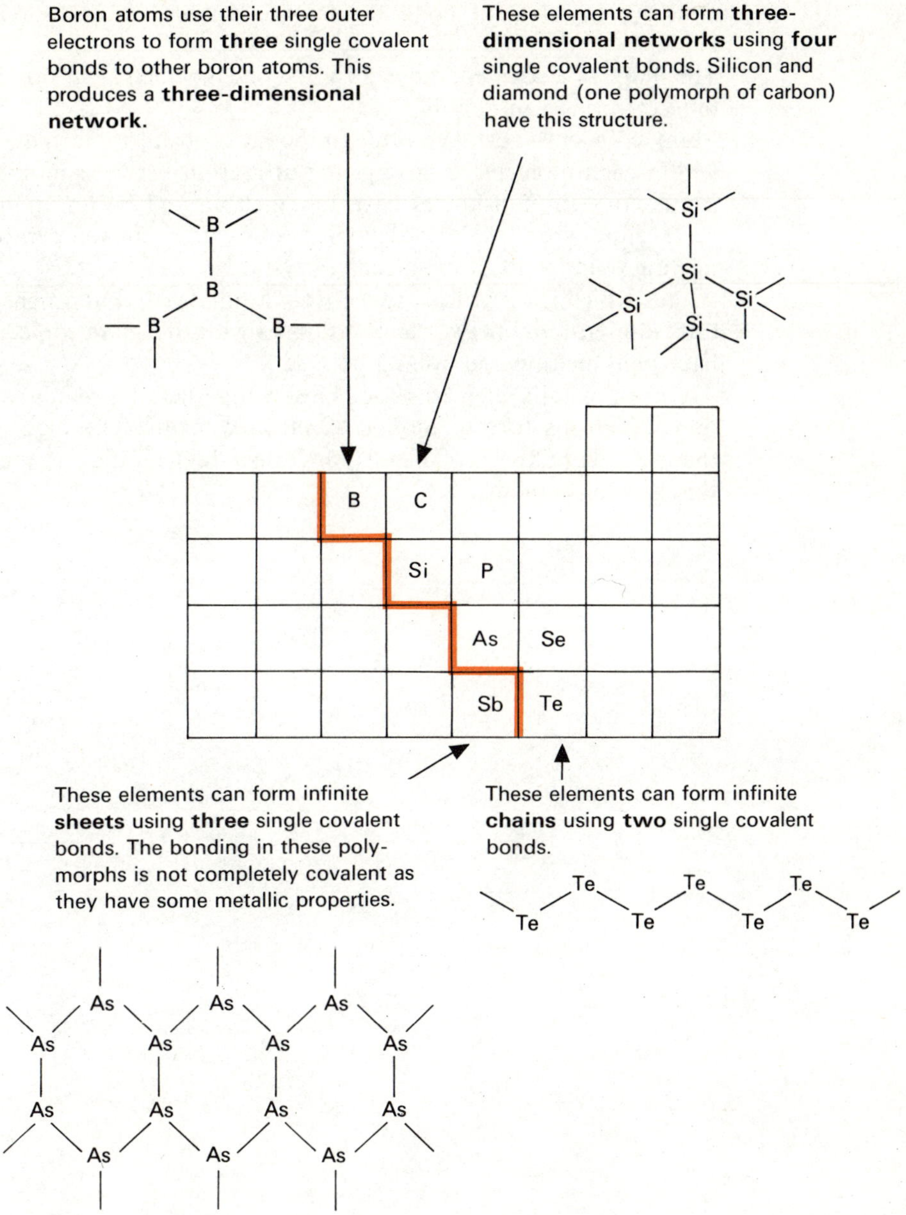

Figure 10.12 Non-metallic elements with a giant structure of atoms

Summary of non-ionic bonding

1. Atoms of non-metallic elements combine with each other by sharing pairs of outer electrons. A shared electron pair is called a covalent bond.
2. The maximum number of covalent bonds that an atom can form is the same as the number of outer electrons. Atoms frequently do not use all their outer

electrons in covalent bond formation. Unused electrons remain as lone pairs.

3. The shape of a covalent molecule is the one in which repulsion between pairs of electrons is a minimum.
4. Many covalent substances are composed of small molecules. The bonding within each molecule is strong, but attraction between molecules is weak. These covalent substances have low melting and boiling points. They are gases, liquids or low melting point solids at room temperature. The low melting point solids are molecular crystals.
5. Atoms which form at least two covalent bonds can link in chains, sheets or three-dimensional networks. Substances with this giant molecular structure have high melting and boiling points.
6. Atoms of metallic elements pack closely together in a regular structure. The outer electrons form a mobile 'electron cement'. The high electrical and thermal conductivity of metals, and their high boiling points, are due to this electron cement.

11 Moles of Molecules and Avogadro's Law

The volume of one mole of gaseous molecules

The volumes of one gramme portions of various gaseous elements and compounds are shown in Table 11.1. These volumes are for room temperature and pressure (r.t.p. is about 19 °C and 760 mmHg).

Table 11.1
Volume Occupied by 1 g of Gas at r.t.p.

Gas	*Volume*	*Gas*	*Volume*
Chlorine, Cl_2	334 cm^3	Hydrogen chloride, HCl	653 cm^3
Carbon dioxide, CO_2	540 cm^3	Oxygen, O_2	755 cm^3
Argon, Ar	600 cm^3	Ammonia, NH_3	1390 cm^3

The volumes in Table 11.1 show a wide spread in values. This is because one gramme portions of each of these substances contain different numbers of molecules. It is more useful to look at the volumes of portions which contain a **constant number of molecules**. One mole of atoms is a term introduced in Chapter 4 to describe portions of various elements which contain a constant number of atoms. Portions of substances which contain this same constant number of molecules are called one mole of molecules.

One mole of atoms of an element contains the same number of atoms as there are carbon atoms in 12 g of carbon.

One mole of molecules of an element or compound contains the same number of molecules as there are carbon atoms in 12 g of carbon.

One mole of molecules is sometimes called one gramme molecule. The mass of one mole of molecules can be called the molar mass or the gramme molecular mass.

One mole of oxygen atoms has a mass of 16 g.

One mole of oxygen molecules, O_2, contains two moles of oxygen atoms and has a mass of 32 g.

One mole of carbon dioxide, CO_2, contains one mole of carbon atoms with mass 12 g and two moles of oxygen atoms with mass 32 g. The mass of one mole of carbon dioxide is 44 g (i.e. 12 g + 32 g).

Table 11.2 shows the volumes of one mole of molecules for various gaseous elements and compounds at room temperature and pressure.

Table 11.2
Volume of One Mole of Gaseous Molecules at r.t.p.

	Mass of 1 mole of molecules	*Volume occupied by 1 g*	*Volume occupied by 1 mole*
Chlorine, Cl_2	71 g	334 cm^3	23 700 cm^3
Carbon dioxide, CO_2	44 g	540 cm^3	23 800 cm^3
Argon, Ar	40 g	600 cm^3	24 000 cm^3
Hydrogen chloride, HCl	36.5 g	653 cm^3	23 900 cm^3
Oxygen, O_2	32 g	755 cm^3	24 200 cm^3
Ammonia, NH_3	17 g	1390 cm^3	23 600 cm^3

The volume occupied by 1 mole of gaseous molecules at r.t.p. is always close to 24 000 cm^3. The volume of a mole of gaseous molecules is sometimes called the gramme molecular volume.

The effect of temperature and pressure changes on the volume of a gas

The volume of a fixed mass of gas changes as its temperature or pressure are altered. The laws of Charles and Boyle describe these volume changes in a quantitative way.

Charles' Law: the volume of a fixed mass of gas at constant pressure is directly proportional to its thermodynamic temperature.
(Thermodynamic temperature = Celsius temperature + 273, e.g. 0 °C = 273 K, 19 °C = 292 K, 100 °C = 373 K.)

Boyle's Law: the volume of a fixed mass of gas at constant temperature is inversely proportional to the pressure.

The laws of Charles and Boyle can be combined in the relation

$$\frac{P_1 V_1}{T_1} = \frac{P_2 V_2}{T_2}$$

This relation describes the change in volume of a gas from V_1 to V_2 when its pressure changes from P_1 to P_2 and its thermodynamic temperature changes from T_1 to T_2. An example of how this equation can be used is given below.

Problem: What will be the volume of one mole of gaseous molecules at 0 °C and 760 mm Hg if its volume is known to be 24 000 cm^3 at 19 °C and 760 mm Hg?

Putting down the values which are known,

P_1 = 760 mmHg, V_1 = 24 000 cm^3, T_1 = 292 K

P_2 = 760 mmHg, V_2 is not known, T_2 = 273 K

$$\frac{760\ \text{mmHg} \times 24\,000\ \text{cm}^3}{292\ \text{K}} = \frac{760\ \text{mmHg} \times V_2}{273\ \text{K}}$$

$$V_2 = 24000\ \text{cm}^3 \times \frac{273\ \text{K}}{292\ \text{K}}$$

$$V_2 = 22400\ \text{cm}^3$$

The volume of 1 mole of gaseous molecules at 0 °C and 760 mmHg is always close to 22400 cm^3. This temperature and pressure is sometimes called standard temperature and pressure or s.t.p.

Avogadro's law

One mole of gaseous molecules occupies 24 000 cm^3 at r.t.p. This volume is almost constant for all gases. A chemist frequently uses volumes of gases rather smaller than 24000 cm^3. The number of moles of molecules in 100 cm^3 of **any** gas at r.t.p., for example, is 0.004 17 (i.e. 100/24 000).

Whatever the gaseous substance, there is a constant number of moles (and a constant number of molecules) in a fixed volume at a fixed temperature and pressure. This is Avogadro's law.

Using Avogadro's law to find the formulae of gaseous molecules

If the volumes of reactants and products in a gaseous reaction are known, it is possible to find the formula of one of the molecules involved. It is, however, necessary to know the formulae of all the other molecules taking part in the reaction.

Problem: Complete combustion of 10 cm^3 of a gaseous hydrocarbon C_xH_y was achieved using 90 cm^3 of oxygen. This was an excess of oxygen. 70 cm^3 of gaseous mixture was left after the reaction, this volume being measured at the original room temperature and pressure. The 70 cm^3 of gas formed was shaken with aqueous sodium hydroxide to dissolve carbon dioxide. The volume then decreased to 40 cm^3.

Solution: The final volume of 40 cm^3 is the oxygen remaining from the original 90 cm^3, so 50 cm^3 of oxygen was used. The decrease in volume on shaking with sodium hydroxide is due to the dissolving of carbon dioxide formed by the combustion. The volume of carbon dioxide is 30 cm^3.

10 cm^3 of $C_xH_y(g)$ + 50 cm^3 of $O_2(g) \rightarrow$ 30 cm^3 of $CO_2(g)$ + liquid water of negligible volume

Avogadro's law says that there is a constant number of molecules in a fixed volume of any gas. If 10 cm^3 of any gas contains n molecules,

n molecules of C_xH_y + $5n$ molecules of $O_2 \rightarrow 3n$ molecules of CO_2 + water

or

1 molecule of C_xH_y + 5 molecules of $O_2 \rightarrow$ 3 molecules of CO_2 + water

3 molecules of CO_2 contain 3 atoms of carbon

$\therefore x = 3$

3 molecules of oxygen are used to form 3 molecules of CO_2
∴ 2 molecules of oxygen are used to form water from the hydrogen present in 1 molecule of C_xH_y

∴ $y = 8$

The molecular formula of the hydrocarbon is C_3H_8. This is propane. The equation for its combustion is

$$C_3H_8(g) + 5O_2 \rightarrow 3CO_2(g) + 4H_2O(l)$$

The bonding in the propane molecule is shown in Figure 11.1.

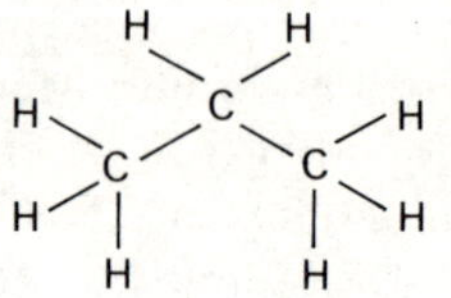

Figure 11.1 Bonding in the propane molecule

Finding the mass of one mole of gaseous molecules

The mass of a known volume of gas at a known temperature and pressure can be determined experimentally in several ways. One simple method involves weighing the gas in a plastic syringe. The measurements can be used to calculate the mass of one mole of the gas.

Problem: 100 cm^3 of ethene (ethylene), C_aH_b, has a mass of 0.117 g at r.t.p. Calculate the mass of 1 mole of ethene and suggest values for *a* and *b*.

Solution:

0.117 g of ethene occupy 100 cm^3 at r.t.p.

∴ $0.117 \times \frac{24\,000}{100}$ g of ethene would occupy 24 000 cm^3 at r.t.p.

this is 1 mole of ethene; its mass is 28 g

but 1 mole of ethene contains *a* moles of carbon atoms and *b* moles of hydrogen atoms

∴ mass of 1 mole of ethene $= a \times 12\text{ g} + b \times 1\text{ g}$

$= (12a + b)\text{ g}$

∴ $12a + b = 28$

a and *b* must both be whole numbers, and *a* cannot be greater than 2

if $a = 1$, $b = 16$

if $a = 2$, $b = 4$

ethene is either CH_{16} or C_2H_4

CH_{16} ↑ Carbon can form only four bonds, and hydrogen forms one bond, so this formula is not possible.

C_2H_4 ↑ It is possible to write a structure for this molecule with carbon and hydrogen forming their correct numbers of bonds. This is shown in Figure 11.2 below.

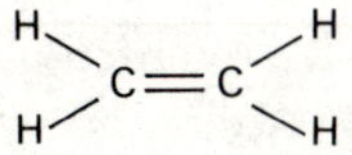

Figure 11.2 Bonding in the ethene molecule

Using the percentage composition by mass of a compound to obtain information about its formula

The percentage composition by mass of ethene, for example, can be found by burning the compound to carbon dioxide and water. Measurement of the masses of carbon dioxide and water produced can be used to calculate the percentage by mass of carbon and hydrogen in this compound.

Ethene, C_aH_b, is found to be composed of 85.7 per cent carbon and 14.3 per cent hydrogen by mass.

100 g of ethene contains 85.7 g of carbon and 14.3 g of hydrogen

100 g of ethene contains $\frac{85.7}{12}$ moles of carbon and $\frac{14.3}{1}$ moles of hydrogen atoms

7.15 moles of C   14.3 moles of H

This shows that in ethene

$$\text{the ratio of moles C:moles H} = 7.15:14.3$$
$$= 1:2$$

The **empirical formula** of ethene is CH_2 (*see* page 29).

The **molecular formula** is $(CH_2)_n$

The mass of 1 mole of ethene is $(12n + 2n)$ g $= 14n$ g

The mass of 1 mole of ethene is known from other measurements to be 28 g (*see* page 110).

$$14n = 28$$
$$n = 2$$

This shows that the molecular formula of ethene is C_2H_4 which is in agreement with the conclusion reached above.

Using Avogadro's law to predict volumes of gases involved in reactions

The problem might be: what volume of air, which is about 20 per cent oxygen, is required for the complete combustion of 10 m^3 of methane, CH_4?

The equation for the reaction must be known if a problem of this type is to be solved.

$$CH_4(g) + 2O_2(g) \rightarrow CO_2(g) + 2H_2O(l)$$

The equation shows that 2 moles of oxygen molecules are required for every mole of methane.

And Avogadro's law indicates that:

2 volumes of oxygen are required for every volume of methane

∴ 20 m^3 of oxygen are required for every 10 m^3 of methane
∴ 100 m^3 of air are required for every 10 m^3 of methane as air is only 20 per cent oxygen.

Using Avogadro's law to investigate equations for gaseous reactions

Avogadro's law can be used to give information about the equations for reactions where gases are involved. The reaction between ammonia and hydrogen chloride, for example, can be investigated by connecting two syringes, each syringe containing a known volume of one of the gases. The tap between the syringes is opened and one gas is pushed into the other syringe. When reaction is over, the volume of any remaining gas is measured. This is illustrated in Figure 11.3.

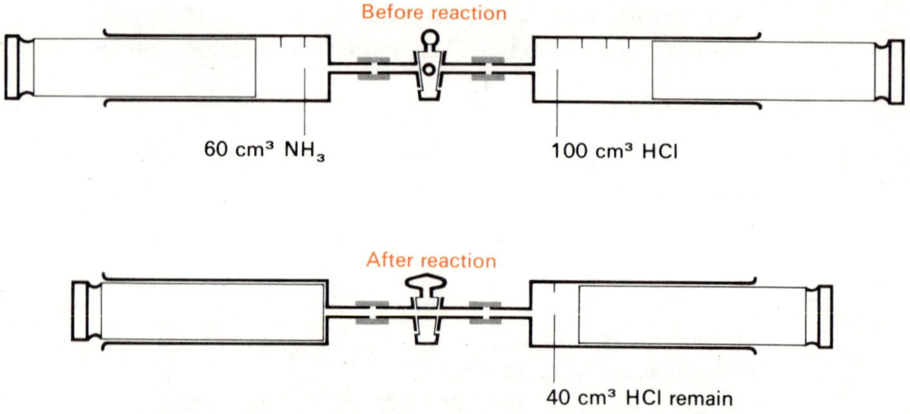

Figure 11.3 The reaction between ammonia and hydrogen chloride

When the gases react they form ammonium chloride which deposits as a fine white solid. In the experiment illustrated, an excess of hydrogen chloride was used. The volume of hydrogen chloride remaining is measured in one syringe. All the ammonia was consumed by reaction with hydrogen chloride.

Volume of ammonia taking part in reaction $= 60 \text{ cm}^3$

Volume of hydrogen chloride used $= 100 \text{ cm}^3 - 40 \text{ cm}^3$

$= 60 \text{ cm}^3$

60 cm^3 of $NH_3(g)$ reacts with 60 cm^3 of $HCl(g) \rightarrow$ ammonium chloride

Let there be *n* molecules in 60 cm^3 of any gas (Avogadro's law).

n molecules of $NH_3(g)$ react with *n* molecules of $HCl(g) \rightarrow$ ammonium chloride

This indicates that the equation for the reaction is

$$NH_3(g) + HCl(g) \rightarrow \text{ammonium chloride}$$

Ammonium chloride is known to be composed of a giant structure of ions with the formula $(NH_4^+Cl^-)$. This makes it possible to write the complete equation.

$$NH_3(g) + HCl(g) \rightarrow (NH_4^+Cl^-)(s)$$

Summary

1. One mole of atoms of an element contains the same number of atoms as there are carbon atoms in 12 g of carbon.

   One mole of molecules of an element or compound contains the same number of molecules as there are carbon atoms in 12 g of carbon.

   The constant number of atoms or molecules in one mole is called the Avogadro constant, represented by the symbol L.

2. The volume occupied by one mole of gaseous molecules at r.t.p. is always close to $24\,000 \text{ cm}^3$. If the volume of a known mass of gas can be measured, the mass of $24\,000 \text{ cm}^3$ of the gas can be calculated. This gives the chemist a method for finding the mass of one mole for a gaseous substance.

3. The volume of a gas changes as its temperature or pressure are altered. The laws of Charles and Boyle describe these changes. The laws can be summarized by the equation

$$\frac{P_1V_1}{T_1} = \frac{P_2V_2}{T_2}$$

4. Avogadro's law: whatever the gaseous substance, there is a constant number of moles (and a constant number of molecules) in a fixed volume at a fixed temperature and pressure.
   Using Avogadro's law it is possible to:
   (a) predict the volumes of gases which will take part in a gaseous reaction if the equation for the reaction is known
   (b) find the equation for a gaseous reaction if the volumes of gases involved can be measured
   (c) find the formula of a gaseous molecule if the volume changes can be measured when this molecule takes part in a gaseous reaction; the formulae of all the other molecules involved must be known.

12 Investigating Chemical Equations

Molecules of substances with a giant structure

Silicon dioxide is a compound with a giant structure of atoms; the empirical formula is SiO_2. Sodium chloride is a compound with a giant structure of ions; the empirical formula is NaCl or (Na^+Cl^-). **One mole of the substance contains the same number of units with stated empirical formula as there are carbon atoms in 12 g of carbon.**

One mole of silicon dioxide, SiO_2, has a mass of 60 g. This mass of silicon dioxide contains 1 mole of silicon atoms (mass 28 g) and 2 moles of oxygen atoms (mass 32 g).

One mole of sodium chloride, (Na^+Cl^-), has a mass of 58.5 g. This mass of sodium chloride contains 1 mole of sodium ions (mass 23 g) and 1 mole of chloride ions (mass 35.5 g).

One mole of calcium carbonate, $(Ca^{2+}CO_3^{2-})$, has a mass of 100 g. This mass of calcium carbonate contains 1 mole of calcium ions (mass 40 g) and 1 mole of carbonate ions (mass 60 g).

One mole of a substance with a giant structure is sometimes called one gramme formula. One gramme formula of silicon dioxide, SiO_2, has a mass of 60 g. It is necessary to state the formula of the substance when using the term gramme formula.

Investigating the equation for the reaction between calcium carbonate and dilute hydrochloric acid

Calcium carbonate can be obtained in a pure crystalline form called calcite. A small crystal of calcite of known mass is added to excess dilute hydrochloric acid in the apparatus shown in Figure 12.1. The calcite reacts rapidly and the volume of carbon dioxide produced is measured using a syringe.

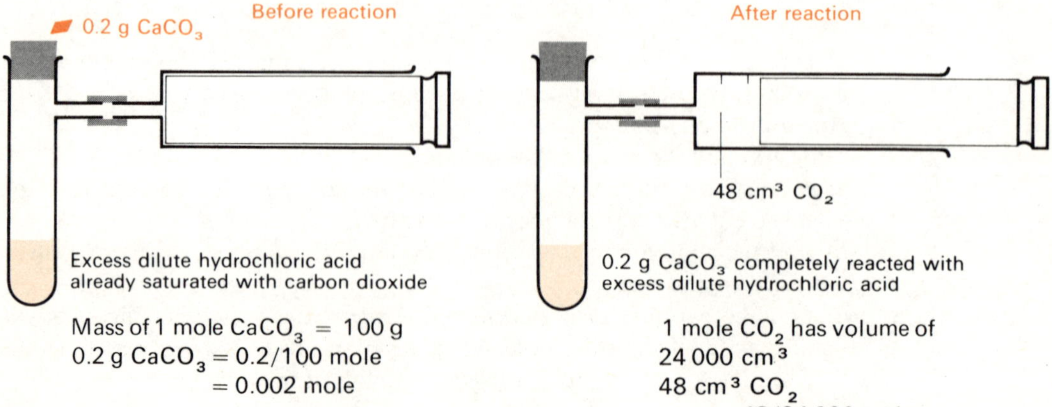

Mass of 1 mole $CaCO_3$ = 100 g
0.2 g $CaCO_3$ = 0.2/100 mole
= 0.002 mole

1 mole CO_2 has volume of 24 000 cm^3
48 cm^3 CO_2
= 48/24 000 mole
= 0.002 mole

Figure 12.1

The experiment outlined in Figure 12.1 shows that

$$0.002 \text{ mole } CaCO_3 \rightarrow 0.002 \text{ mole } CO_2$$

$$\therefore 1 \text{ mole } CaCO_3 \rightarrow 1 \text{ mole } CO_2$$

The equation for the reaction is of the form

$$CaCO_3(s) + n\ HCl(aq) \rightarrow CO_2(g) +$$

The experiment only gives information about the relative numbers of moles of $CaCO_3$ and CO_2. The complete equation cannot be written from the evidence of this experiment alone.

The balanced equation which is in agreement with this experiment is

$$CaCO_3(s) + 2HCl(aq) \rightarrow CO_2(g) + H_2O(l) + CaCl_2(aq)$$

The information summarized by the equation can be put into the following words: 1 mole of solid calcium carbonate reacts with 2 moles of aqueous hydrochloric acid. The reaction produces 1 mole of gaseous carbon dioxide, 1 mole of liquid water and 1 mole of aqueous calcium chloride.

Molar solutions

A **molar** solution of a soluble substance contains 1 mole of that substance in 1000 cm^3 (one litre) of aqueous solution.

The number of moles of a substance contained in 1000 cm^3 of aqueous solution is called the **molarity** of that solution.

	Molarity
58.5 g of NaCl (1 mole of NaCl) in 1000 cm^3 of solution	1 M
5.85 g of NaCl (0.1 mole of NaCl) in 1000 cm^3 of solution	0.1 M
5.85 g of NaCl (0.1 mole of NaCl) in 100 cm^3 of solution	1 M
7.3 g of HCl (0.2 mole of HCl) in 1000 cm^3 of solution	0.2 M
7.3 g of HCl (0.2 mole of HCl) in 100 cm^3 of solution	2 M

A chemist frequently uses solutions of known molarity both to investigate equations for reactions and also to find the molarities of other solutions. These experiments use volumes of solutions considerably smaller than 1000 cm^3. This makes it necessary for a chemist to be able to calculate the number of moles of a substance contained in a few cm^3 of solution. Some examples of these calculations are given below.

1000 cm^3 of 2 M HCl contain 2 moles of HCl

$$1 \text{ cm}^3 \text{ of 2 M HCl contains } 2 \times \frac{1}{1000} = 0.002 \text{ mole of HCl}$$

$$20 \text{ cm}^3 \text{ of 2 M HCl contain } 2 \times \frac{20}{1000} = 0.04 \text{ mole of HCl}$$

$$15 \text{ cm}^3 \text{ of 0.1 M NaOH contain } 0.1 \times \frac{15}{1000} = 0.0015 \text{ mole of NaOH}$$

Using solutions of known molarity to investigate equations

The reaction between calcium carbonate and dilute hydrochloric acid

A lump of calcium carbonate (calcite) of known mass is added to a known volume of known molarity hydrochloric acid. The quantity of calcium carbonate can be chosen so that it consumes all the hydrochloric acid, leaving a small lump of solid which is taken out, washed, dried and weighed. The experiment is illustrated in Figure 12.2.

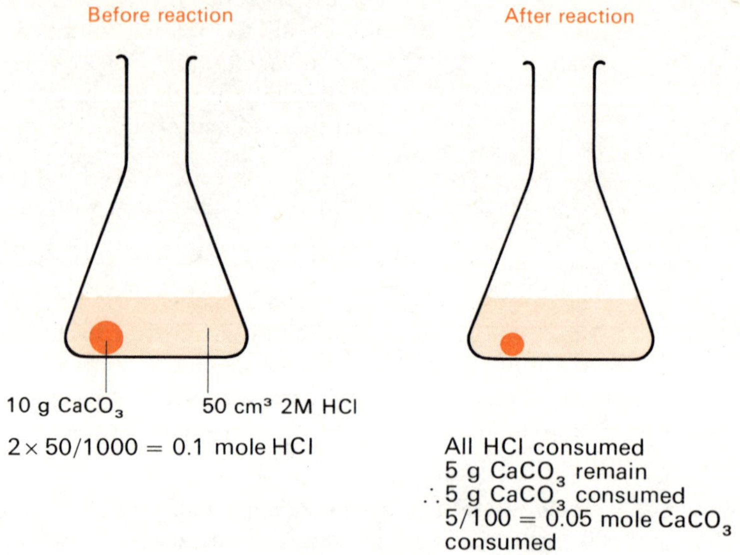

Figure 12.2

The experiment shows that

0.05 mole of $CaCO_3$ react with 0.1 mole of HCl

∴ 1 mole of $CaCO_3$ reacts with 2 moles of HCl

The equation for the reaction is of the form

$$CaCO_3(s) + 2HCl(aq) \rightarrow$$

This experiment alone gives no further information about the equation. Further investigation would show that the complete equation is

$$CaCO_3(s) + 2HCl(aq) \rightarrow CO_2(g) + H_2O(l) + CaCl_2(aq)$$

This experiment should be compared with the investigation of the same reaction described on pages 114–5.

The reaction between lead nitrate and potassium iodide

Lead nitrate and potassium iodide react rapidly in aqueous solution. A deep yellow precipitate of lead iodide is formed. The result of an investigation of this reaction is illustrated in Figure 12.3. 8 cm^3 of molar potassium iodide were measured out into a test-tube. 1 cm^3 of molar lead nitrate was added. The mixture was stirred and the test-tube was then placed in a centrifuge for a few seconds to settle the precipitate. The height of the precipitate was recorded.

Further 1 cm^3 portions of lead nitrate were added, the mixture being centrifuged after each addition.

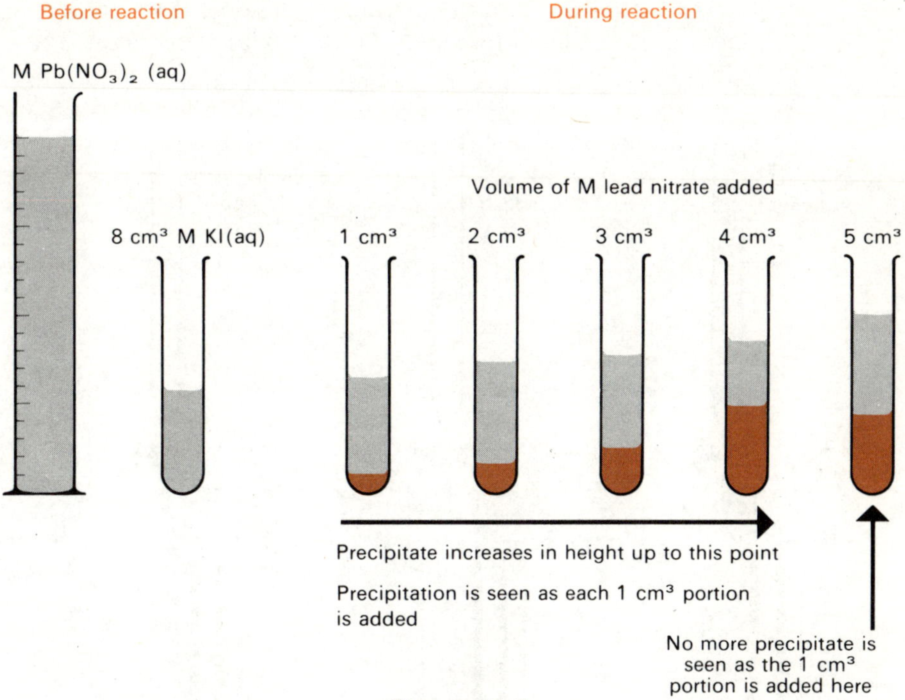

Figure 12.3

The precipitate was found to increase in height until 4 cm^3 of lead nitrate had been added. Further addition of lead nitrate produced no more precipitate. The height of the precipitate then dropped as further centrifuging made it more compact.

This experiment shows that

$$4\ cm^3\ M\ Pb(NO_3)_2 \text{ react with } 8\ cm^3\ M\ KI$$

$$4\ cm^3\ M\ Pb(NO_3)_2 \text{ contain } 1\times\frac{4}{1000} = 0.004 \text{ mole } Pb(NO_3)_2$$

$$8\ cm^3\ M\ KI \qquad \text{contain } 1\times\frac{8}{1000} = 0.008 \text{ mole KI}$$

$$0.004 \text{ mole } Pb(NO_3)_2 \text{ react with } 0.008 \text{ mole KI}$$

$$\therefore\ 1 \text{ mole } Pb(NO_3)_2 \text{ reacts with } 2 \text{ moles KI}$$

These results suggest the equation

$$Pb(NO_3)_2(aq) + 2KI(aq) \rightarrow PbI_2(s) + 2KNO_3(aq)$$

The equation written in ionic form is

$$Pb^{2+}(aq) + 2I^-(aq) \rightarrow (Pb^{2+}2I^-)(s)$$

The reaction between aqueous barium hydroxide and dilute sulphuric acid

An experiment to investigate this reaction is shown in Figure 12.4. Dilute sulphuric acid of known molarity is added from a burette to a measured volume of known molarity barium hydroxide. Two carbon electrodes are placed in the reaction mixture. The electrodes are connected to a battery and bulb.

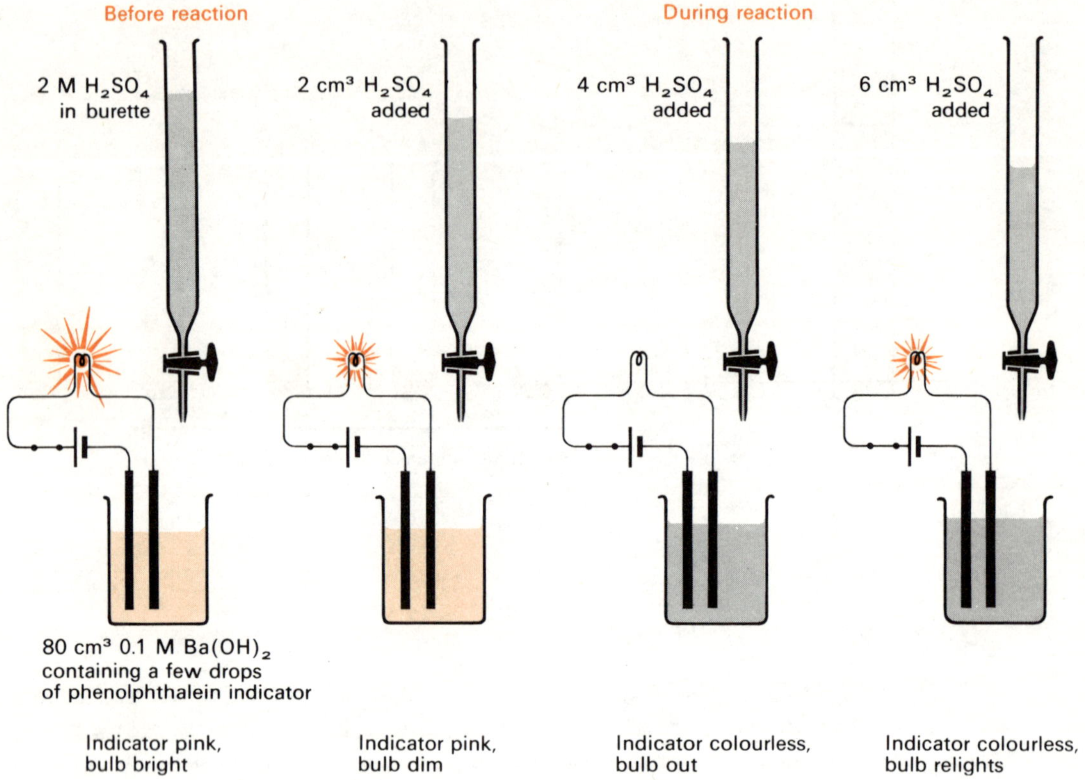

Figure 12.4

As the sulphuric acid is added, the reaction mixture becomes less conducting and the bulb decreases in brightness. When 4 cm^3 of acid have been added, the bulb goes out. The bulb relights as more acid is added. The reaction mixture also contains an indicator which is pink in alkaline solution and colourless in acidic solution. The indicator changes from pink to colourless when 4 cm^3 of acid have been added.

This method for following a reaction is called a **titration**. The **end-point** in the titration occurs when just enough sulphuric acid has been added to completely consume the barium hydroxide. The end-point is shown by a sudden change in the properties of the reaction mixture. In this reaction the end-point is when the conductivity of the mixture reaches a minimum, and when the indicator changes colour.

4 cm^3 of 2 M H_2SO_4 react with 80 cm^3 of 0.1 M $Ba(OH)_2$

Number of moles of H_2SO_4 in 4 cm^3 of 2 M $= 4 \times \frac{2}{1000} = 0.008$

Number of moles of $Ba(OH)_2$ in 80 cm^3 of 0.1 M $= 0.1 \times \frac{80}{1000} = 0.008$

0.008 mole of H_2SO_4 react with 0.008 mole of $Ba(OH)_2$

$\therefore$ 1 mole of H_2SO_4 reacts with 1 mole of $Ba(OH)_2$

The equation for the reaction is of the form

$$H_2SO_4(aq) + Ba(OH)_2(aq) \rightarrow$$

The titration alone gives no further information about the equation. During the reaction, however, a white precipitate is seen to form. This could be identified as barium sulphate, $BaSO_4$. If the reaction is recognized to be one in which an acid neutralizes a base to give a salt and water, the complete equation can be written.

$$H_2SO_4(aq) + Ba(OH)_2(aq) \rightarrow BaSO_4(s) + 2H_2O(l)$$

The ionic form of the equation is

$$2H^+(aq) + SO_4^{2-}(aq) + Ba^{2+}(aq) + 2OH^-(aq) \rightarrow (Ba^{2+}SO_4^{2-})(s) + 2H_2O(l)$$

Ions are removed from solution as this reaction takes place. Here is the reason for the bulb going out at the end-point: the bulb relights after the end-point as an excess of ionized sulphuric acid is then present.

Using equations for reactions to find molarities of solutions

An experiment to find the molarity of aqueous sodium hydroxide is illustrated in Figure 12.5. A known volume of sodium hydroxide is placed in the insulated

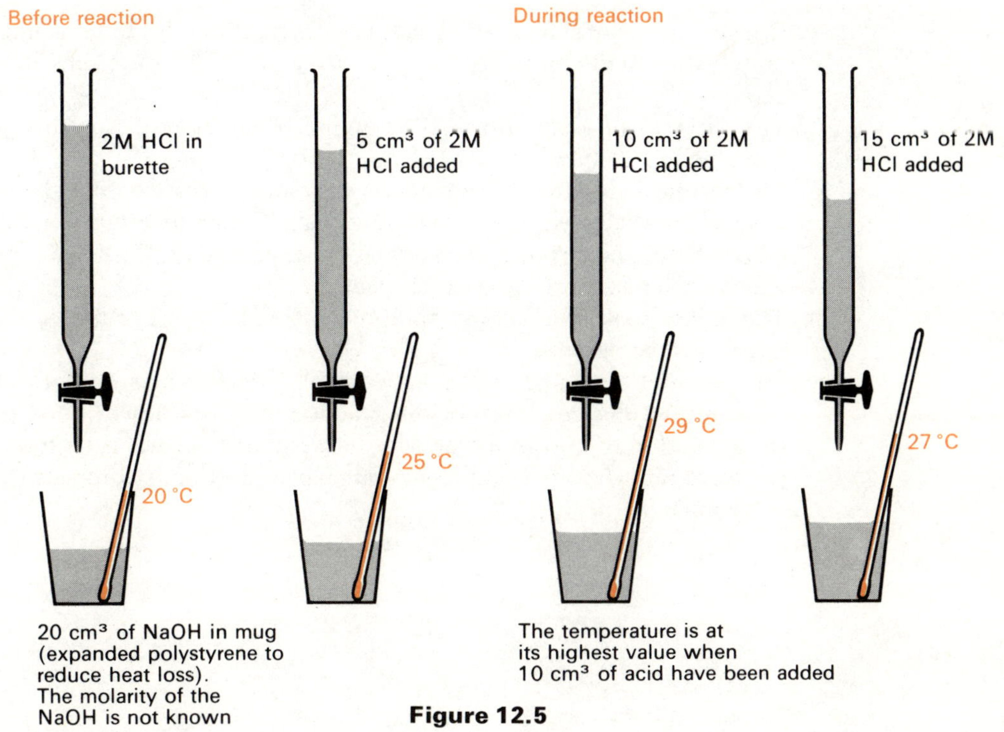

Figure 12.5

container, and its temperature is found. Measured volumes of known molarity acid are added. The temperature of the mixture is recorded after each addition of acid. A graph could be drawn of temperature of mixture plotted against volume of acid added. The temperature increases at first as the acid is added. The temperature increases because of the heat produced by the neutralization reaction. After sufficient acid has been added for neutralization, the temperature begins to decrease. The temperature drops because colder acid is being added and because heat is gradually lost to the surroundings.

The end-point in the titration illustrated in Figure 12.5 occurs when 10 cm^3 of acid have been added.

10 cm^3 of 2 M HCl are needed to neutralize 20 cm^3 of NaOH

10 cm^3 of 2 M HCl contain $2 \times \frac{10}{1000} = 0.02$ mole of HCl

The equation for the neutralization reaction is

$$HCl(aq) + NaOH(aq) \longrightarrow NaCl(aq) + H_2O(l)$$

The equation shows that 1 mole of HCl reacts with 1 mole of NaOH or 0.02 mole of HCl react with 0.02 mole of NaOH.

The solution of NaOH used must contain 0.02 mole of NaOH in 20 cm^3 or $0.02 \times 1000/20 = 1.0$ mole in 1000 cm^3. Therefore, the solution of NaOH is 1.0 M.

Summary

1. One mole of a substance with a giant structure contains the same number of units with stated empirical formula as there are carbon atoms in 12 g of carbon.
2. The molarity of a solution is the number of moles of dissolved substance contained in 1000 cm^3 (one litre) of solution.
3. The chemical equation for a reaction can only be written after the reaction has been investigated in a quantitative way. The relative numbers of moles of reactants and products are found by experiments. The equation summarizes the results of these experiments.
4. The molarity of an aqueous solution can frequently be found by titration. For this to be possible:
   (a) a second substance must be available which reacts with the first in solution. The equation for this reaction must be known
   (b) a solution of known molarity of the second substance is required
   (c) there must be a method for finding when reaction is complete (the end-point)

13 Energy Changes
1: Heat energy and electrical energy

The heat of reaction

As a chemical reaction takes place there usually is a change in the temperature of the reaction mixture. The temperature sometimes drops, but more frequently it increases. The temperature changes during the reaction between zinc and aqueous copper(II) sulphate are illustrated in Figure 13.1. This reaction can be represented by the equation

$$Zn(s) + CuSO_4(aq) \rightarrow Cu(s) + ZnSO_4(aq)$$

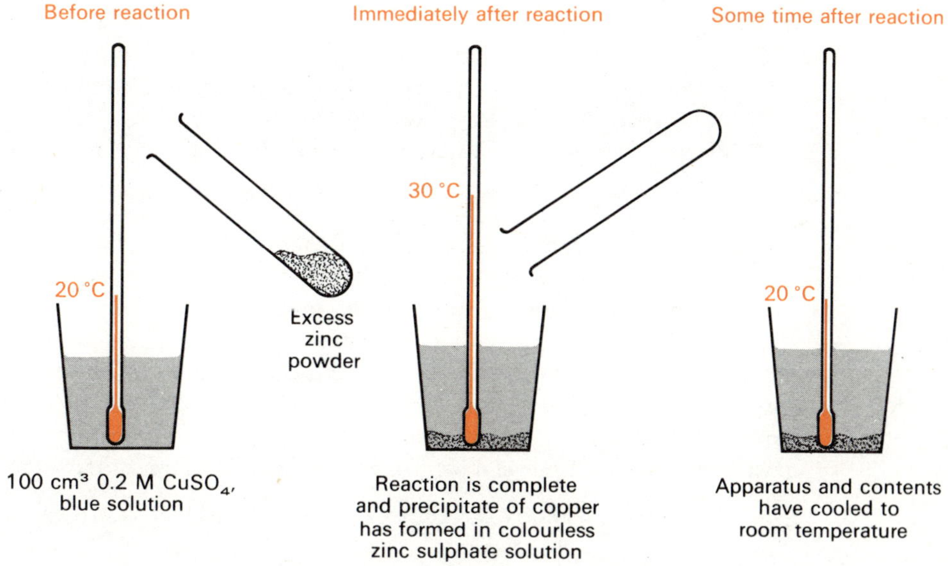

Figure 13.1 Temperature changes in the reaction between zinc and aqueous copper(II) sulphate

When the reaction between zinc and aqueous copper(II) sulphate is complete, the temperature of the mixture has increased by about 10 °C. The mixture then gradually loses heat energy to the surroundings and the temperature drops back to room temperature. The products of the reaction at room temperature (copper and aqueous zinc sulphate) contain less energy than the original reactants at the same temperature (zinc and aqueous copper(II) sulphate).

This energy has not disappeared; it has been transferred from the chemicals to their surroundings. The energy transferred to the surroundings in the experiment illustrated in Figure 13.1 can be calculated in the following way.

For water and for dilute aqueous solutions it is known that about 4.2 joules are transferred to the surroundings when the temperature of 1 cm^3 decreases by 1 °C.

4.2 J are transferred when 1 cm^3 decreases by 1 °C

4.2×100 J are transferred when 100 cm^3 decreases by 1 °C

$4.2 \times 100 \times 10$ J are transferred when 100 cm^3 decreases by 10 °C. This is 4200 J, or 4.2 kJ.

When excess zinc reacts with 100 cm^3 of 0.2 M $CuSO_4$, 4.2 kJ are transferred to the surroundings.

$$100 \text{ cm}^3 \text{ of } 0.2 \text{ M } CuSO_4 \text{ contain } 0.2 \times \frac{100}{1000} = 0.02 \text{ mole of } CuSO_4.$$

When excess zinc reacts with **one mole** of aqueous copper(II) sulphate, the energy transferred to the surroundings will be $4.2 \times 1/0.02 = 210$ kJ.

There are several sources of error in this simple experiment to find the energy transferred to the surroundings during a chemical reaction. The sources of error include the following.

- Some energy is transferred to the surroundings before the maximum temperature has been reached. If this energy transfer could be prevented, the temperature rise would be a little more than 10 °C.
- Some energy is transferred to the container and thermometer as their temperature rises by 10 °C. This energy was not considered in the calculation.
- The energy transferred to the surroundings when the temperature of 1 cm^3 of an aqueous solution drops by 1 °C is not exactly 4.2 J. The energy varies slightly from solution to solution.

These sources of error can be eliminated or allowed for in accurate work. It is then found that one mole of copper and one mole of aqueous zinc sulphate (products) contain 216 kJ less energy than 1 mole of zinc and 1 mole of aqueous copper(II) sulphate (reactants) at the same temperature. This information about the energy change can be given in the chemical equation for the reaction.

$$Zn(s) + CuSO_4(aq) \rightarrow Cu(s) + ZnSO_4(aq) \quad \Delta H = -216 \text{ kJ/mole}$$

↑ This is called the **heat of reaction** (or enthalpy change). The energy change is for molar quantities as represented by the equation for the reaction. The value of ΔH is negative as energy is transferred to the surroundings, i.e. lost by the chemicals. A reaction in which energy is transferred to the surroundings is said to be **exothermic**.

The energy change in the reaction can be shown on an **energy level diagram.**

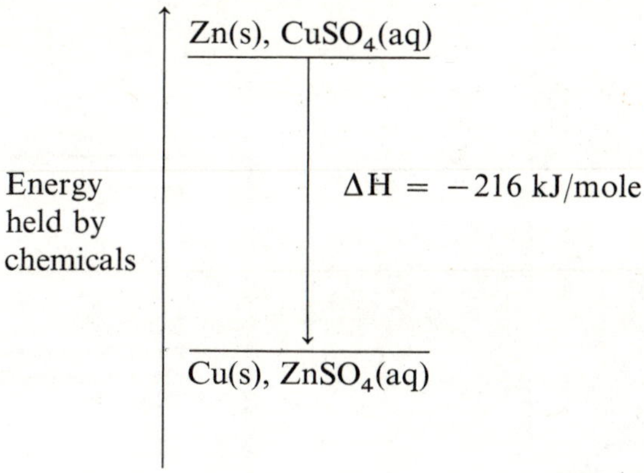

The reactants in this reaction are on the higher energy level.

Energy is transferred to the surroundings in this reaction which is exothermic.

The products in this reaction are on the lower energy level.

Obtaining electrical energy from a chemical reaction

The reaction between zinc and copper(II) sulphate takes place in aqueous solution. Aqueous copper(II) sulphate is composed of separate positive and negative ions. Aqueous zinc sulphate, one of the reaction products, is also composed of separate ions. The equation for the reaction can be written in ionic form.

$$Zn(s) + Cu^{2+}(aq) + SO_4^{2-}(aq) \rightarrow Cu(s) + Zn^{2+}(aq) + SO_4^{2-}(aq)$$

The sulphate ions are spectator ions. They take no part in the reaction and can be excluded from the equation, which becomes

$$Zn(s) + Cu^{2+}(aq) \rightarrow Cu(s) + Zn^{2+}(aq)$$

When the equation is written in this way it can be seen that the reaction is one in which zinc is ionizing, while copper is deionizing.

$$Zn(s) \rightarrow Zn^{2+}(aq) + 2e$$

$$\downarrow$$

$$2e + Cu^{2+}(aq) \rightarrow Cu(s)$$

The electrons released when zinc atoms ionize are transferred to copper ions. The neutralization of the positive charge on copper ions produces unionized copper atoms. This change is called **electron transfer**, and is discussed more fully in Chapter 17. It is possible to construct a simple cell, called a **Daniell cell**, which produces electrical energy from this electron transfer reaction. The cell is illustrated in Figure 13.2. Electrical energy is produced as the electrons move through the external circuit from the zinc rod to the aqueous copper sulphate. The voltage of the cell is 1.1 volt. This is the maximum voltage of the cell, which is the value when no current is flowing. If the bulb is switched into the circuit, current flows, the bulb lights and the voltage drops below 1.1 volt.

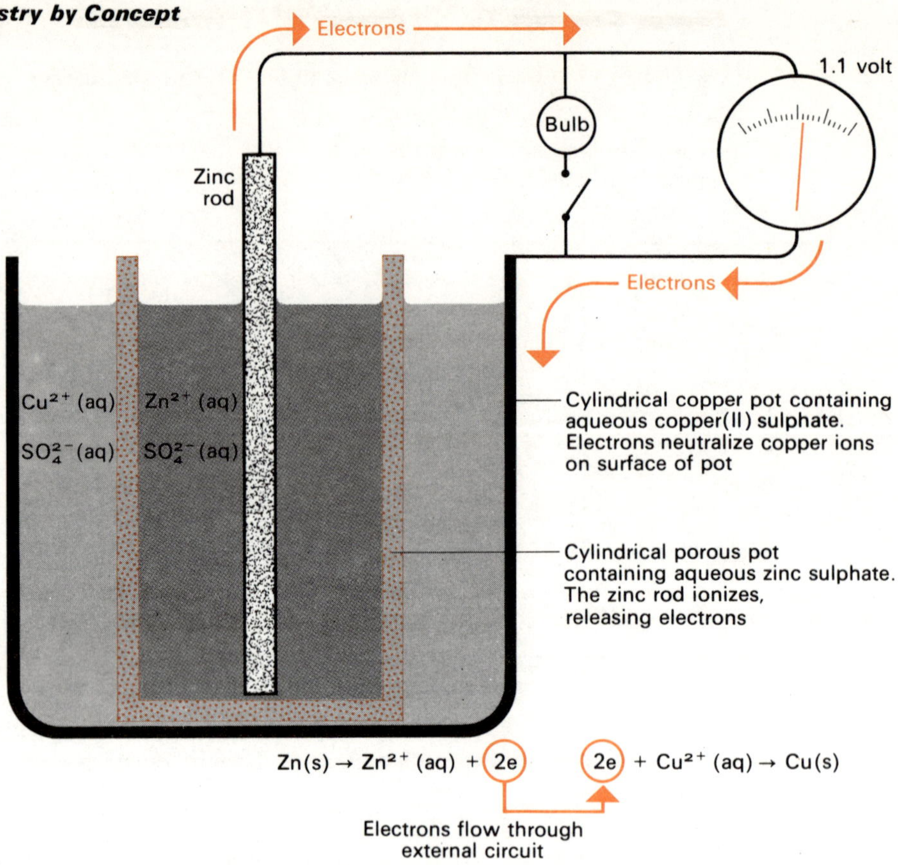

Figure 13.2 Electron transfer in the Daniell cell

When one mole of zinc reacts with one mole of aqueous copper(II) ions, two moles of electrons are transferred from zinc atoms to copper ions.

2 moles of electrons = 2 Faradays = 2 × 96 500 coulombs

energy change when 1 coulomb is transferred at 1 volt = 1 joule (by definition)

∴ energy change when 2 × 96 500 coulombs are transferred at 1.1 volt = 2 × 96 500 × 1.1 J

= 212 kJ

For the reaction

$$Zn(s) + Cu^{2+}(aq) \rightarrow Cu(s) + Zn^{2+}(aq)$$

Maximum electrical energy which can be obtained	= **212** kJ/mole	Transferred from chemicals in form of electrical energy
Heat of reaction (enthalpy change)	= **216** kJ/mole	Transferred from chemicals to surroundings in form of heat energy

The electrical energy that can be obtained from a chemical reaction does not have the same value as the heat of reaction.

The Principle of Conservation of Energy

The total energy change in the reaction between zinc and aqueous copper(II) sulphate is always 216 kJ/mole. This energy can appear in many ways.

- All the energy can appear as heat energy which is transferred to the surroundings.
- Some of the energy (but never more than 212 kJ/mole) can appear as electrical energy. When this happens, some of the energy (but never less than 4 kJ/mole) always appears as heat energy. The temperature of the Daniell cell increases slightly and heat energy is transferred to the surroundings.
- The electrical energy (never more than 212 kJ/mole) can be converted to other forms of energy such as light, sound, heat and work.

Some of the ways in which the energy of 216 kJ/mole can appear are shown in Figure 13.3.

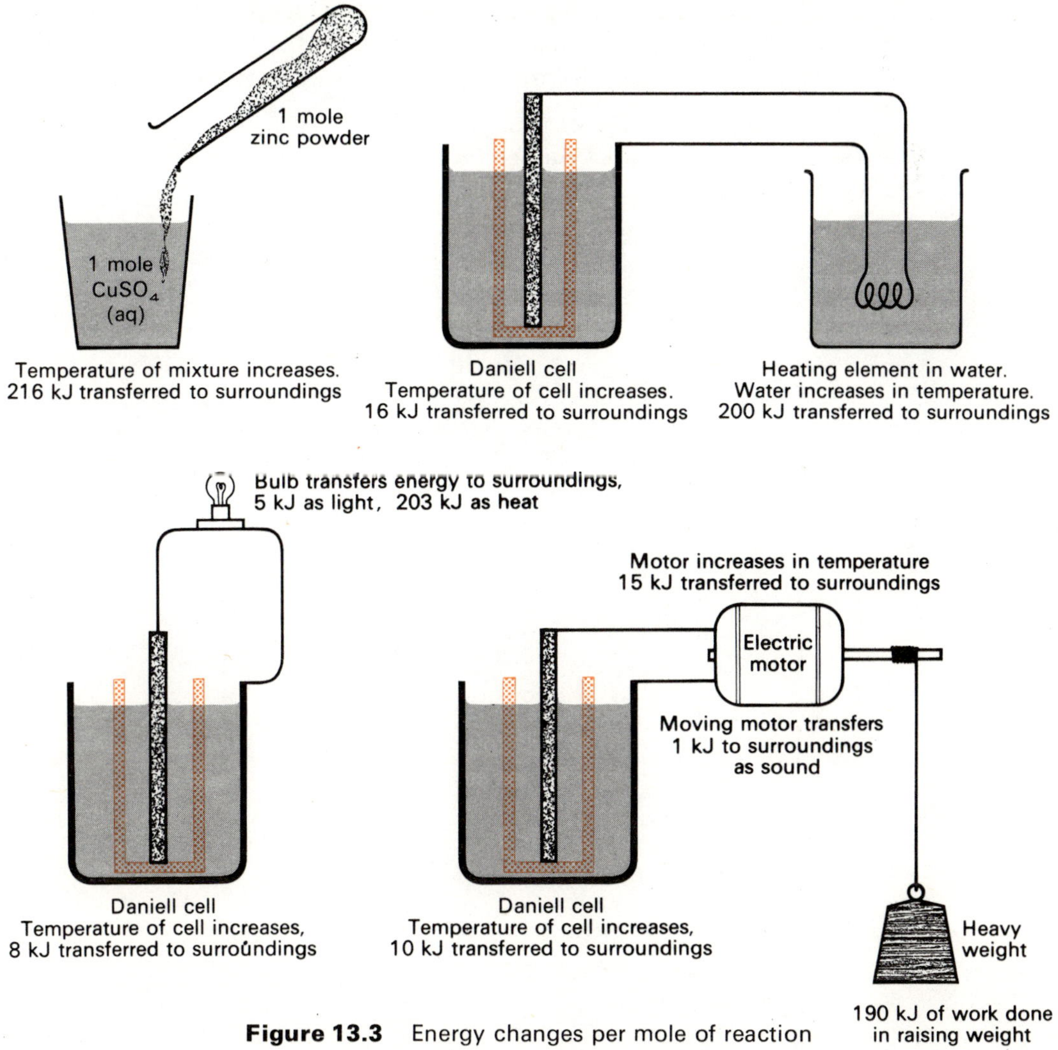

Figure 13.3 Energy changes per mole of reaction
$Zn(s) + Cu^{2+}(aq) \rightarrow Cu(s) + Zn^{2+}(aq)$

Looking at the four situations outlined in Figure 13.3 it will be seen that the total energy produced by the chemical reaction is always 216 kJ/mole. The Principle of Conservation of Energy is at work here. Imagine the laboratory in which the experiments are being carried out. On a bench stand molar quantities of zinc and aqueous copper(II) sulphate. These chemicals contain 216 kJ more energy than will be contained in the copper and aqueous zinc sulphate which in a moment they are going to form. The chemicals are then allowed to react, either directly or in a Daniell cell.

When the zinc and aqueous copper(II) sulphate react directly, 216 kJ of energy stored in the original chemicals become 216 kJ of heat energy stored in the hot aqueous solution of zinc sulphate. The hot solution cools down and the 216 kJ is dispersed, first into the air of the room and then into all the objects in the room.

When the zinc and aqueous copper(II) react in a Daniell cell, the possible ways of dispersing the energy stored in the chemicals are endless. Three such ways are illustrated in Figure 13.3. All these ways, however, have one common feature. If the various forms of energy which have been produced are gathered together, the total is always 216 kJ. The total is never more or less than this value.

Energy can be changed from one form to another. Energy cannot be created or destroyed.

This is one way of stating the Principle of Conservation of Energy.

Summary for Chapters 13 and 14

1. The products of an exothermic reaction contain less energy than the reactants at the same temperature. This energy is lost to the surroundings.
2. The products of an endothermic reaction contain more energy than the reactants at the same temperature. This energy is gained from the surroundings.
3. The heat of reaction is the energy change on using the numbers of moles shown in the chemical equation.
4. The heat of combustion is the energy lost on complete combustion of one mole of the substance.
5. The heat of solution is the energy change on dissolving one mole of the substance.
6. Neutralization reactions are exothermic. The heat of reaction (heat of neutralization) is frequently close to -58 kJ/mole. The neutralization reaction for fully ionized acids and bases is always the same.

$$H^+(aq) + OH^-(aq) \rightarrow H_2O(l) \quad \Delta H = -58 \text{ kJ/mole}$$

7. Energy level diagrams can be used to show the energy changes in reactions. In an exothermic reaction the products are at a lower energy level than the reactants. In an endothermic reaction the products are at a higher energy level than the reactants.
8. The Principle of Conservation of Energy states: energy can be changed from one form to another; energy cannot be created or destroyed.
9. Electrical energy can be obtained from some chemical reactions in suitable cells. The electrical energy is not the same as the heat of reaction.
10. The energy change in a chemical reaction comes from the forming and breaking of bonds of different strength. In an exothermic reaction, the bonds formed are stronger than the bonds broken. In an endothermic reaction, the bonds formed are weaker than the bonds broken.

14 Energy Changes
2: Combustion, solution and neutralization

Heat of combustion

Reactions involving combustion can be investigated readily in simple laboratory experiments. A volatile flammable liquid is burnt in a spirit lamp. The change in mass of the lamp shows the mass of the liquid which has been burnt. The energy lost by the chemicals during the reaction can be used to increase the temperature of a measured volume of water held in a suitable container. One method for investigating the energy change on combustion is illustrated in Figure 14.1.

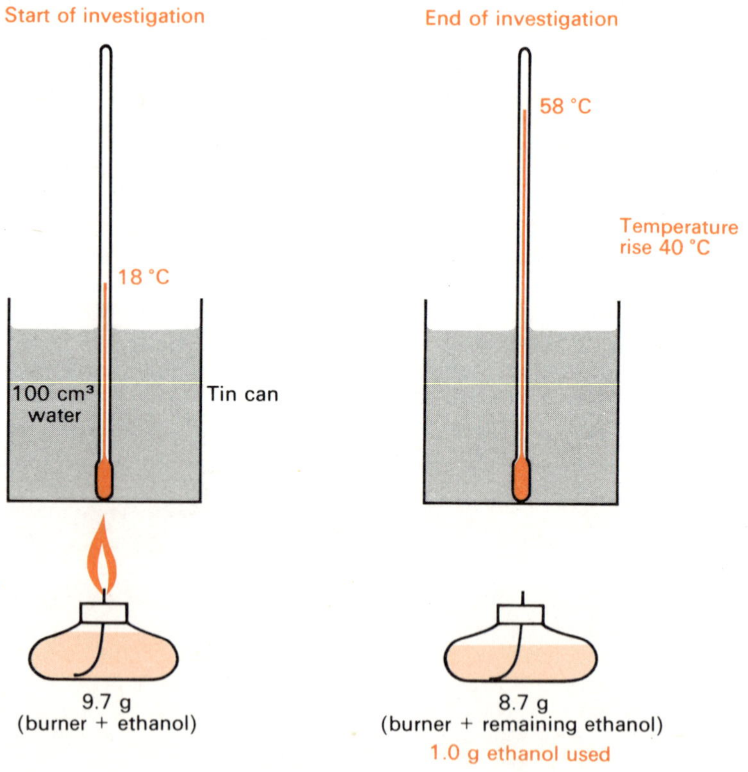

Figure 14.1 Investigation of heat of combustion of ethanol

The experiment illustrated in Figure 14.1 shows that when 1 g of ethanol burns, the energy transferred to the water increases the temperature of 100 cm^3 by 40 °C. It is known that 4.2 J must be transferred to 1 cm^3 of water to increase the temperature by 1 °C.

4.2 J are transferred to the water when 1 cm^3 increases in temperature by 1 °C

$\therefore$ $4.2 \times 100 \times 40$ J are transferred when 100 cm^3 increases in temperature by 40 °C. This is 16800 J, or 16.8 kJ.

Energy lost by ethanol (C_2H_5OH) when 1 g burns = 16.8 kJ

$\therefore$ Energy lost by ethanol when 1 mole (46 g) burns $= 16.8 \times 46$ kJ

$= 773$ kJ

More accurate experiments, however, show that the total energy lost by ethanol when 1 mole burns is 1230 kJ. In the simple experiment illustrated in Figure 14.1 only about one-half of the energy lost by the ethanol is transferred to the water of the container. A large proportion of the energy is lost straight to the atmosphere. The information about the total energy change obtained from accurate experiments can be included in the equation for the reaction.

$$C_2H_5OH(l) + 3O_2(g) \rightarrow 2CO_2(g) + 3H_2O(g) \quad \Delta H = -1230 \text{ kJ/mole}$$

The heat of reaction is negative because the products of the reaction, carbon dioxide and steam, contain less energy than the original ethanol and oxygen. **The energy lost by one mole of a substance on combustion is called the heat of combustion.** The energy level diagram for this reaction is shown in Figure 14.2.

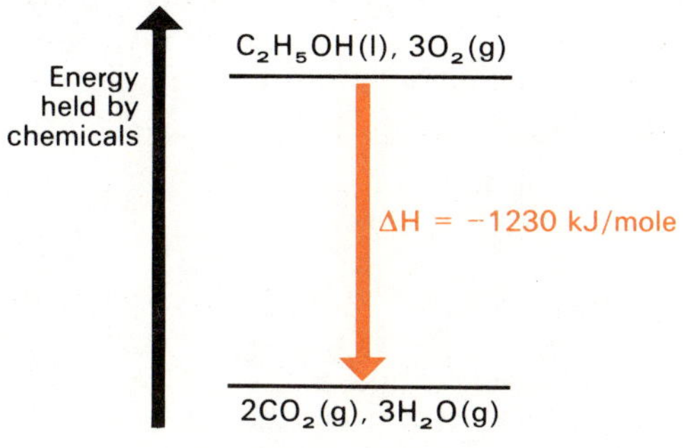

Figure 14.2 Energy level diagram for combustion of ethanol

Heat of solution

There is usually an energy change when a solid dissolves in water. **The energy change when one mole of solid is dissolved is called the heat of solution.** An experiment to find the heat of solution of copper(II) sulphate pentahydrate is illustrated in Figure 14.3.

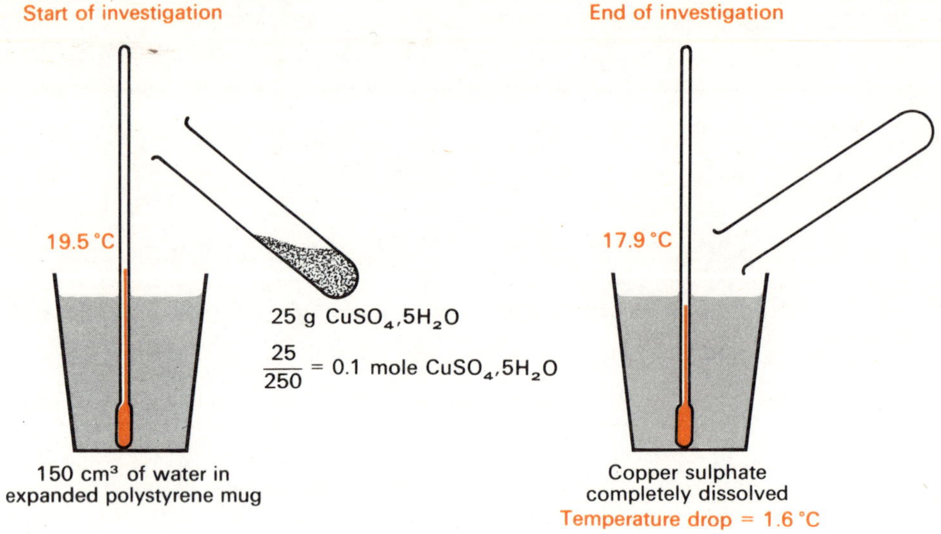

Figure 14.3 Investigation of heat of solution of copper(II) sulphate pentahydrate

The temperature drops during the experiment illustrated in Figure 14.3. The aqueous solution of copper(II) sulphate then gradually warms up again to room temperature. An aqueous solution of copper(II) sulphate contains more energy at room temperature than is contained in the solid copper(II) sulphate and water from which it is formed. **A reaction in which the chemicals contain more energy after the reaction has taken place is said to be endothermic.** The energy absorbed by the aqueous copper(II) sulphate when its temperature increases from 17.9 °C to 19.5 °C is calculated as shown below.

1 cm^3 of aqueous solution absorbs about 4.2 J when the temperature increases by 1 °C

∴ 150 cm^3 of aqueous solution absorbs $150 \times 1.6 \times 4.2$ J when the temperature increases by 1.6 °C. This is very close to 1000 J, or 1 kJ

The energy absorbed from the surroundings when 0.1 mole of $CuSO_4,5H_2O$ dissolves is 1 kJ. When 1 mole of the compound is used, the energy absorbed is 10 kJ. This information can be included in the equation for the reaction.

$$CuSO_4,5H_2O(s) + aq \rightarrow CuSO_4(aq) \quad \Delta H = +10 \text{ kJ/mole}$$

The heat of solution of $CuSO_4,5H_2O$ in water is +10 kJ/mole. The energy level diagram for this reaction is shown in Figure 14.4.

The heat of solution of anhydrous copper(II) sulphate is very different from that of the pentahydrate. An experiment to find this heat of solution is illustrated in Figure 14.5.

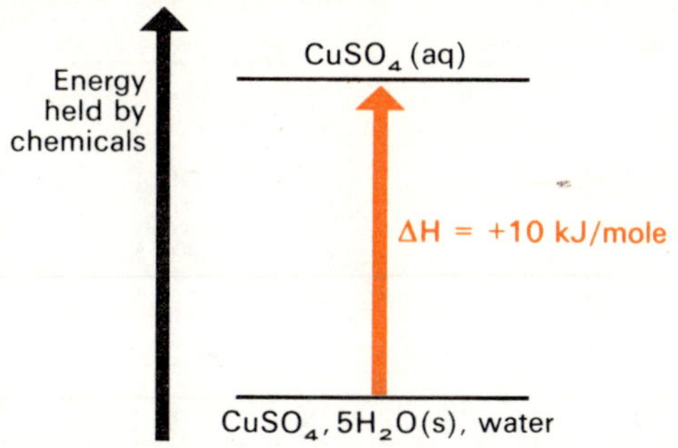

Figure 14.4 Energy level diagram for copper(II) sulphate pentahydrate dissolving in water

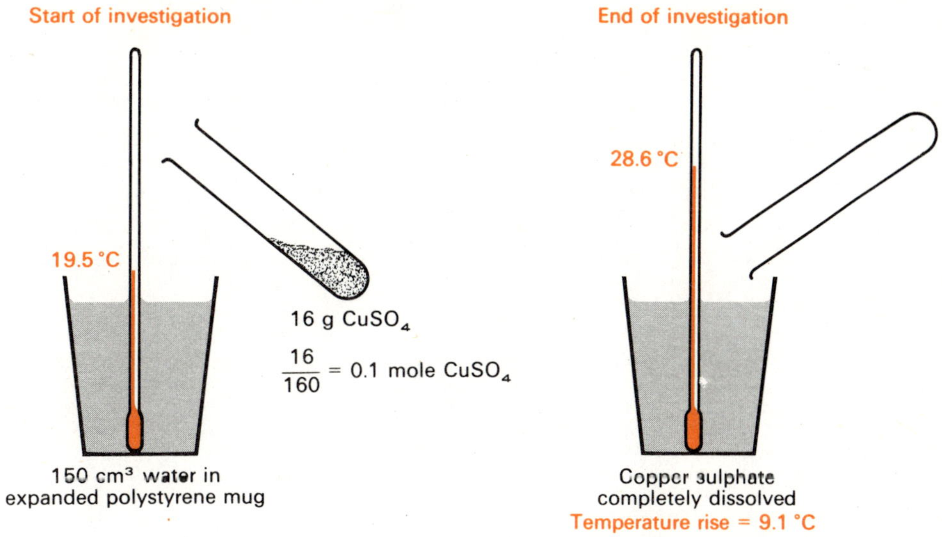

Fig. 14.5 Investigation of heat of solution of anhydrous copper(II) sulphate

The dissolving of anhydrous copper(II) sulphate is an exothermic change. When the solution has cooled down again to room temperature, energy will have been lost to the surroundings. ΔH is negative for this reaction.

$$\text{Energy lost by solution on cooling by } 9.1\,^{\circ}\text{C} = 150 \times 9.1 \times 4.2\ \text{J}$$

$$= 5700\ \text{J, or } 5.7\ \text{kJ}$$

This is the energy lost to the surroundings when 0.1 mole of $CuSO_4$ is used. When 1 mole of the compound is used, the energy lost is 57 kJ. This information can be included in the equation for the reaction.

$$CuSO_4(s) + aq \rightarrow CuSO_4(aq) \quad \Delta H = -57\ \text{kJ/mole}$$

The heats of solution of anhydrous and hydrated copper(II) sulphate have been put on the same energy level diagram in Figure 14.6.

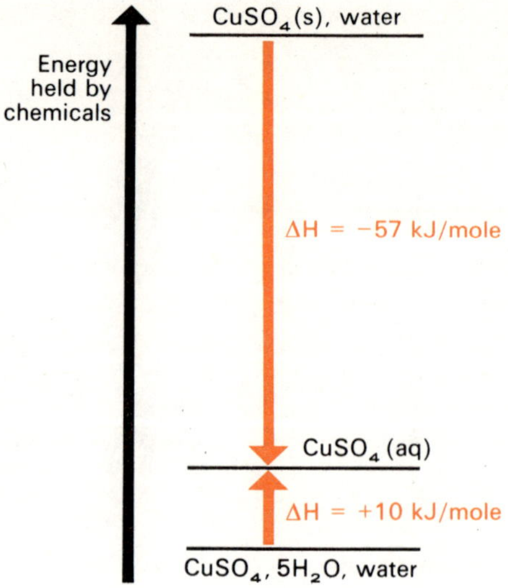

Figure 14.6 Energy level diagram for anhydrous and hydrated copper(II) sulphate dissolving in water

The energy changes for the two reactions shown in Figure 14.6 can be used to **predict** the energy change in a third reaction. Anhydrous copper(II) sulphate contains 57 kJ/mole less than is contained in aqueous copper(II) sulphate. Aqueous copper(II) sulphate contains 10 kJ/mole more energy than is contained

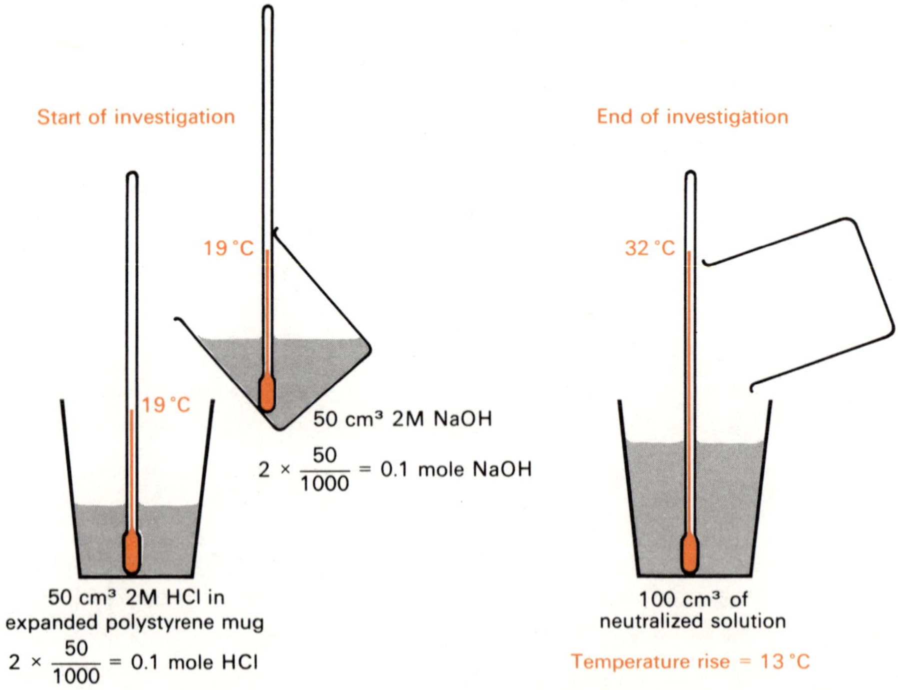

Figure 14.7 Investigation of heat of neutralization of hydrochloric acid by aqueous sodium hydroxide

in hydrated copper(II) sulphate. This shows that anhydrous copper(II) sulphate contains 67 kJ/mole more energy than is contained in the hydrated salt.

$$CuSO_4(s) + 5H_2O(l) \rightarrow CuSO_4{,}5H_2O(s) \quad \Delta H = -67 \text{ kJ/mole}$$

Heat of neutralization

When aqueous solutions of acids and alkalis are mixed, there is a rise in temperature. **Neutralization reactions are exothermic.** The investigation of a neutralization reaction is illustrated in Figure 14.7.

The neutralization illustrated in Figure 14.7 produces hot aqueous sodium chloride. As this solution cools to room temperature, energy is transferred to the surroundings.

Energy transferred to surroundings when 100 cm^3 of aqueous solution drops in temperature by 13 °C $= 100 \times 13 \times 4.2$ J

$$= 5500 \text{ J}$$

$$= 5.5 \text{ kJ}$$

The energy transferred to the surroundings when 0.1 mole of HCl reacts with 0.1 mole of NaOH is 5.5 kJ. When 1 mole of HCl reacts with 1 mole of NaOH, 55 kJ is transferred to the surroundings. More accurate experiments which allow for all sources of error find the value for the energy lost to be slightly greater than this at 58 kJ.

$$HCl(aq) + NaOH(aq) \rightarrow NaCl(aq) + H_2O(l) \quad \Delta H = -58 \text{ kJ/mole}$$

This energy change is called the **heat of neutralization**. The energy changes in various neutralization reactions are given in Table 14.1.

Table 14.1

Neutralization reaction	*Heat of neutralization*
$HCl(aq) + NaOH(aq) \rightarrow NaCl(aq) + H_2O(l)$	−58 kJ/mole
$HBr(aq) + NaOH(aq) \rightarrow NaBr(aq) + H_2O(l)$	−58 kJ/mole
$HNO_3(aq) + NaOH(aq) \rightarrow NaNO_3(aq) + H_2O(l)$	−58 kJ/mole
$HCl(aq) + KOH(aq) \rightarrow KCl(aq) + H_2O(l)$	−58 kJ/mole
$HNO_3(aq) + KOH(aq) \rightarrow KNO_3(aq) + H_2O(l)$	−58 kJ/mole
$HBr(aq) + KOH(aq) \rightarrow KBr(aq) + H_2O(l)$	−58 kJ/mole

The heats of neutralization in Table 14.1 are all the same because the same reaction is taking place in every case. This can be seen if the equations are written in ionic form. The equation for the reaction which forms sodium chloride is, in ionic form

$$H^+(aq) + Cl^-(aq) + Na^+(aq) + OH^-(aq) \rightarrow Na^+(aq) + Cl^-(aq) + H_2O(l)$$

The ions $Na^+(aq)$ and $Cl^-(aq)$ are unchanged by the reaction. They can be omitted from the equation, and are sometimes called spectator ions. The equation then becomes

$$H^+(aq) + OH^-(aq) \rightarrow H_2O(l)$$

When the spectator ions are omitted from the other equations in Table 14.1 the same simple equation remains each time. **All neutralization reactions produce unionized water from aqueous hydrogen ions and hydroxide ions.** The energy change in this reaction is −58 kJ/mole.

$$H^+(aq) + OH^-(aq) \rightarrow H_2O(l) \quad \Delta H = -58 \text{ kJ/mole}$$

The energy level diagram for neutralization reactions is shown in Figure 14.8.

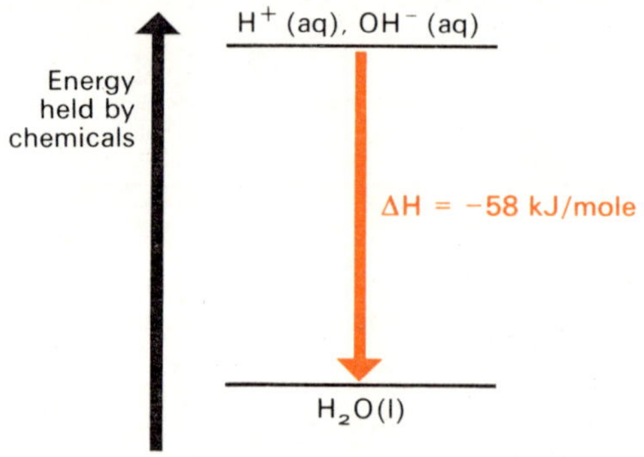

Figure 14.8 Energy level diagram for neutralization reactions

Bond breaking and bond forming

The atoms in a covalent molecule, and the ions in an ionic lattice, are held together by strong electrostatic forces of attraction (*see* Chapter 10). Considerable energy is needed to overcome these forces of attraction and break the bonds. The values for some bond-breaking energies are given in Table 14.2. In these reactions, gaseous molecules are broken into separate gaseous atoms.

Table 14.2

Bond-breaking reaction	*Energy change*
H–H(g) → 2H(g)	+435 kJ/mole
H–Cl(g) → H(g) + Cl(g)	+430 kJ/mole
H–Br(g) → H(g) + Br(g)	+365 kJ/mole
H–I(g) → H(g) + I(g)	+295 kJ/mole
Cl–Cl(g) → 2Cl(g)	+240 kJ/mole
Br–Br(g) → 2Br(g)	+190 kJ/mole
I–I(g) → 2I(g)	+150 kJ/mole

It will be seen from Table 14.2 that bond-breaking energies have positive values. Energy must be supplied to the molecules if the bonds are to be broken. **The breaking of bonds is an endothermic change. The forming of bonds is an exothermic change.**

Bond-breaking: $HCl(g) \rightarrow H(g) + Cl(g) \quad \Delta H = +430$ kJ/mole

Bond-forming: $H(g) + Cl(g) \rightarrow HCl(g) \quad \Delta H = -430$ kJ/mole

The energy level diagram for bond-breaking and bond-forming in hydrogen chloride is shown in Figure 14.9.

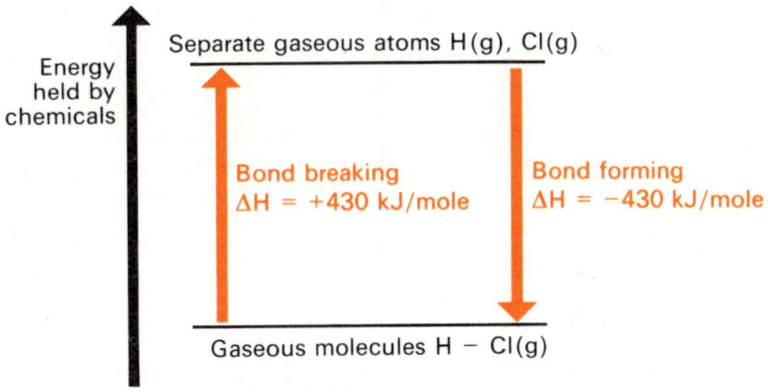

Figure 14.9 Energy level diagram for bond breaking and bond forming in hydrogen chloride

In a chemical reaction, atoms rearrange their bonding. This can be seen clearly in the formation of hydrogen chloride from hydrogen and chlorine.

$$H_2(g) + Cl_2(g) \rightarrow 2HCl(g) \quad \Delta H = -185 \text{ kJ/mole}$$

During this reaction the bonds between hydrogen atoms in the H_2 molecule must be broken (energy required = 435 kJ/mole). The bonds between chlorine atoms in the Cl_2 molecule must also be broken (energy required = 240 kJ/mole). The total bond-breaking energy which must be supplied is 675 kJ/mole. When the reaction is complete, new bonds between hydrogen atoms and chlorine atoms have been formed. The forming of these new bonds is an exothermic change. The energy released is $2 \times 430 = 860$ kJ when 2 moles of HCl are formed from their gaseous atoms. These energy changes are shown on an energy level diagram in Figure 14.10 on page 136.

In the reaction between hydrogen and chlorine, the forming of new bonds releases more energy than is required to break the original bonds. This reaction is therefore exothermic. **The energy change in a chemical reaction comes from the forming and breaking of bonds of different strengths.**

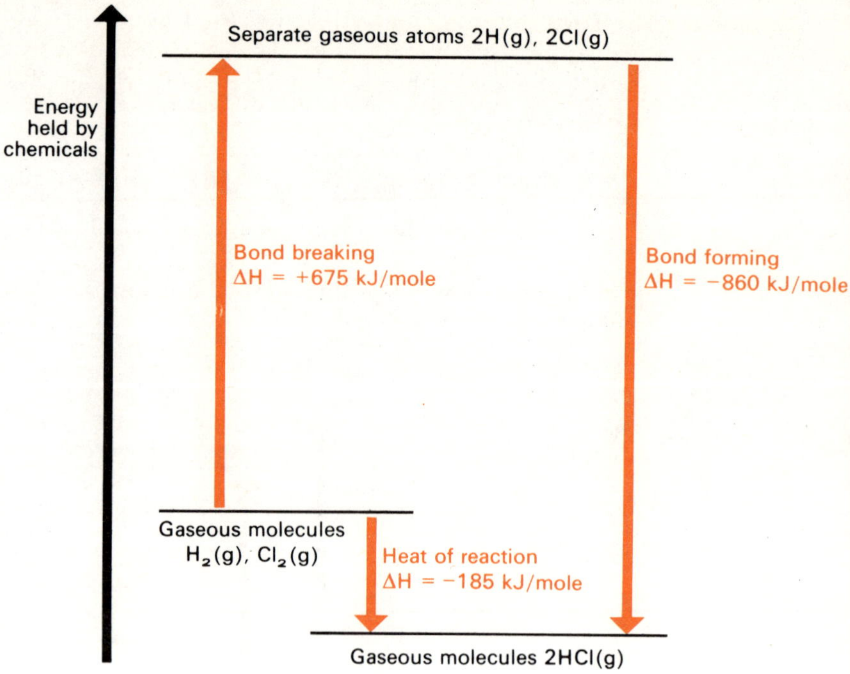

Figure 14.10 Energy level diagram for the reaction
$H_2(g) + Cl_2(g) \rightarrow 2HCl(g) \quad \Delta H = -185$ kJ/mole

Summary for Chapters 13 and 14

1. The products of an exothermic reaction contain less energy than the reactants at the same temperature. This energy is lost to the surroundings.

2. The products of an endothermic reaction contain more energy than the reactants at the same temperature. This energy is gained from the surroundings.

3. The heat of reaction is the energy change on using the numbers of moles shown in the chemical equation.

4. The heat of combustion is the energy lost on complete combustion of one mole of the substance.

5. The heat of solution is the energy change on dissolving one mole of the substance.

6. Neutralization reactions are exothermic. The heat of reaction (heat of neutralization) is frequently close to −58 kJ/mole. The neutralization reaction for fully ionized acids and bases is always the same.

$$H^+(aq) + OH^-(aq) \rightarrow H_2O(l) \quad \Delta H = -58 \text{ kJ/mole}$$

7. Energy level diagrams can be used to show the energy changes in reactions. In an exothermic reaction the products are at a lower energy level than the reactants. In an endothermic reaction the products are at a higher energy level than the reactants.

8. The Principle of Conservation of Energy states: energy can be changed from one form to another; energy cannot be created or destroyed.

9. Electrical energy can be obtained from some chemical reactions in suitable cells. The electrical energy is not the same as the heat of reaction.

10. The energy change in a chemical reaction comes from the forming and breaking of bonds of different strength. In an exothermic reaction, the bonds formed are stronger than the bonds broken. In an endothermic reaction, the bonds formed are weaker than the bonds broken.

15 Rates of Reaction

The rate at which magnesium ribbon reacts with dilute hydrochloric acid can be investigated using the apparatus shown in Figure 15.1. This method can be used to study many reactions where a gas is one of the products.

$$Mg(s) + 2H^+(aq) \rightarrow Mg^{2+}(aq) + H_2(g)$$

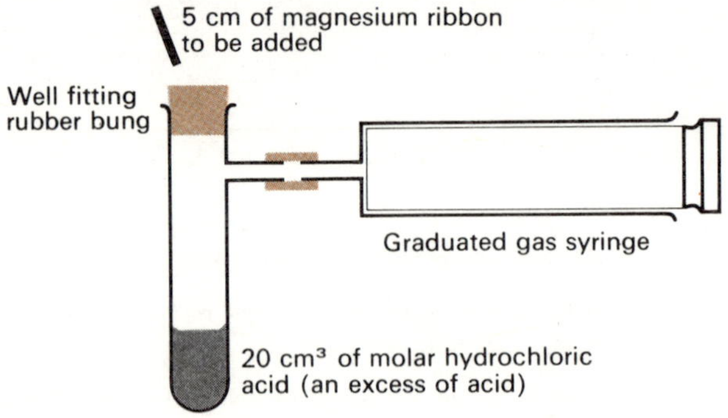

Figure 15.1 Apparatus for investigating rate at which gas is produced during a chemical reaction

To start the reaction, the rubber bung is removed for a moment and the magnesium ribbon is dropped in. The bung is replaced rapidly. As the reaction progresses, the volume of hydrogen evolved is measured by the volume of gas collected in the syringe. The reaction is fairly rapid and it is necessary to record the volume of gas at 5 second or 10 second intervals. The results of an experiment at room temperature are shown in graphical form in Figure 15.2.

It will be seen from Figure 15.2 that during the first 50 seconds of reaction the hydrogen is being produced at a steady rate. This rate is about 8 cm^3 every 10 seconds. The rate of reaction begins to decrease after about 50 seconds and eventually becomes zero. No more hydrogen is produced after a certain time as all the magnesium has reacted.

There are several ways of altering the steady rate at which hydrogen is produced at the beginning of the reaction. These ways will now be discussed.

The influence of concentration on rate of reaction

The experiment is repeated with only one alteration. The concentration of the acid is made 2 M instead of M. The results of this experiment are shown in Figure 15.3.

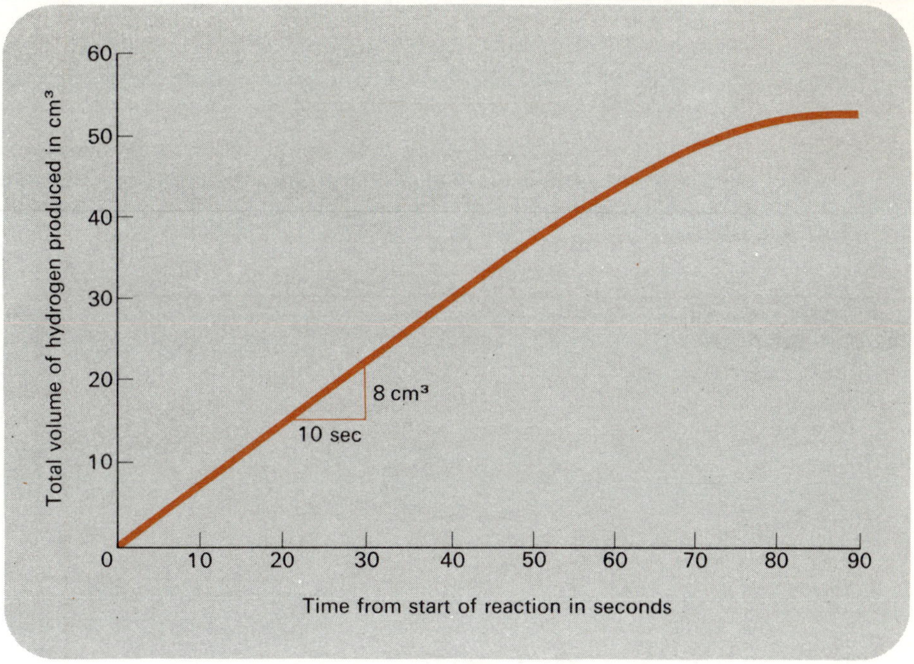

Figure 15.2 The rate of reaction of magnesium with M HCl at 18 °C

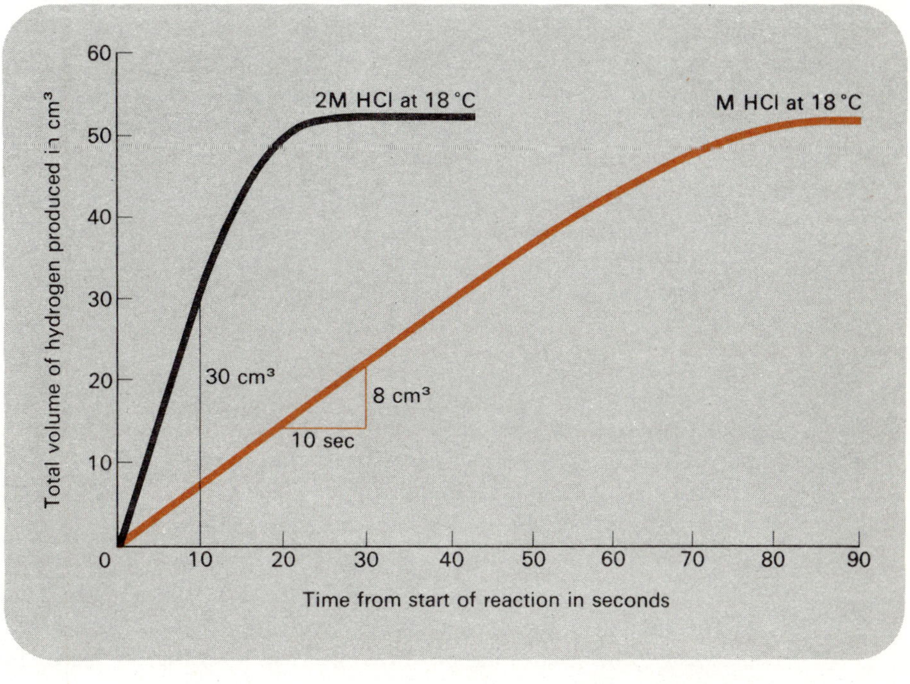

Figure 15.3 The influence of acid concentration on the rate of reaction of magnesium with HCl at 18 °C

Hydrogen is still produced at a steady rate at the beginning of the experiment, but the rate has increased to about 30 cm^3 in every 10 seconds. The rate of reaction has been increased about four times by increasing the molarity of the hydrochloric acid from M to 2 M.

The rate of a chemical reaction is usually increased by increasing the concentrations of the reactants (and decreased by decreasing the concentrations of the reactants).

Refer back to Figure 15.2. It will be seen that the rate of reaction of magnesium with M HCl begins to slow down after about 50 seconds. One reason for this is that the concentration of the acid decreases during the reaction as the H^+(aq) ions are consumed. As the concentration of acid decreases, so does the rate of reaction.

The increase in rate of reaction with increasing concentration can be explained on a simple kinetic model. The H^+(aq) ions in an aqueous solution of hydrochloric acid are in rapid chaotic motion. Reaction can take place when these ions collide with the surface of the magnesium ribbon. The concentration of H^+(aq) ions is increased by changing from M HCl to 2 M HCl. This increases both the rate at which H^+(aq) ions collide with the magnesium surface and the rate of reaction.

The influence of temperature on rate of reaction

The first investigation (*see* Figure 15.2), using 5 cm of magnesium ribbon and 20 cm^3 of M HCl at 18 °C, is repeated with one alteration. The temperature of the acid is increased to 36 °C. The results of the experiment at a higher temperature are shown in Figure 15.4. It will be seen from Figure 15.4 that hydrogen is once again produced at a steady rate at the beginning of the reaction, but the rate has now increased to 11 cm^3 in every 10 seconds.

The rate of a chemical reaction is increased by increasing the temperature of the reactants (and decreased by decreasing the temperature of the reactants).

The increase in rate of reaction with temperature can be explained on the kinetic model. Increase in temperature increases the energy of the reacting particles. This makes more energy available for the breaking of bonds which must take place if chemical reaction is to occur. Bond breaking is more frequent at higher temperature and the rate of reaction increases.

The influence of surface area on rate of reaction

The first investigation (*see* Figure 15.2) is repeated with one alteration. The 5 cm length of magnesium ribbon is cut into two 2.5 cm portions which are then glued face to face. This reduces the surface area of the magnesium to about half compared with the original experiment. The results of the experiment using magnesium with reduced surface area are shown in Figure 15.5. It will be seen that hydrogen is once again produced at a steady rate, but the rate has now decreased to about 4 cm^3 in every 10 seconds.

The rate of a chemical reaction involving a solid reactant is decreased by decreasing the surface area of the solid (and increased by increasing the surface area of the solid).

In the early stages of the reaction between magnesium and dilute hydrochloric

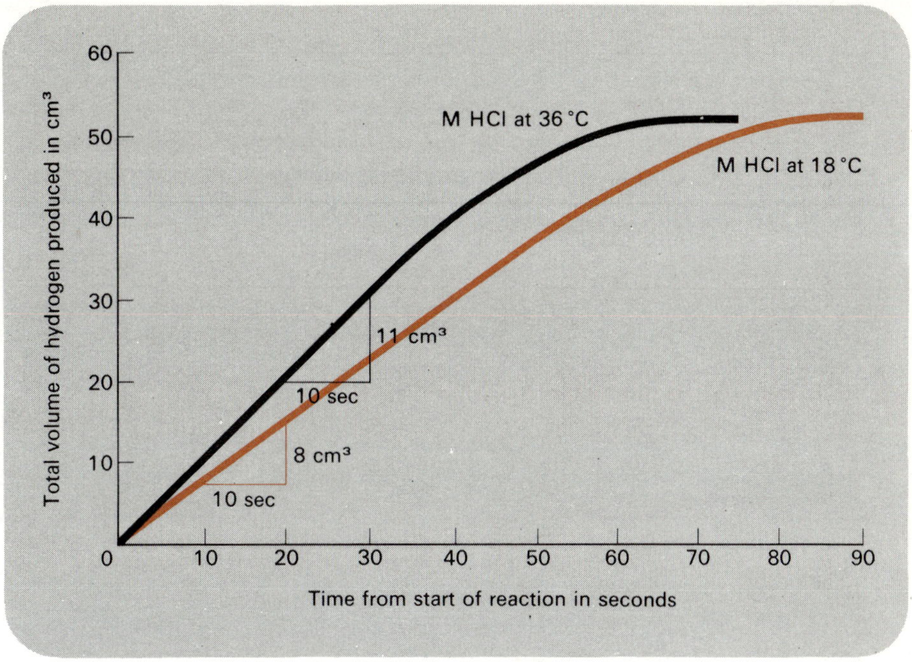

Figure 15.4 The influence of temperature on the rate of reaction of magnesium with HCl

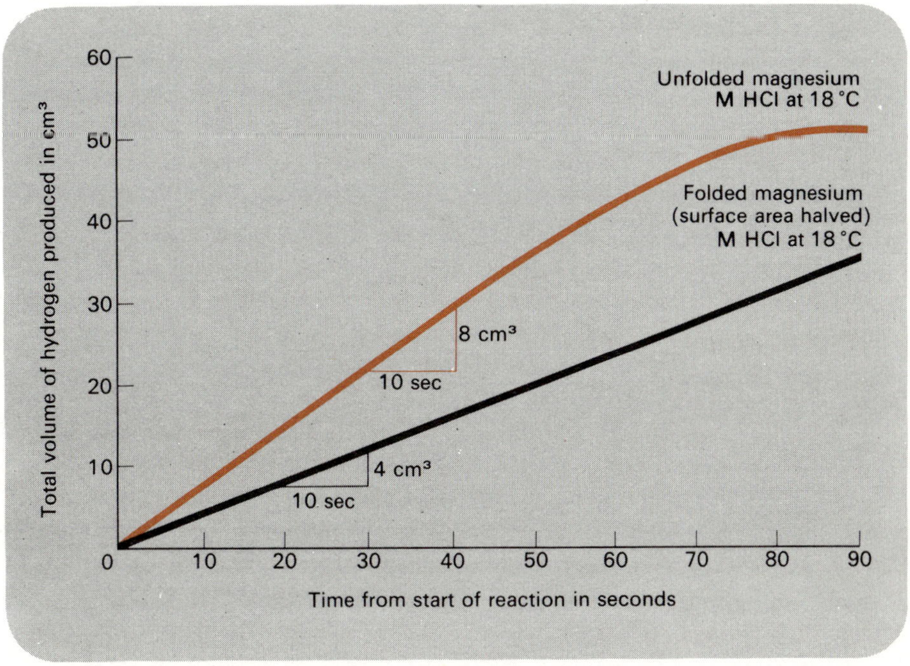

Figure 15.5 The influence of surface area on the rate of reaction of magnesium with M HCl at 18 °C

acid, the rate of reaction is steady. After a certain point, however, the rate begins to decrease. One reason for this decrease is that $H^+(aq)$ ions are consumed during the reaction. A more important reason for the decrease in rate is that the strip of magnesium becomes smaller during the reaction. The surface area of the magnesium decreases and so does the rate of reaction.

The change in rate of reaction with change in surface area can be explained on the kinetic model. Reaction can take place when the solid surface is hit by the moving particles of the other reactant. If the surface area is smaller, the chance of collisions with the surface is reduced and so is the rate of reaction. If the surface area is made larger, the chance of collisions with the surface is increased and so is the rate of reaction.

The influence of catalysts on rate of reaction

The rate of many reactions can be altered by the addition of various substances which do not appear in the chemical equation for the reaction. These substances take part in the reaction but they are unchanged in amount at the end. They are called **catalysts**.

The rate of reaction between magnesium and dilute hydrochloric acid is changed if the acid contains dissolved salts such as sodium chloride. The results of an experiment in which the hydrochloric acid is saturated with sodium chloride are shown in Figure 15.6.

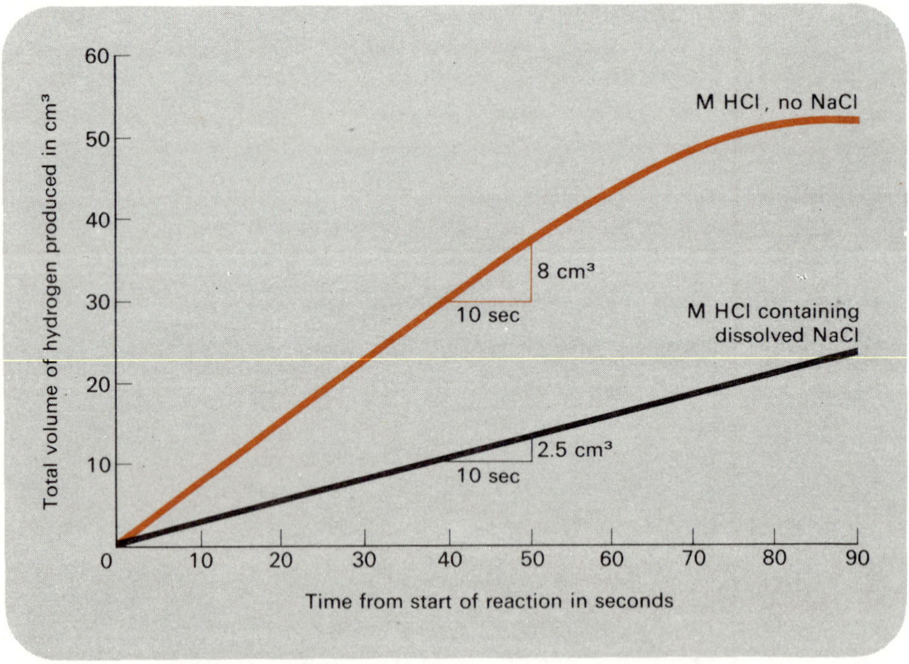

Fig. 15.6 The influence of sodium chloride on the rate of reaction of magnesium with M HCl at 18 °C

The steady rate of production of hydrogen is 8 cm^3 in every 10 seconds when sodium chloride is absent. When the acid is saturated with sodium chloride, the rate of production of hydrogen drops to only 2.5 cm^3 in every 10 seconds. Sodium chloride could be called an **inhibitor** for the reaction rather than a catalyst. The term catalyst is frequently used to describe only those substances which increase the rate of reaction.

The kinetic model can be used to explain the way in which sodium chloride inhibits this reaction. As the reaction progresses, $Mg^{2+}(aq)$ ions have to escape into the solution to make room for more $H^+(aq)$ ions to attack the metal surface. A high concentration of $Na^+(aq)$ and $Cl^-(aq)$ ions hinders either the escape of $Mg^{2+}(aq)$ ions or the approach of more $H^+(aq)$ ions. This reduces the rate of reaction.

Sodium chloride has to be present in high concentration to have a significant influence on the rate of reaction between magnesium and dilute hydrochloric acid. Many catalysts produce a considerable change in rate of reaction when present in only small amounts. This can be shown by adding very small quantities of various metals and metal oxides to aqueous hydrogen peroxide. This compound can decompose to water and oxygen.

$$2H_2O_2(aq) \rightarrow 2H_2O(l) + O_2(g)$$

The decomposition of hydrogen peroxide is very slow unless a suitable catalyst is present. Figure 15.7 shows the effect of adding 0.1 g samples of

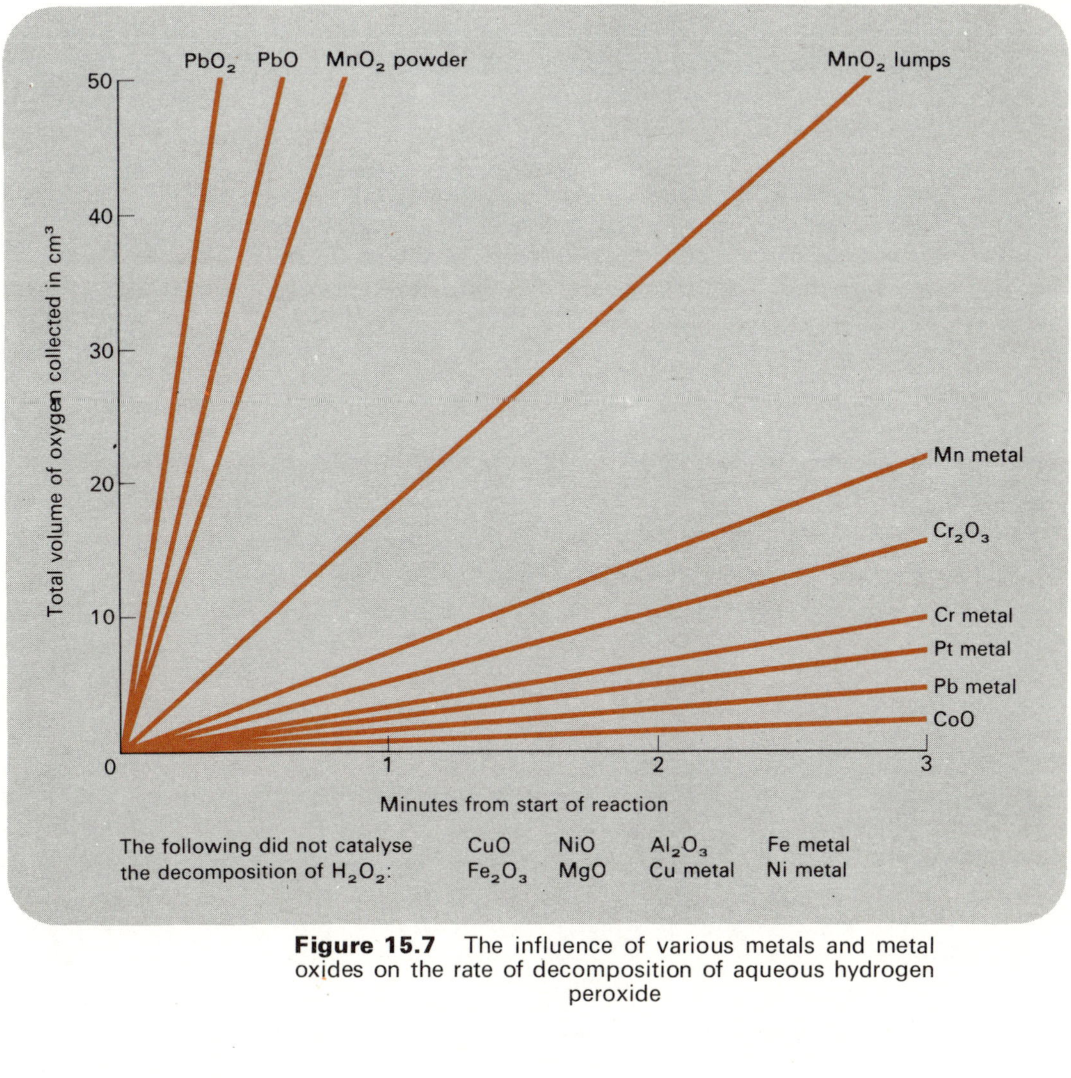

Figure 15.7 The influence of various metals and metal oxides on the rate of decomposition of aqueous hydrogen peroxide

metals and metal oxides to 20 cm^3 of hydrogen peroxide. The reaction produces a gas, so it can be followed by the same method used for investigating the reaction between magnesium and dilute hydrochloric acid.

Inspecting Figure 15.7 the following points can be seen.

- Very small amounts of some metals and metal oxides have a considerable catalytic effect.
- Powdered manganese(IV) oxide is a better catalyst than the same compound in lump form. This is due to the greater surface area of the powder. Reaction takes place on the surface.
- Catalysts vary in their effect. Some are much better than others. Some metals and metal oxides do not catalyse the reaction at all.
- The reaction is catalysed by certain transition metals and their oxides. Transition metals and their compounds are good catalysts for many reactions. Lead and its oxides also catalyse this reaction. Lead is not a transition metal.

Summary

The rate of a chemical reaction can be increased by increasing the concentration, temperature and surface area of the reactants. Addition of a suitable catalyst also increases the rate. These points are illustrated in Figure 15.8.

A chemist is frequently interested in influencing the rate of a chemical reaction. Usually he wants to increase the rate, but in some cases he may want to moderate the rate to make the reaction more manageable. The Haber process for the production of ammonia, the Contact process for the production of sulphuric acid and the oxidation of ammonia in the production of nitric acid are important examples of catalysed reactions (*see* Figures 24.2, 24.4).

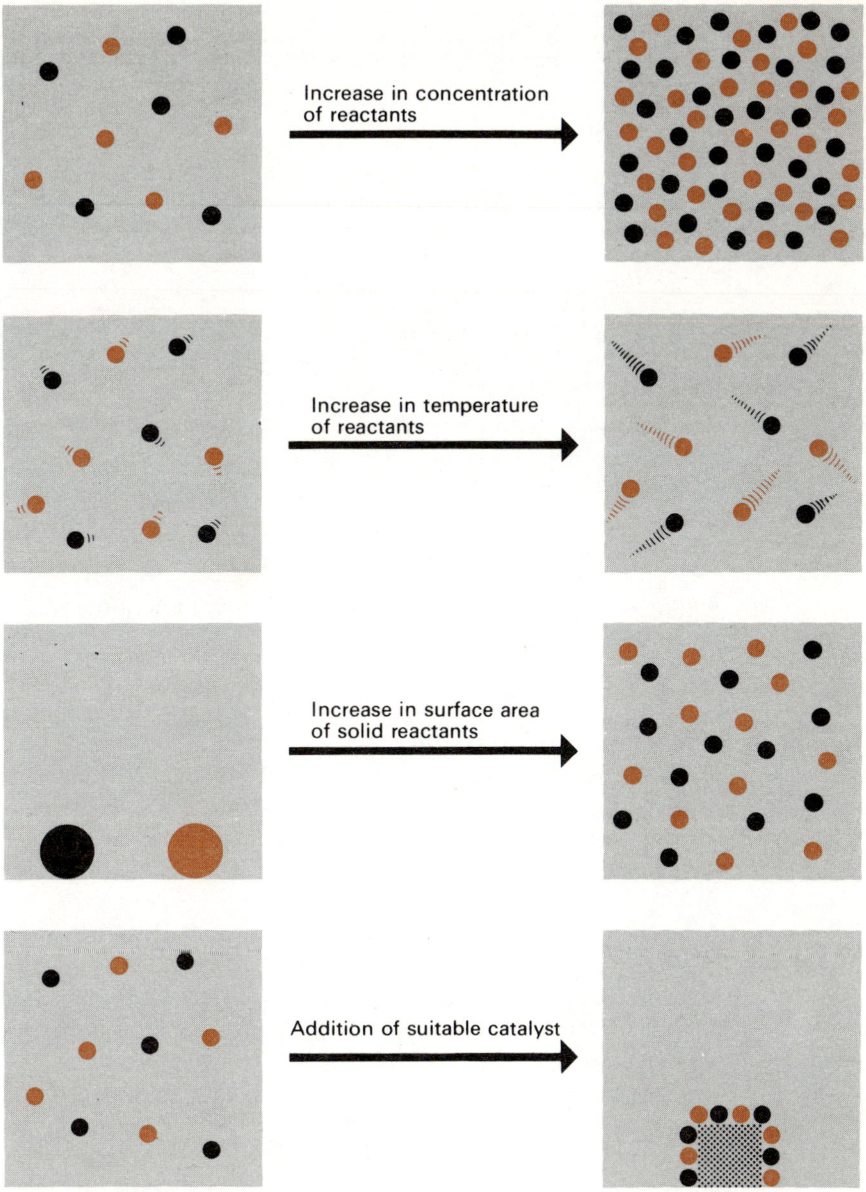

Figure 15.8 Ways in which the rate of a chemical reaction can be increased

16 Dynamic Equilibrium

When magnesium is added to dilute hydrochloric acid, reaction continues rapidly until either all the magnesium or all the acid has been used up. If hydrogen is bubbled through an aqueous solution of magnesium ions, no reaction takes place. The reaction only goes in one direction, and goes to completion. **This reaction is irreversible.**

$$Mg(s) + 2H^+(aq) \rightarrow Mg^{2+}(aq) + H_2(g)$$

$$\xrightarrow{\hspace{4cm}}$$

Reaction only goes this way

Many reactions, however, do not go to completion. An example of such a reaction is the dissolving of solid iodine in water. A brown solution of iodine molecules is formed.

$$I_2(s) + aq \rightarrow I_2(aq)$$

This reaction can be investigated by steadily shaking about 10 g of iodine with a large volume of water (about 1000 cm^3). The colour of the water darkens as the iodine goes into solution. Small portions of the solution are removed at various times during the reaction. The concentration of iodine in these portions is then found by titration. The results of an investigation are shown in Figure 16.1.

Inspection of Figure 16.1 shows that at the beginning of the reaction iodine is going into solution at a steady rate. After shaking for about 18 minutes, however, the concentration of iodine in solution reaches a constant value of 0.001 mole/1000 cm^3. Further shaking produces no increase in concentration although most of the solid iodine remains undissolved.

If 20 g of iodine is used instead of 10 g, the rate at which iodine dissolves at the start is more rapid. The final constant concentration of the aqueous solution is still, however, 0.001 mole/1000 cm^3. Similarly, if the iodine is more finely powdered or if more water is used, the rate of dissolving is changed but not the final constant concentration. This is shown in Figure 16.2 on page 148.

The iodine concentration in aqueous solution always reaches the same constant value of 0.001 mole/1000 cm^3. At this point **reaction has not stopped**. It can be shown that reaction has not stopped by treating the mixture of solid iodine and its aqueous solution in the following ways.

- A small quantity of solid iodine containing radioactive iodine is added. The solution rapidly becomes radioactive. This shows that solid iodine is still dissolving.
- A few drops of aqueous iodine containing radioactive iodine are added. The solid iodine rapidly becomes radioactive. This shows that iodine is precipitating from solution.

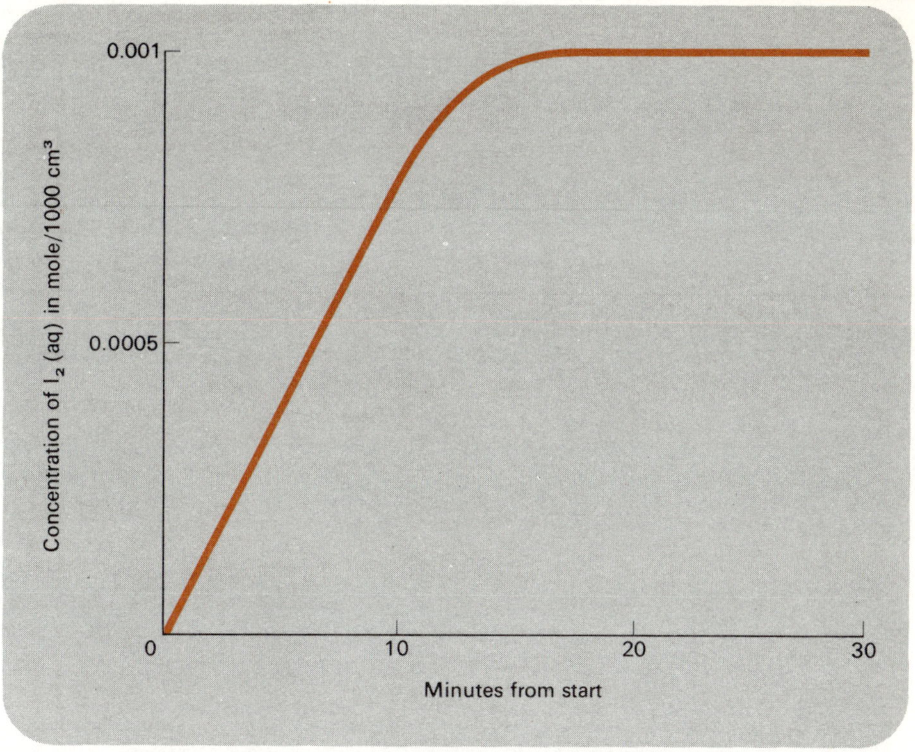

Figure 16.1 An investigation of the rate at which 10 g of iodine dissolves in 1000 cm³ of water

These experiments show that solid iodine and its aqueous solution have reached a state of dynamic equilibrium. **Iodine molecules are dissolving and precipitating at equal rates when dynamic equilibrium has been reached.** This is the reason for the concentration of iodine in solution reaching a constant value. To show that the reaction reaches a state of dynamic equilibrium, the equation is written

$$I_2(s) + aq \rightleftharpoons I_2(aq)$$

The kinetic model of dynamic equilibrium is illustrated in Figure 16.3 on page 148.

The dissociation of dinitrogen tetroxide, N_2O_4

You will probably have met the gas dinitrogen tetroxide when heating certain metal nitrates, such as lead nitrate.

$$2Pb(NO_3)_2(s) \rightarrow 2PbO(s) + N_2O_4(g) + O_2(g)$$

Dinitrogen tetroxide can also be obtained by mixing the gases nitrogen monoxide and oxygen in the correct volume ratio. The gases react rapidly and there is an immediate decrease in volume.

$$\begin{array}{lll} 2NO(g) & +O_2(g) & \rightarrow N_2O_4(g) \\ 2\text{ volumes} & +1\text{ volume} & \rightarrow 1\text{ volume} \\ 80\text{ cm}^3 & +40\text{ cm}^3 & \rightarrow 40\text{ cm}^3 \end{array}$$

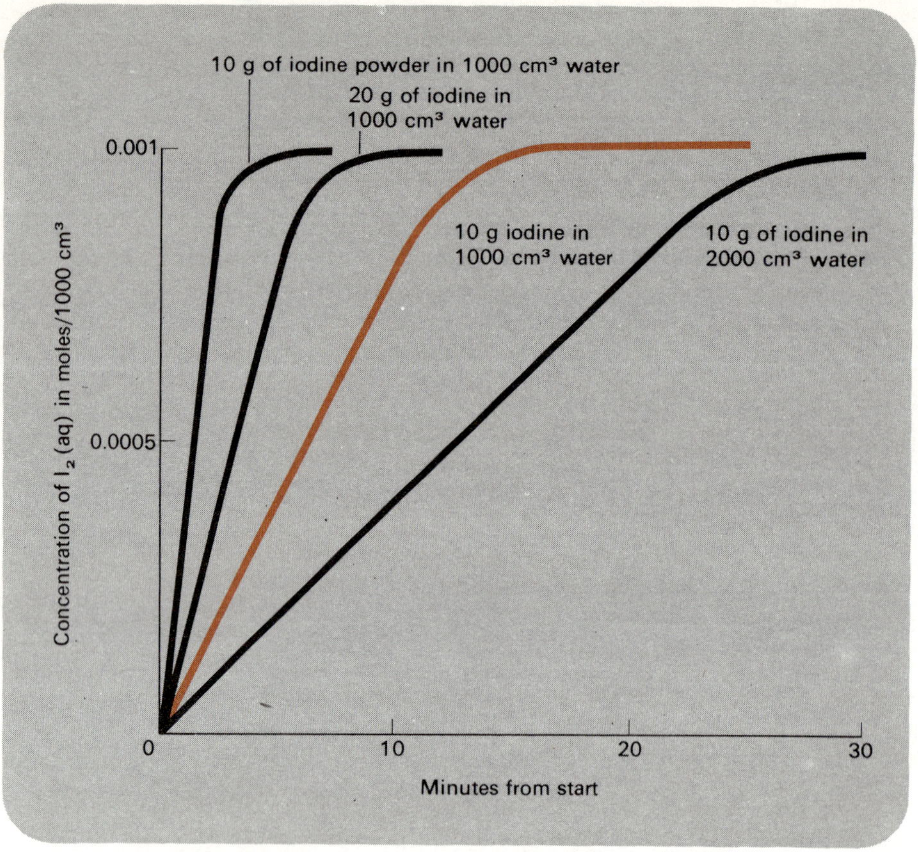

Figure 16.2 A further investigation of the dissolving of iodine in water. The iodine concentration in aqueous solution always reaches the same final constant value

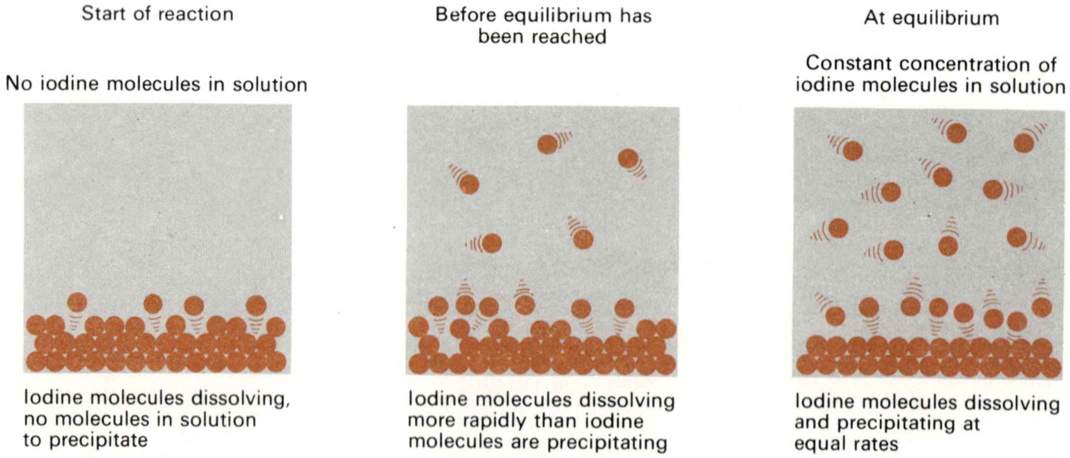

Figure 16.3 The dissolving of solid iodine in water. The kinetic model of dynamic equilibrium

This reaction can be investigated using two connected gas syringes. The apparatus is shown in Figure 16.4.

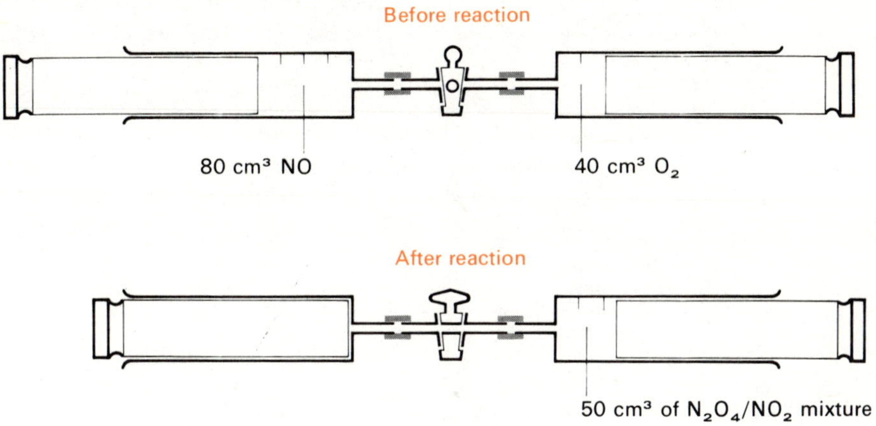

Figure 16.4 An investigation of the gaseous reaction between nitrogen monoxide and oxygen

The volume of gas produced in this experiment should be 40 cm^3 if the equation for the reaction used above is correct and complete. The volume found by experiment is almost 50 cm^3. The experimental volume of gas is higher than predicted because dinitrogen tetroxide **dissociates** into nitrogen dioxide, NO_2. If this dissociation is complete, the volume of gas would be 80 cm^3.

$$N_2O_4(g) \rightarrow 2NO_2(g)$$

$$1 \text{ mole} \rightarrow 2 \text{ moles}$$

$$\therefore 1 \text{ volume} \rightarrow 2 \text{ volumes}$$

$$\therefore 40 \text{ cm}^3 \rightarrow 80 \text{ cm}^3$$

If there were no dissociation, the volume of gas would be 40 cm^3. The experimental volume of 50 cm^3 shows that the dinitrogen tetroxide is only partly dissociated at room temperature. The experimental volume is much closer to 40 cm^3 than to 80 cm^3 because only about 25 per cent of the N_2O_4 molecules are dissociated at room temperature. Molecules of N_2O_4 and NO_2 are in dynamic equilibrium and this can be shown in the equation for the dissociation.

$$N_2O_4(g) \rightleftharpoons 2NO_2(g) \quad \Delta H = +58 \text{ kJ/mole}$$

Molecules of N_2O_4 are dissociating into molecules of NO_2 all the time, while molecules of NO_2 are recombining all the time. The rates of dissociation and recombination are equal when equilibrium has been reached, so the volume remains steady at 50 cm^3. The dissociation is endothermic because bonds are broken. Energy has to be supplied from the surroundings to break these bonds.

Influence of temperature on the N_2O_4/NO_2 equilibrium

The temperature of the equilibrium mixture of N_2O_4 and NO_2 can be easily altered. The syringe containing 50 cm^3 of gas is sealed with a rubber cap. The syringe is then lowered into a beaker of boiling water. This is illustrated in Figure 16.5.

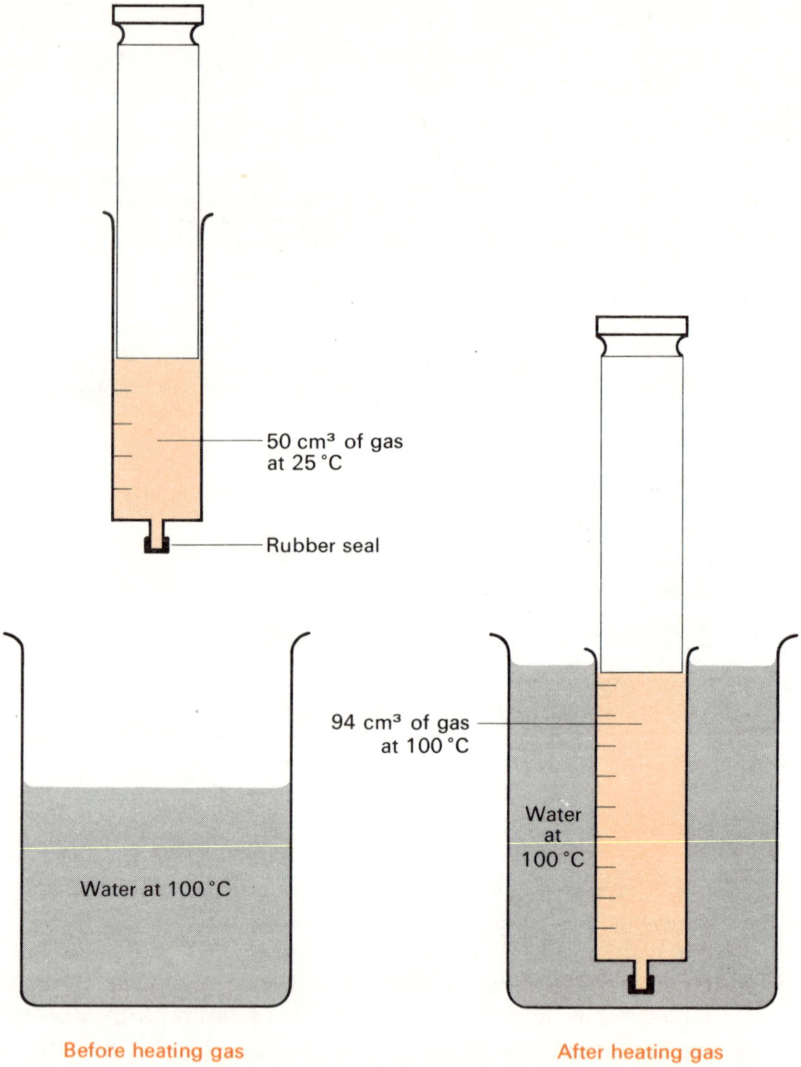

Figure 16.5 The change in volume with temperature fo the equilibrium mixture of N_2O_4 and NO_2

The volume of gas almost doubles on raising the temperature from 25 °C to 100 °C, and its colour changes from pale brown to deep brown.

$$50\ cm^3 \text{ of } N_2O_4/NO_2 \text{ at } 25\ °C \rightarrow 94\ cm^3 \text{ of } N_2O_4/NO_2 \text{ at } 100\ °C$$

What volume change would be expected in this experiment? The change can be predicted using the equation which combines the laws of Boyle and Charles (*see* page 108).

$$\frac{P_1V_1}{T_1} = \frac{P_2V_2}{T_2}$$

$P_1 = 1$ atm $\quad P_2 = 1$ atm $\quad$ (1 atmosphere = 760 mmHg)

$V_1 = 50\ cm^3 \quad V_2$ is to be predicted

$T_1 = 298$ K $\quad T_2 = 373$ K

$$\frac{1\ atm \times 50\ cm^3}{298\ K} = \frac{1\ atm \times V_2}{373\ K}$$

$$V_2 = 50\ cm^3 \times \frac{373\ K}{298\ K}$$

$$V_2 = 63\ cm^3$$

The experimental volume of the equilibrium mixture at 100 °C is very much greater than predicted. This is because the dissociation of N_2O_4 into NO_2 is increased by raising the temperature (*see* Table 16.1).

Table 16.1

Low temperature	*High temperature*
$N_2O_4(g) \rightleftharpoons 2NO_2(g)$ $\Delta H = +58$ kJ/mole	$N_2O_4(g) \rightleftharpoons 2NO_2(g)$ $\Delta H = +58$ kJ/mole
Colourless molecules → Brown molecules	Colourless molecules → Brown molecules
Only slight dissociation at 25 °C (about 25 per cent). Equilibrium mixture is pale brown	Far more dissociation at 100 °C (about 90 per cent). Equilibrium mixture is deep brown due to high concentration of brown molecules

When the temperature of an equilibrium mixture is increased, the equilibrium position shifts in the direction which absorbs energy from the surroundings (the endothermic direction).

The dissociation of N_2O_4 is increased by increasing the temperature because dissociation is endothermic. The shift in equilibrium position with change in temperature is important in the Haber process for the production of ammonia from nitrogen and hydrogen (*see* Figure 24.2, page 220).

$$N_2(g) + 3H_2(g) \rightleftharpoons 2NH_3(g) \quad \Delta H = -92\ \text{kJ/mole}$$

← Endothermic direction $\qquad$ Exothermic direction →

← Equilibrium shifts this way with increase in temperature

Influence of pressure on the N_2O_4/NO_2 equilibrium

The influence of pressure on this equilibrium can be shown by resting a large mass on top of the syringe plunger while the syringe is tightly clamped in hot water. The experiment is illustrated in Figure 16.6.

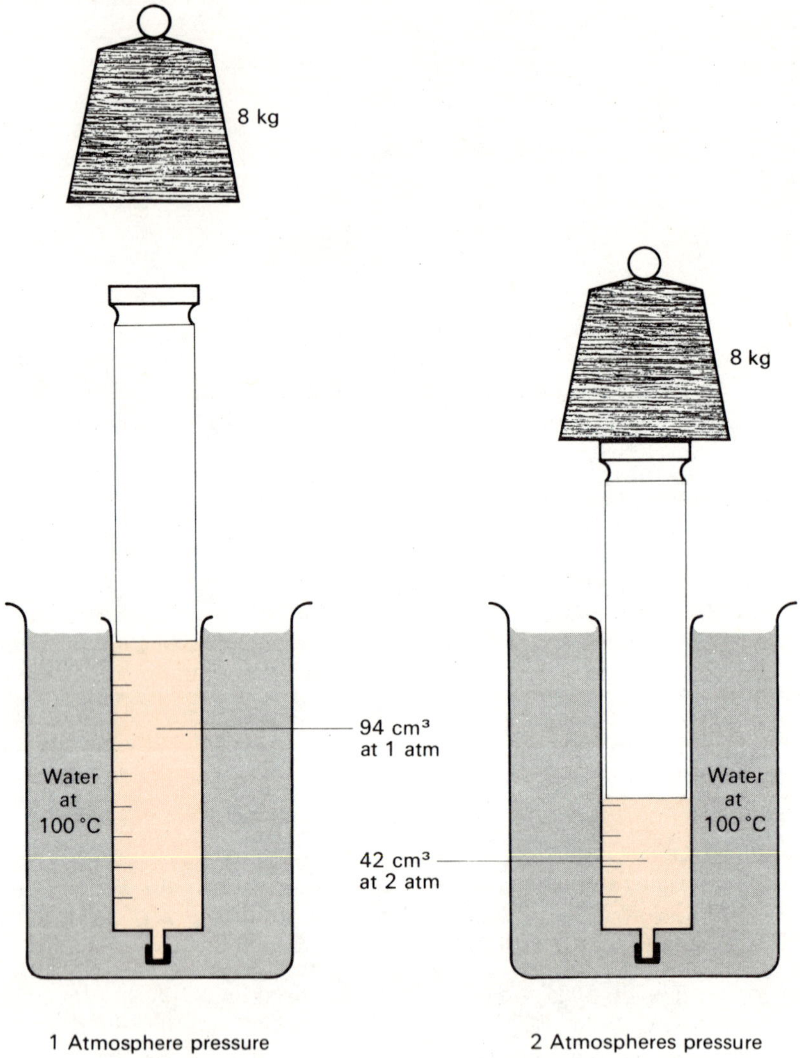

Figure 16.6 The change in volume with pressure for the equilibrium mixture of N_2O_4 and NO_2

When the gaseous equilibrium mixture has a volume of 94 cm^3, it is at a pressure of one atmosphere. When the 8 kg mass is in place the pressure is increased to two atmospheres. (The new pressure can be calculated if the diameter of the syringe piston is known.) The volume of the gaseous equilibrium mixture has decreased to 42 cm^3 on doubling the pressure. From Boyle's law it would be expected that doubling the pressure would halve the volume to 47 cm^3. In this experiment, however, the volume has decreased considerably more than expected. The extra decrease in volume is caused by some NO_2 molecules recombining to form N_2O_4 molecules.

$$N_2O_4(g) \rightleftharpoons 2NO_2(g)$$

1 mole of gaseous molecules	2 moles of gaseous molecules

← Equilibrium position shifts this way with increase in pressure

When the pressure is increased, the equilibrium position shifts in the direction with the smaller number of gaseous molecules.

The shift in equilibrium position with change in pressure is important in the Haber process for the production of ammonia from nitrogen and hydrogen (*see* Figure 24.2, page 220).

$$N_2(g) + 3H_2(g) \rightleftharpoons 2NH_3(g)$$

4 moles of gaseous molecules	2 moles of gaseous molecules

→ Equilibrium position shifts this way with increase in pressure

Influence of concentration on equilibrium position

When a few drops of aqueous sodium hydroxide are added to bromine water, the brown colour of the bromine disappears. The reaction can be represented by the equation

$$Br_2(aq) + OH^-(aq) \rightarrow Br^-(aq) + OBr^-(aq) + H^+(aq)$$

$Br_2(aq)$	$OH^-(aq)$	$Br^-(aq)$	$OBr^-(aq)$	$H^+(aq)$
Brown bromine molecules	Colourless hydroxide ions	Colourless bromide ions	Colourless hypobromite ions	

When reaction has taken place, the concentration of bromide ions in the mixture can be increased by addition of sodium bromide. The mixture then becomes brown due to the reformation of bromine molecules. The reaction is reversible. A dynamic equilibrium is present and the equilibrium position is altered by a high concentration of bromide ions.

$$Br_2(aq) + OH^-(aq) \rightleftharpoons Br^-(aq) + OBr^-(aq) + H^+(aq)$$

← Equilibrium position shifts this way as $Br^-(aq)$ ions are added

When a component of an equilibrium mixture is added, the equilibrium position shifts in the direction which removes this component.

The equilibrium position in this reaction can also be altered by removing hydroxide ions from the equilibrium mixture. When hydroxide ions are added to bromine water, the brown colour disappears as described earlier. If a dilute acid is then added to the colourless mixture, the brown colour of bromine is reformed as the acid combines with hydroxide ions. The brown colour can then be removed again by the addition of more hydroxide ions.

$$Br_2(aq) + OH^-(aq) \rightleftharpoons Br^-(aq) + OBr^-(aq) + H^+(aq)$$

Brown Colourless Colourless Colourless

$\longleftarrow$ Equilibrium position shifts this way as $OH^-(aq)$ ions are removed by addition of acid $[OH^-(aq) + H^+(aq) \rightarrow H_2O(1)]$

$\longrightarrow$ Equilibrium position shifts this was as $OH^-(aq)$ ions are added using sodium hydroxide

When a component of an equilibrium mixture is removed, the equilibrium position shifts in the direction which forms this component.

Influence of catalysts on equilibria

A suitable catalyst increases the rate at which equilibrium is reached. A catalyst does not influence the final equilibrium position.

Summary

Many reactions do not go to completion. They reach a position of dynamic equilibrium. At equilibrium, reaction is taking place at equal rates in both directions. The equilibrium position can be altered by changing certain experimental conditions (Table 16.2).

Table 16.2

Changes in conditions	*Shift in equilibrium position*
Increase in temperature	In endothermic direction
Decrease in temperature	In exothermic direction
Increase in pressure	In direction with smaller number of gaseous molecules
Decrease in pressure	In direction with larger number of gaseous molecules
Addition of reaction component	In direction which removes component
Removal of reaction component	In direction which forms more component
Addition of catalyst to increase rate of reaching equilibrium	No change in equilibrium position

17 Electron Transfer

Atoms and ions

Atoms are neutral particles. The negatively charged electrons outside the nucleus balance the positively charged protons in the nucleus. The numbers of electrons and protons are each the same as the atomic number of the element. Atoms can change their electron configuration by losing or gaining electrons. Loss or gain of electrons produces ions.

A sodium atom can lose one electron and become a positive sodium ion.

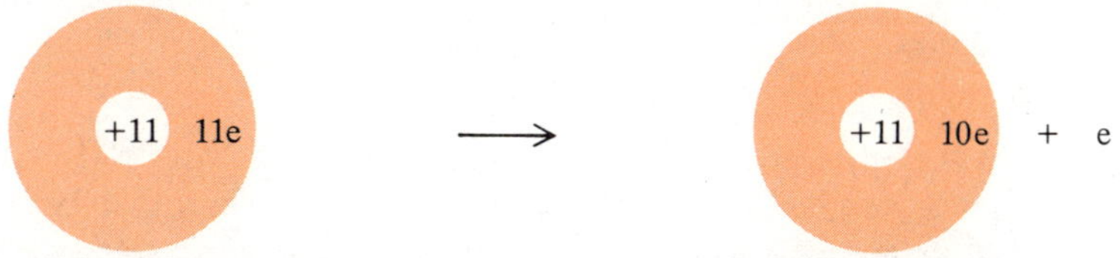

Neutral sodium atom | Positive sodium ion

The reaction would be written:

$$Na \rightarrow Na^{+} + e$$

A chlorine atom can gain one electron and become a negative chloride ion.

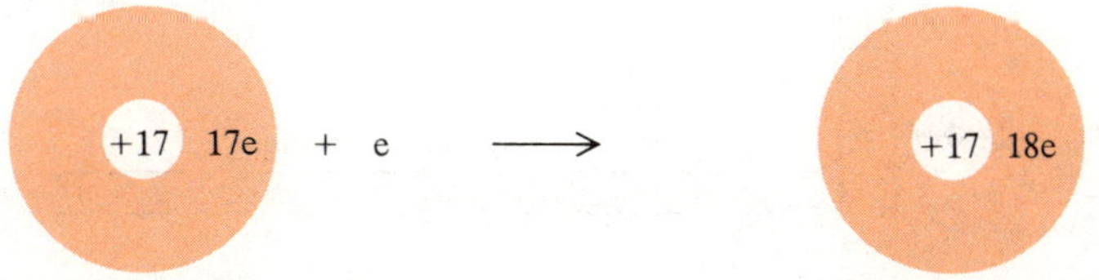

Neutral chlorine atom | Negative chloride ion

The reaction would be written:

$$Cl + e \rightarrow Cl^{-}$$

Ions can also change their electron configuration. In doing so they may become atoms, or they may become ions with a charge different from the original.

A negative chloride ion can lose one electron and become a neutral chlorine atom.

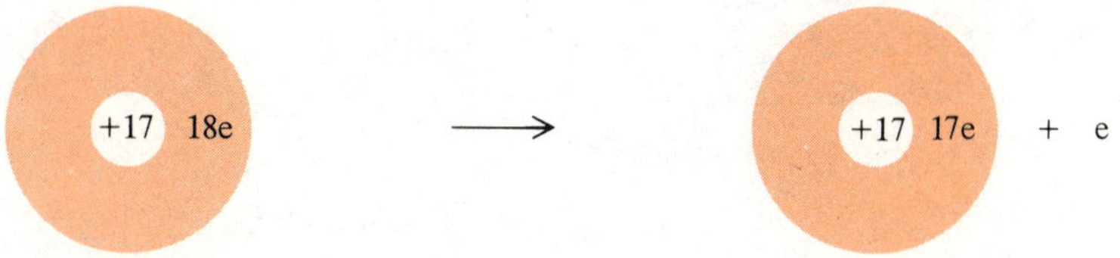

The reaction would be written:

$$Cl^- \rightarrow Cl + e$$

Some other examples of electron loss and gain are given in Table 17.1.

Table 17.1
Some Reactions where Electrons are Lost or Gained

Electron loss	*Electron gain*
$Fe^{2+} \rightarrow Fe^{3+} + e$	$Fe^{3+} + e \rightarrow Fe^{2+}$
$Mg \rightarrow Mg^{2+} + 2e$	$Cu^{2+} + 2e \rightarrow Cu$
$Fe \rightarrow Fe^{2+} + 2e$	$Na^+ + e \rightarrow Na$
$Br^- \rightarrow Br + e$	$O_2 + 4e \rightarrow 2O^{2-}$
$Li \rightarrow Li^+ + e$	$2H^+ + 2e \rightarrow H_2$

In many chemical reactions one of the reactants loses electrons while the other reactant gains electrons. **Electron transfer** is taking place.

Investigating electron transfer

In the apparatus of Figure 17.1 a solution of copper(II) sulphate in one beaker has a weighed copper rod dipping into it. The other beaker contains iron(II) sulphate and a weighed iron rod. The solutions are connected ('bridged') by a tube containing agar jelly and salt, which acts as an electrical conductor. The metal rods are connected by wires through a meter measuring electric current.

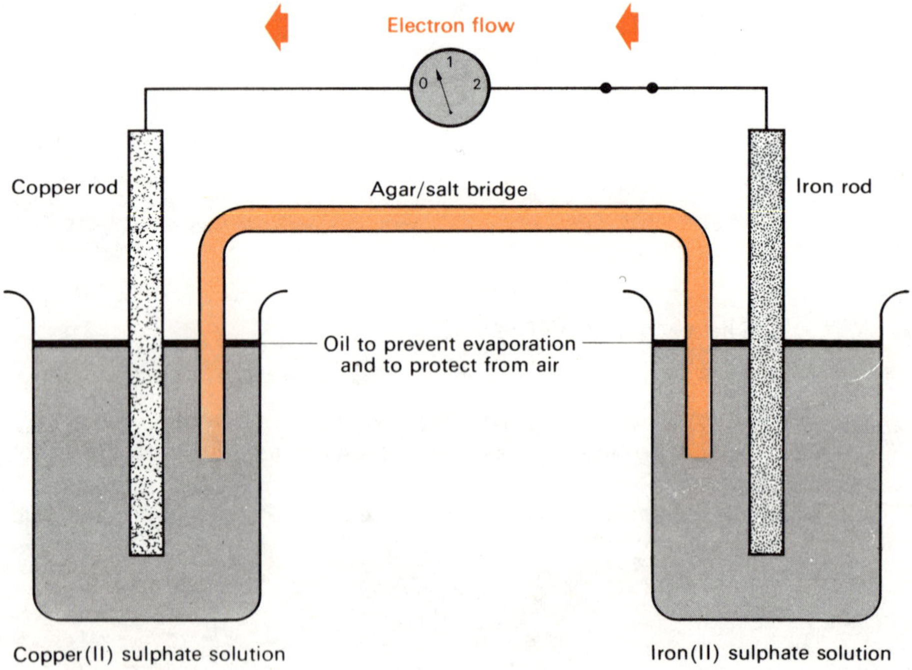

Figure 17.1

This apparatus is a form of electric cell. Current flows through the meter when the switch is closed. If current is allowed to flow for some time several observations can be made.

- The copper rod gains in mass. A deposit of fresh copper can be seen on it.
- The copper(II) sulphate solution becomes paler, showing that the concentration of copper(II) ions has decreased.
- The iron rod loses mass.

In an actual experiment, tests could be made to see whether the concentration of Fe^{2+} ions (iron(II)) had changed. It is found to increase.

The chemical changes may be written:

$$\text{Cu ions} \rightarrow \text{Cu atoms}$$

$$\text{Fe atoms} \rightarrow \text{Fe ions}$$

For every mole of electrons transferred from iron atoms to copper ions the increase in mass of the copper rod is half a mole of copper. The decrease in mass of the iron rod is half a mole of iron. One mole of copper is deposited if two moles of electrons are transferred. The electrons discharge the copper ions, which are positively charged:

$$Cu^{2+}(aq) + 2e \rightarrow Cu(s)$$

This equation represents a **half-reaction** since it takes no account of the other reacting substance. The half-reaction in the other beaker can be represented:

$$Fe(s) \rightarrow Fe^{2+}(aq) + 2e$$

The copper ions cannot be discharged unless there is an **electron donor** (reducing agent) to provide the electrons. Similarly, the iron cannot ionize unless the electrons can be donated to an **electron acceptor** (oxidizing agent). The complete reaction between copper ions and iron atoms is the sum of the two half-reactions:

$$Cu^{2+}(aq) + 2e \rightarrow Cu(s)$$

$$Fe(s) \rightarrow Fe^{2+}(aq) + 2e$$

Adding gives the equation

$$Cu^{2+}(aq) + Fe(s) \rightarrow Cu(s) + Fe^{2+}(aq)$$

Metals and acids

Electron transfer takes place when a metal reacts with an acid to produce hydrogen. Acids contain $H^+(aq)$ ions in solution. H^+ ions accept electrons to form hydrogen gas:

$$2H^+(aq) + 2e \rightarrow H_2(g)$$

The electrons could have come from a reactive metal such as magnesium:

$$Mg(s) \rightarrow Mg^{2+}(aq) + 2e$$

The full reaction would be represented by the equation:

$$Mg(s) + 2H^+(aq) \rightarrow Mg^{2+}(aq) + H_2(g)$$

Acids in aqueous solution contain not only $H^+(aq)$ ions but also negatively charged ions, sometimes called the acid radicals. Since the solution as a whole is electrically neutral the total positive charge is equal to the total negative charge. Hydrochloric acid contains $Cl^-(aq)$ ions (chloride ions); sulphuric acid contains $SO_4^{2-}(aq)$ ions (sulphate ions). When a metal reacts with an acid, the metal attacks only the $H^+(aq)$ ions, leaving the negative ions unchanged. Ions which are not changed in a reaction are called **spectator ions.**

The product containing the metal ions and the spectator ions is a salt. It is obtainable as a solid if the water is evaporated from the solution. Otherwise metal and spectator ions remain in electrical balance in solution. The full equations of reaction between hydrochloric or sulphuric acids and magnesium are:

$$Mg(s) + 2(H^+Cl^-)(aq) \rightarrow (Mg^{2+}2Cl^-)(aq) + H_2(g)$$

$$Mg(s) + (2H^+SO_4^{2-})(aq) \rightarrow (Mg^{2+}SO_4^{2-})(aq) + H_2(g)$$

Reactions such as these are exothermic.

Some metals do not react with acids to give hydrogen. They are called the **noble metals**. They are not as easily oxidized as the metals which react with acids to give hydrogen. Examples are copper, platinum, gold and silver. Copper does not react with H^+ ions to give hydrogen, but it is an exception amongst noble metals in being comparatively easily oxidized. Metals which transfer electrons to hydrogen ions are called **base metals**. Familiar base metals and noble metals are shown in Table 17.2.

Table 17.2

Familiar base metals		*Familiar noble metals*	
Sodium Calcium Magnesium Aluminium Zinc Iron	These metals react with $H^+(aq)$ ions easily and usually oxidize easily	Copper Silver Gold Mercury Platinum	These metals do not react with $H^+(aq)$ ions and are resistant to oxidation

Group 1 metals such as sodium transfer electrons to hydrogen ions so readily that it is dangerous to put them with acid. They react fast with the low concentration of $H^+(aq)$ ions in water in strongly exothermic reactions.

Electronic configuration of ions

Metal atoms have small numbers of electrons in their outer shells. These electrons can be lost relatively easily, so metals are electron donors (reducing agents). Table 17.3 shows electron configuration and charge on metal ions in the sodium and magnesium groups.

Ions of **transition** metals do **not** have noble gas electron configuration. For example, Cu^{2+} has the configuration 2.8.17. and Fe^{2+} the configuration 2.8.14.

Table 17.3

Element	*Group*	*Ion*	*First shell*	*Second shell*	*Third shell*
Li	1	Li^+	2		
Na	1	Na^+	2	8	
Mg	2	Mg^{2+}	2	8	
K	1	K^+	2	8	8
Ca	2	Ca^{2+}	2	8	8

These electron configurations are the same as those of He(Li^+) or Ne(Na^+, Mg^{2+}) or Ar(K^+, Ca^{2+})

The atoms of non-metals have larger numbers of electrons in their outer shells. They do not lose electrons but they can accept them and so can form negative ions. They can behave as oxidizing agents. The electron configuration of each negative ion is the same as that of the nearest noble gas. Table 17.4 gives the configuration of ions in Groups 6 and 7.

Table 17.4

Element	*Group*	*Ion*	*First shell*	*Second shell*	*Third shell*	*Fourth shell*
O	6	O^{2-}	2	8		
F	7	F^-	2	8		
S	6	S^{2-}	2	8	8	
Cl	7	Cl^-	2	8	8	
Br	7	Br^-	2	8	18	8

These electron configurations are the same as those of Ne(F^-, O^{2-}) or Ar(Cl^-, S^{2-}) or Kr(Br^-)

Some electron transfer equations

Equations of electron transfer reactions between sodium group elements and chlorine group elements are all of the same pattern: each sodium group element transfers one electron per atom and since the molecules of the chlorine group elements are diatomic, two atoms of the sodium group metal are required for each chlorine group molecule.

$$2Na(s)+Cl_2(g) \rightarrow 2(Na^+Cl^-)(s)$$

$$2K(s)+Br_2(l) \rightarrow 2(K^+Br^-)(s)$$

$$2Li(s)+I_2(s) \rightarrow 2(Li^+I^-)(s)$$

$$2Na(s)+Br_2(l) \rightarrow 2(Na^+Br^-)(s)$$

Reactions between sodium group elements and sulphur group elements also follow a pattern: since 2 moles of electrons are required to form 1 mole of sulphur group ions, 2 moles of sodium group atoms are needed:

$$2Na(s)+S(s) \rightarrow (2Na^+S^{2-})(s)$$

$$2K(s)+S(s) \rightarrow (2K^+S^{2-})(s)$$

In the case of oxygen the molecule is diatomic and 4 moles of sodium group atoms are needed:

$$4Na(s) + O_2(g) \rightarrow 2(2Na^+O^{2-})(s)$$

$$4Li(s) + O_2(g) \rightarrow 2(2Li^+O^{2-})(s)$$

Examples of equations of reaction between magnesium group (Group 2) elements and chlorine group (Group 7) or sulphur group (Group 6) elements are given below. Students should practise writing equations of reaction between any Group 1 or 2 element and any Group 6 or 7 element.

$$Mg(s) + Cl_2(g) \rightarrow (Mg^{2+}2Cl^-)(s)$$

$$Ca(s) + Br_2(l) \rightarrow (Ca^{2+}2Br^-)(s)$$

$$2Mg(s) + O_2(g) \rightarrow 2(Mg^{2+}O^{2-})(s)$$

$$Mg(s) + S(s) \rightarrow (Mg^{2+}S^{2-})(s)$$

When writing these equations, it may be found easier to begin with the half-reactions showing electron transfer. The equations for the two half-reactions can then be added together. For example, in the reaction between magnesium and oxygen, 4 moles of electrons are required since the oxygen molecule is diatomic:

$$2Mg(s) \rightarrow 2Mg^{2+}(s) + 4e$$

$$O_2(g) + 4e \rightarrow 2O^{2-}(g)$$

$$2Mg(s) + O_2(g) \rightarrow 2(Mg^{2+}O^{2-})(s)$$

Summary

1. Metal atoms can donate electrons. They can act as reducing agents.
2. Metal ions can accept electrons and act as oxidizing agents.
3. Non-metal atoms can accept electrons. They can act as oxidizing agents.
4. Non-metal ions can donate electrons and act as reducing agents.
5. $H^+(aq)$ ions act like metal ions in being electron acceptors.
6. The total number of electrons donated by electron donors in a reaction is equal to the total number accepted by electron acceptors.
7. Acids contain hydrogen ions in solution, $H^+(aq)$, which can act as acceptors of electrons from base metals.
8. Hydrogen ions in solution do not accept electrons from noble metals.
9. Ions of normal elements which are formed from single atoms often have the electron configuration of the nearest noble gas.
10. Ions of transition metals do not have noble gas electron configurations.
11. Tin is a normal metal but the Sn^{2+} ion does not have a noble gas electron configuration. It has the configuration 2.8.18.18.2. In^+ and Sb^{3+} have the same configuration. Note the position of these metals in the Periodic Table.

18 The Electrochemical Series

The electron transfer reaction between iron atoms and aqueous copper(II) ions in an electrochemical cell is discussed in Chapter 17.

$$\begin{array}{l} Fe(s) \rightarrow Fe^{2+}(aq) + 2e \\ \qquad\qquad\qquad\qquad\downarrow \\ \qquad\qquad\qquad\quad 2e + Cu^{2+}(aq) \rightarrow Cu(s) \end{array}$$

The reaction can be further investigated in a simple calorimeter such as that shown in Figure 18.1.

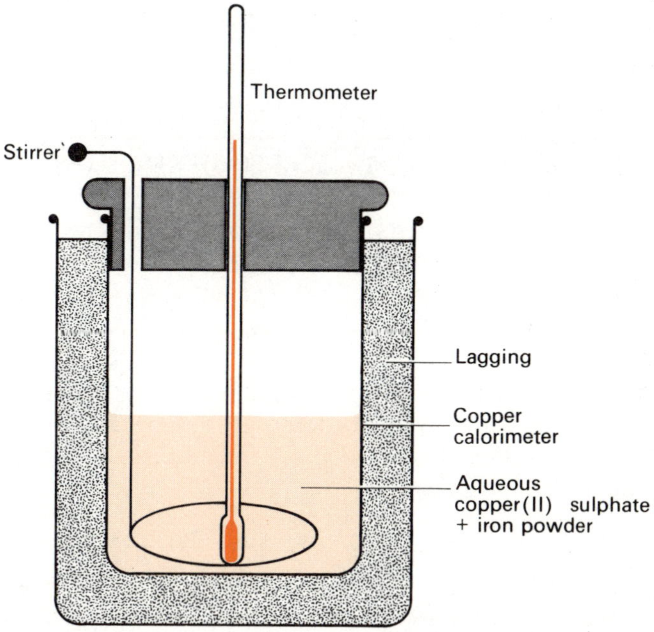

Figure 18.1 A calorimeter for investigating the heat of reaction between iron and aqueous copper(II) sulphate

Excess iron powder is added quickly to a known volume of known molarity aqueous copper(II) sulphate. The reaction is exothermic and the temperature rises. The mixture is stirred and the temperature is recorded at the beginning and when it has reached its highest value. As the mixture cools back to room temperature, heat energy is lost to the surroundings. The heat energy lost when molar quantities of reactants are used can be calculated (*see* page 122 for a specimen calculation). The sign of ΔH is negative because the products of the reaction contain less energy than the reactants at the same temperature.

$$Fe(s)+Cu^{2+}(aq) \rightarrow Fe^{2+}(aq)+Cu(s) \quad \Delta H = -152 \text{ kJ/mole}$$

If the experiment is repeated using zinc instead of iron, the heat energy change is different (*see* page 122).

$$Zn(s)+Cu^{2+}(aq) \rightarrow Zn^{2+}(aq)+Cu(s) \quad \Delta H = -216 \text{ kJ/mole}$$

In both these reactions two moles of electrons (2 Faradays) are transferred per mole of $Cu^{2+}(aq)$ ions discharged. The heat energy lost to the surroundings is, however, considerably greater in the second case. Table 18.1 shows the heat energy lost to the surroundings when the two moles of electrons are obtained from various metals.

Table 18.1

Metal providing electrons	*Equation for electron transfer reaction*	*Heat energy lost to surroundings per mole of $Cu^{2+}(aq)$ discharged by metal*
Magnesium	$Mg(s) + Cu^{2+}(aq) \rightarrow Mg^{2+}(aq) + Cu(s)$	526 kJ
Aluminium	$2\,Al(s) + 3\,Cu^{2+}(aq) \rightarrow 2\,Al^{3+}(aq) + 3\,Cu(s)$	414 kJ
Zinc	$Zn(s) + Cu^{2+}(aq) \rightarrow Zn^{2+}(aq) + Cu(s)$	216 kJ
Iron	$Fe(s) + Cu^{2+}(aq) \rightarrow Fe^{2+}(aq) + Cu(s)$	152 kJ
Lead	$Pb(s) + Cu^{2+}(aq) \rightarrow Pb^{2+}(aq) + Cu(s)$	63 kJ

All the reactions in Table 18.1 involve the same half-reaction: the transfer of two moles of electrons to $Cu^{2+}(aq)$ ions. The heat energy change in this half-reaction makes a constant contribution to the total heat energy change in each of the five reactions.

$$2e+Cu^{2+}(aq) \rightarrow Cu(s)$$

The heat energy changes in the complete reactions vary because the other half-reaction is different in each case. The complete reaction using magnesium is the most exothermic because the ionization of magnesium is more exothermic than the ionizations of the other four metals.

$$Mg(s) \rightarrow Mg^{2+}(aq)+2e$$

$$Al(s) \rightarrow Al^{3+}(aq)+3e$$

$$Zn(s) \rightarrow Zn^{2+}(aq)+2e$$

$$Fe(s) \rightarrow Fe^{2+}(aq)+2e$$

$$Pb(s) \rightarrow Pb^{2+}(aq)+2e$$

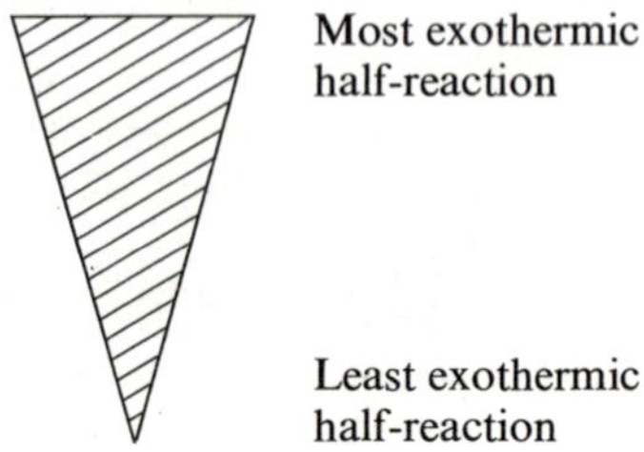

It is possible to obtain **electrical energy** from reactions where electrons are transferred. This is discussed in Chapter 13 for the reaction between zinc and aqueous copper(II) ions in the Daniell cell. A more general apparatus for investigating this electrical energy is shown in Figure 18.2.

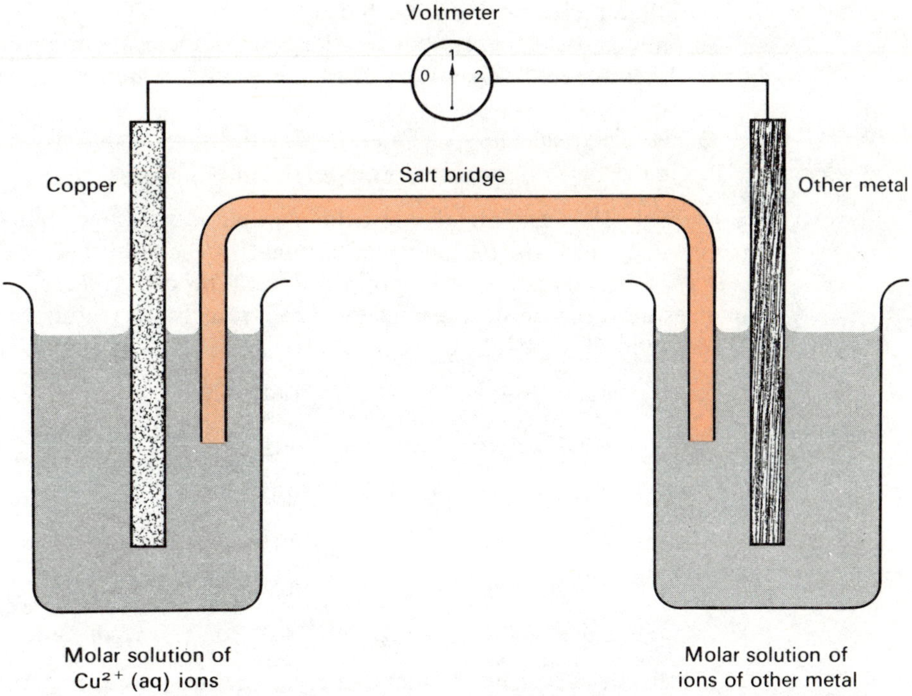

Figure 18.2 An apparatus for investigating electrical energy changes in electron transfer reactions

The voltage of the cell is a measure of the maximum electrical energy which can be obtained from the electron transfer reaction taking place in the cell. The results of the investigation are set out in Table 18.2, together with the heat energy changes already listed in Table 18.1.

Table 18.2

Metal being tested against copper electrode	*Maximum voltage of cell*	*Maximum electrical energy which can be obtained per mole of $Cu^{2+}(aq)$ discharged by metal*	*Heat energy lost to surroundings per mole of $Cu^{2+}(aq)$ discharged by metal*
Magnesium	−2.71 V	505 kJ	526 kJ
Aluminium	−2.00 V	374 kJ	414 kJ
Zinc	−1.10 V	212 kJ	216 kJ
Iron	−0.78 V	146 kJ	152 kJ
Lead	−0.47 V	89 kJ	63 kJ

Points of interest in this table which summarizes the experimental results include the following.

- The five metals all form the negative terminal of the cell when tested against copper. This is why the cell voltage is shown with a negative value.
- Each cell reaction gives a different voltage, which corresponds to a different electrical energy change.
- The electrical energy change is large when the heat energy change is large. When the electrical energy change is small, the heat energy change is also small.
- The electrical energy change is not the same value as the heat energy change. This is discussed in more detail in Chapter 13.

In Table 18.2, five metals are being compared for their ability to lose electrons. They all form the negative terminal of their cell because they all lose electrons more readily than copper does. The cell voltage is highest with magnesium because magnesium loses electrons more readily than any of the other metals.

$Mg(s)$	$\longrightarrow$	$Mg^{2+}(aq) + 2e$	Electrons most readily lost
$Al(s)$	$\longrightarrow$	$Al^{3+}(aq) + 3e$	
$Zn(s)$	$\longrightarrow$	$Zn^{2+}(aq) + 2e$	
$Fe(s)$	$\longrightarrow$	$Fe^{2+}(aq) + 2e$	
$Pb(s)$	$\longrightarrow$	$Pb^{2+}(aq) + 2e$	
$Cu(s)$	$\rightarrow$	$Cu^{2+}(aq) + 2e$	Electrons least readily lost

The electrons lost by the metal (e.g. magnesium) are transferred to aqueous copper(II) ions, which are discharged.

$$Mg(s) \longrightarrow Mg^{2+}(aq) + 2e$$
$$Cu(s) \leftarrow Cu^{2+}(aq) + 2e$$

Electrons are transferred to $Cu^{2+}(aq)$, forming copper

Copper in molar aqueous copper(II) ions is a **comparison electrode**; the other reactions are being standardized against this electrode. A more useful comparison electrode for accurate work is the **hydrogen electrode** which is illustrated in Figure 18.3. The hydrogen electrode is a platinum sheet, coated with platinum black, immersed in molar hydrogen ions. Hydrogen gas is bubbled round the platinum sheet. This electrode compares the ability of various reactions to give electrons to aqueous hydrogen ions.

$$H_2(g) \longleftarrow 2H^+(aq) + 2e$$

When the hydrogen electrode is used as the standard of comparison, the voltage of the complete cell is called the **standard electrode potential**. Table 18.3 gives the values of some standard electrode potentials.

The reactions of Table 18.3 form part of a series called the **Electrochemical Series**, frequently abbreviated to ECS. The Electrochemical Series can be used to make predictions about electron transfer reactions in aqueous solution. Three types of example of these predictions are given below.

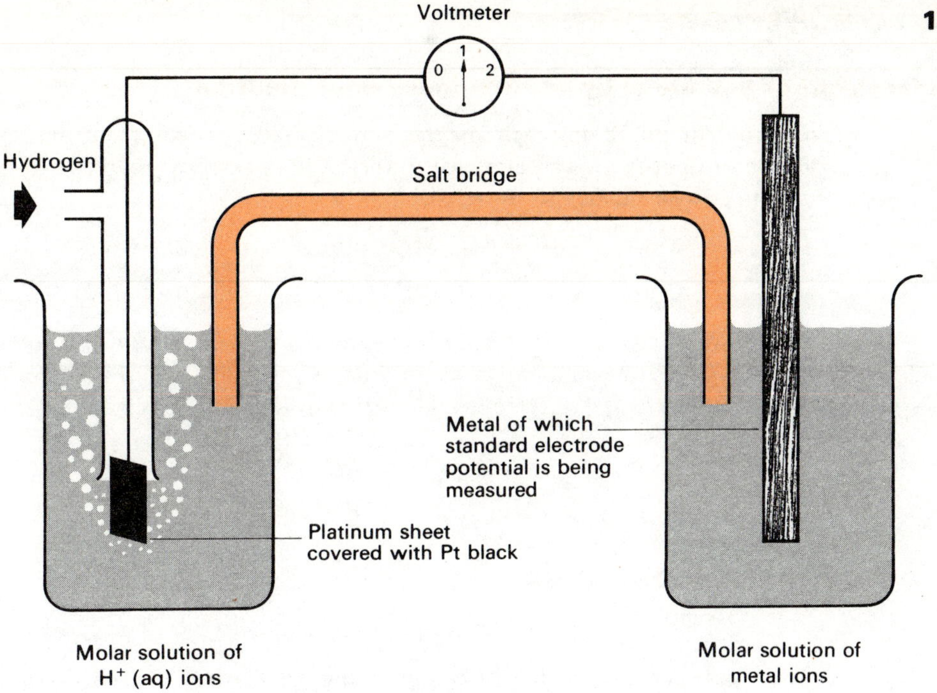

Figure 18.3 A hydrogen electrode

Table 18.3

Element	*Reaction giving electrons*	*Standard electrode potential (volts)*	
Lithium	$Li(s) \longrightarrow Li^+(aq) + e$	−3.05	↑
Potassium	$K(s) \longrightarrow K^+(aq) + 3$	−2.92	
Calcium	$Ca(s) \longrightarrow Ca^{2+}(aq) + 2e$	−2.87	Electrons
Sodium	$Na(s) \longrightarrow Na^+(aq) + e$	−2.71	most
Magnesium	$Mg(s) \longrightarrow Mg^{2+}(aq) + 2e$	−2.37	readily
Aluminium	$Al(s) \longrightarrow Al^{3+}(aq) + 3e$	−1.66	lost
Zinc	$Zn(s) \longrightarrow Zn^{2+}(aq) + 2e$	−0.76	
Iron	$Fe(s) \longrightarrow Fe^{2+}(aq) + 2e$	−0.44	
Lead	$Pb(s) \longrightarrow Pb^{2+}(aq) + 2e$	−0.13	
HYDROGEN	$H_2(g) \longrightarrow 2H^+(aq) + 2e$	0.00	
Copper	$Cu(s) \longrightarrow Cu^{2+}(aq) + 2e$	+0.34	
Oxygen	$4OH^-(aq) \longrightarrow O_2(g) + 2H_2O(1) + 4e$	+0.40	
Iodine	$2I^-(aq) \longrightarrow I_2(aq) + 2e$	+0.54	
Silver	$Ag(s) \longrightarrow Ag^+(aq) + e$	+0.80	
Nitrate ions	$NO(g) + 2H_2O(l) \longrightarrow NO_3^-(aq) + 4H^+(aq) + 3e$	+0.96	Electrons
Bromine	$2Br^-(aq) \longrightarrow Br_2(aq) + 2e$	+1.09	least
Chlorine	$2Cl^-(aq) \longrightarrow Cl_2(aq) + 2e$	+1.36	readily
Sulphate ions	$2SO_4^{2-}(aq) \longrightarrow S_2O_8^{2-}(aq) + 2e$	+2.01	lost
Fluorine	$2F^-(aq) \longrightarrow F_2(g) + 2e$	+2.85	↓

Displacement of one metal by another in aqueous solution

The problem might be: will magnesium displace copper from an aqueous solution of copper ions? Reference to the ECS shows how readily magnesium and copper lose electrons.

Mg(s)	$\longrightarrow$	$Mg^{2+}(aq) + 2e$	-2.37 volt
Cu(s)	$\rightarrow$	$Cu^{2+}(aq) + 2e$	$+0.34$ volt

Magnesium atoms lose electrons more readily than copper atoms lose electrons. This shows that when magnesium is put into a solution of $Cu^{2+}(aq)$, magnesium atoms lose electrons and $Cu^{2+}(aq)$ ions gain electrons.

$$\begin{array}{lll} Mg(s) & \longrightarrow & Mg^{2+}(aq) + 2e \;\urcorner \\ Cu(s) & \longleftarrow & Cu^{2+}(aq) + 2e \;\leftarrow\!\lrcorner \end{array}$$

The chemical equation is

$$Mg(s) + Cu^{2+}(aq) \rightarrow Mg^{2+}(aq) + Cu(s)$$

The reaction higher in the ECS goes to the right while the reaction lower in the ECS goes to the left.

Will lead displace zinc from an aqueous solution of zinc ions? Reference to the ECS shows how readily atoms of lead and zinc lose electrons.

$$\begin{array}{lll} Zn(s) & \longrightarrow & Zn^{2+}(aq) + 2e \;\urcorner \\ Pb(s) & \longleftarrow & Pb^{2+}(aq) + 2e \;\leftarrow\!\lrcorner \end{array}$$

Higher reaction goes to the right

Lower reaction goes to the left

Lead atoms do not lose electrons as readily as zinc atoms lose electrons. This means that lead will not displace zinc from aqueous solution. The relative positions in the ECS shows that the reverse of this reaction will take place. Zinc will displace lead ions from aqueous solution.

$$Zn(s) + Pb^{2+}(aq) \rightarrow Zn^{2+}(aq) + Pb(s)$$

Metals reacting with acids in aqueous solution to produce hydrogen

Metals above hydrogen in the ECS can displace hydrogen from acids.

$$\begin{array}{lll} Mg(s) & \longrightarrow & Mg^{2+}(aq) + 2e \;\urcorner \\ H_2(g) & \longleftarrow & 2H^{+}(aq) + 2e \;\leftarrow\!\lrcorner \end{array}$$

Higher reaction goes to the right; metal dissolves as ions

Lower reaction goes to the left; hydrogen gas produced

$$Mg(s) + 2H^{+}(aq) \rightarrow Mg^{2+}(aq) + H_2(g)$$

If the metal is below hydrogen in the ECS, this type of reaction cannot take place. **Metals below hydrogen in the ECS do not react with acids to give hydrogen.** These metals are called **noble metals** (*see* page 158).

Displacement of one halogen by another in aqueous solution

What will happen when aqueous chlorine is added separately to aqueous solutions of a bromide and an iodide? Reference to the ECS shows how readily halide ions lose electrons.

$$2I^-(aq) \longrightarrow I_2(aq) + 2e \quad \text{Electrons most readily lost}$$

$$2Br^-(aq) \longrightarrow Br_2(aq) + 2e$$

$$2Cl^-(aq) \longrightarrow Cl_2(aq) + 2e \quad \text{Electrons least readily lost}$$

The reaction higher in the ECS goes to the right while the reaction lower in the ECS goes to the left. When chlorine is added to a bromide or iodide, chlorine will form chloride ions (lower reaction going to the left). The bromide and iodide ions will lose electrons and form halogen molecules (higher reaction going to the right). The equations for the reactions are:

$$Cl_2(aq) + 2Br^-(aq) \rightarrow 2Cl^-(aq) + Br_2(aq)$$

$$Cl_2(aq) + 2I^-(aq) \rightarrow 2Cl^-(aq) + I_2(aq)$$

Chlorine will displace bromine from a bromide and iodine from an iodide.

It can be predicted in the same way that bromine will displace iodine from an iodide, but cannot displace chlorine from a chloride.

$$2I^-(aq) \longrightarrow I_2(aq) + 2e \quad \text{Higher reaction goes to the right}$$

$$2Br^-(aq) \longleftarrow Br_2(aq) + 2e \quad \text{Lower reaction goes to the left}$$

$$Br_2(aq) + 2I^-(aq) \rightarrow 2Br^-(aq) + I_2(aq)$$

The Electrochemical Series and chemical properties

Chlorine reacts more readily with sodium than with magnesium. Water also reacts more readily with sodium than with magnesium. Table 18.3 shows that sodium is higher than magnesium in the ECS. Position in the ECS is not the only factor which determines reactivity. Physical state, temperature, concentration and solubility of reaction products also influence reactivity. In a general way, however, the reactivity of an element is related to its position in the ECS. The name **Reactivity Series** is sometimes used instead of Electrochemical Series for this reason. The order of important elements in the ECS or Reactivity Series is shown in Table 18.4 on page 168.

To illustrate the relation between chemical reactivity and position in the ECS, the behaviour of some elements with water is shown in Tables 18.5 and 18.6. Look at these tables and decide to what extent chemical reactivity is related to the position of the elements in the ECS.

Table 18.4

Electrochemical Series of Important Elements

Lithium	↑	
Potassium		*Metals:*
Calcium		Increasing readiness to
Sodium		lose electrons and form
Magnesium		positive ions such as Na^+
Aluminium		
Zinc		Increasingly powerful
Iron		reducing agents
Lead		
Hydrogen		
Copper		*Non-metals:*
Oxygen		Increasing readiness to
Iodine		gain electrons and form
Silver		negative ions such as Cl^-
Bromine		
Chlorine		Increasingly powerful
Fluorine	↓	oxidizing agents

Table 18.5

Order of metal in ECS	*Action of metal with water to displace hydrogen*
Lithium	Moderate. Strongly alkaline solution formed $2Li(s) + 2H_2O(l) \rightarrow 2Li^+(aq) + 2OH^-(aq) + H_2(g)$
Potassium	Fast. Always catches fire. Strongly alkaline solution formed $2K(s) + 2H_2O(l) \rightarrow 2K^+(aq) + 2OH^-(aq) + H_2(g)$
Calcium	Moderate. Alkaline solution formed $Ca(s) + 2H_2O(l) \rightarrow Ca^{2+}(aq) + 2OH^-(aq) + H_2(g)$
Sodium	Fast. Sometimes catches fire. Strongly alkaline solution formed $2Na(s) + 2H_2O(l) \rightarrow 2Na^+(aq) + 2OH^-(aq) + H_2(g)$
Magnesium	Very slow. Slightly alkaline solution formed $Mg(s) + 2H_2O(l) \rightarrow Mg^{2+}(aq) + 2OH^-(aq) + H_2(g)$
Metals lower in ECS	No reaction with water at room temperature

Table 18.6

Order of halogen in ECS	*Action of halogen with water to displace oxygen*
Iodine	Dissolves slightly. No further reaction
Bromine	Dissolves. Water decomposed very slowly $2Br_2(aq) + 2H_2O(l) \rightarrow 4H^+(aq) + 4Br^-(aq) + O_2(g)$
Chlorine	Dissolves. Water decomposed slowly $2Cl_2(aq) + 2H_2O(l) \rightarrow 4H^+(aq) + 4Cl^-(aq) + O_2(g)$
Fluorine	Vigorous reaction. Water decomposed immediately $2F_2(g) + 2H_2O(l) \rightarrow 4H^+(aq) + 4F^-(aq) + O_2(g)$

Summary

1. Many reactions in aqueous solution involve electron transfer. An electron transfer reaction can be regarded as the sum of two half-reactions. One half-reaction provides electrons, the other half-reaction accepts these electrons.

$$Mg(s) \longrightarrow Mg^{2+}(aq) + 2e$$

Half-reaction providing electrons

$$Cu(s) \longleftarrow Cu^{2+}(aq) + 2e$$

these electrons are transferred to $Cu^{2+}(aq)$ the electron acceptor

2. The readiness with which a half-reaction provides electrons is shown by the position of the reaction in the Electrochemical Series (ECS). The higher the reaction is in the ECS, the more readily are electrons provided.

$$Mg(s) \longrightarrow Mg^{2+}(aq) + 2e$$

Reaction higher in ECS. Electrons more readily provided than in any reaction lower in ECS

$$Cu(s) \longrightarrow Cu^{2+}(aq) + 2e$$

Reaction lower in ECS. Electrons less readily provided than in any reaction higher in ECS

3. The ECS can be used to make predictions about electron transfer reactions in aqueous solution by applying the following general principle.

   The reaction higher in the ECS goes to the right while the reaction lower in the ECS goes to the left.

$$Mg(s) \longrightarrow Mg^{2+}(aq) + 2e$$

Higher reaction goes to the right

$$Cu(s) \longleftarrow Cu^{2+}(aq) + 2e$$

Lower reaction goes to the left

4. Metals higher in the ECS displace metals lower in the ECS from aqueous solutions of their ions.

5. Metals above hydrogen in the ECS can displace hydrogen from aqueous acids.

6. Halogens lower in the ECS displace halogens higher in the ECS from aqueous solutions of their halide ions.

7. Position in the ECS is a guide to general chemical reactivity. The most reactive metals are found near the top of the ECS. The most reactive non-metals are found near the bottom of the ECS.

19 Electrolysis

Many ions containing transition elements are coloured, the colour showing very clearly when the ions are in aqueous solution. Examples of coloured aqueous ions are the copper(II) ion, $Cu^{2+}(aq)$, which is blue, the dichromate ion, $Cr_2O_7^{2-}(aq)$, which is orange and the chromate ion, CrO_4^{2-}, which is yellow. Aqueous ions containing only normal elements are colourless. Examples of colourless aqueous ions are the potassium ion, $K^+(aq)$, and the sulphate ion, $SO_4^{2-}(aq)$. If an aqueous solution contains coloured ions, interesting studies can be made on the effect of passing an electric current through the solution.

When aqueous solutions of copper(II) sulphate and potassium dichromate are mixed, a pale murky green solution is formed. This solution contains the following ions:

$Cu^{2+}(aq)$	$SO_4^{2-}(aq)$	$K^+(aq)$	$Cr_2O_7^{2-}(aq)$
Pale blue	Colourless	Colourless	Orange

The mixture of pale blue and orange gives the solution its green colour.

If ammonia solution is then added, the solution becomes a brilliant green colour. This solution contains the following ions:

$Cu^{2+}(aq)$	$SO_4^{2-}(aq)$	$K^+(aq)$	$CrO_4^{2-}(aq)$	$NH_4^+(aq)$
Deep blue when ammonia is present	Colourless	Colourless	Yellow	Colourless

The mixture of deep blue and yellow gives the solution its brilliant green colour.

A rectangle of filter paper is rolled and slipped into a glass tube (Figure 19.1). The paper is soaked in the brilliant green solution and graphite rods are pushed into each end of the tube to make electrical contact with the paper. The graphite rods are connected to a battery of about 45 volts and current is passed for some time.

At first the filter paper is uniformly green. After a while, colour changes are seen to be taking place. Near the negative cathode the paper turns deep blue, and near the positive anode the paper turns yellow. The region near the middle remains green. A dark deposit forms where the rods touch the paper. After some time all the copper ions have moved near the cathode and that end of the paper is deep blue. The yellow chromate ions have all moved near the anode, and the region in the middle of the paper is white.

This experiment vividly illustrates what happens during electrolysis. As current passes, the positive ions move towards the negative cathode and the negative ions move towards the positive anode. When the ions reach the electrode surface they are discharged and a dark deposit of metal is formed.

Figure 19.2 illustrates what happens when electricity is passed through any

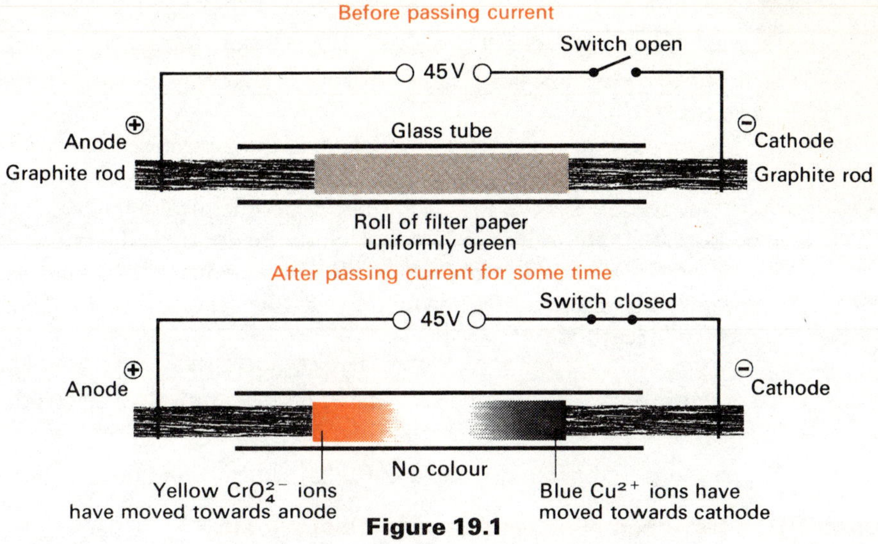

Figure 19.1

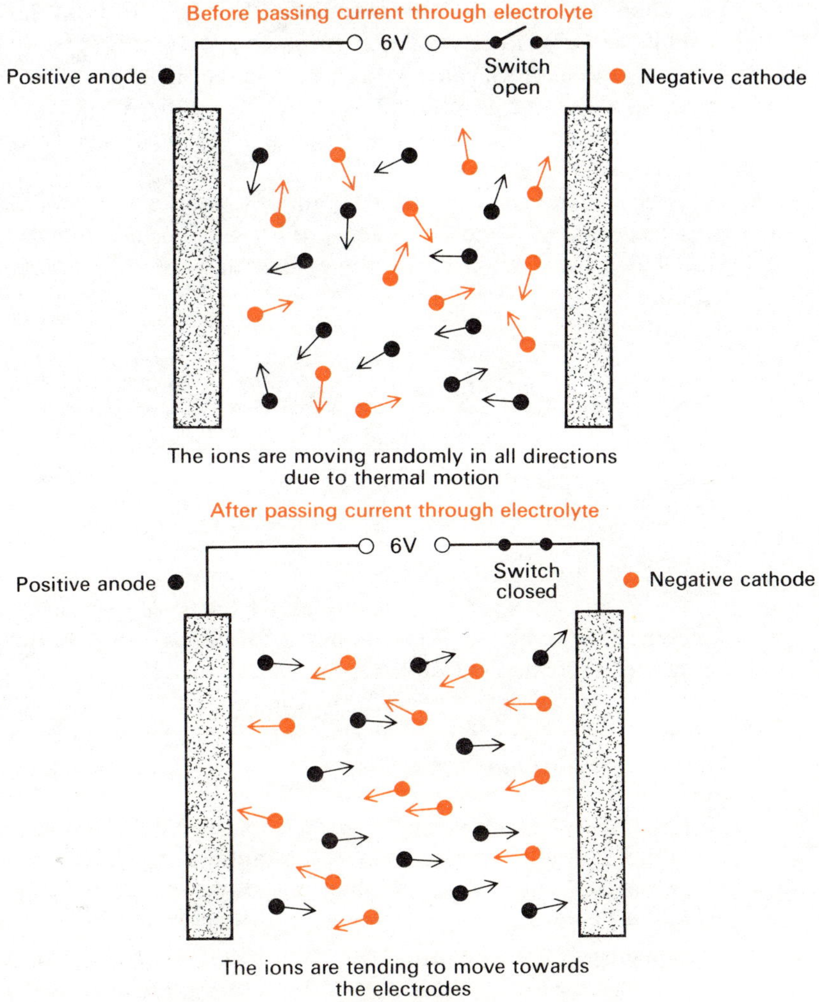

Figure 19.2

liquid containing ions. The process is called **electrolysis** and the liquid is called an **electrolyte**. The apparatus in which electrolysis takes place is called a **voltameter**.

An electrolyte is a substance which undergoes chemical change when electricity is passed through it. Its ions are discharged at the electrodes. Compare this with conduction by a metal in which there is no chemical change.

Aqueous solutions of ionic compounds are not the only liquids which conduct electricity. Molten ionic compounds also conduct. If graphite electrodes are dipped into molten lead bromide, metallic lead is liberated at the negative electrode (cathode).

$$Pb^{2+}(l) + 2e \rightarrow Pb(l)$$

Brown bromine gas can be seen rising from the positive electrode (anode).

$$2Br^{-}(l) \rightarrow Br_2(g) + 2e$$

Lead ions and bromide ions are discharged by gain and loss of electrons.

Competition between reactions during electrolysis

In molten lead bromide, only lead ions, $Pb^{2+}(l)$, and bromide ions, $Br^{-}(l)$, are present. Since neither lead nor bromine attack graphite, the only possible products of electrolysis are lead and bromine. If molten lead oxide is electrolysed, however, the graphite of the anode is oxidized to carbon monoxide and carbon dioxide. The products of electrolysis may therefore depend on the material of the electrodes.

In the electrolysis of aqueous solutions there is a complication due to the presence of $H^{+}(aq)$ and $OH^{-}(aq)$ ions formed by slight ionization of the water. Under certain circumstances these ions may be discharged. If an aqueous solution of copper(II) chloride is electrolysed with inert electrodes, copper is deposited at the cathode because the reaction

$$Cu^{2+}(aq) + 2e \rightarrow Cu(s)$$

takes place in preference to the reaction

$$2H^{+}(aq) + 2e \rightarrow H_2(g)$$

At the anode oxygen is released. The reaction

$$4OH^{-}(aq) \rightarrow O_2(g) + 2H_2O(l) + 4e$$

takes place in preference to the reaction

$$2Cl^{-}(aq) \rightarrow Cl_2(aq) + 2e$$

Reference to the Electrochemical Series on page 165 shows that this illustrates an important principle (Table 19.1).

Table 19.1

At the cathode ⊖	*At the anode* ⊕
Metal ions (positive ions) low in the ECS are discharged in preference to metal ions higher in the ECS.	Non-metal ions (negative ions) high in the ECS are discharged in preference to non-metal ions lower in the ECS.
Metal ions gain electrons. This is easier for a reaction lower in the ECS.	Non-metal ions lose electrons. This is easier for a reaction higher in the ECS.

This principle cannot be applied with certainty if the concentrations of the ions which could be discharged are very different. In this situation the ions present in high concentration may be discharged even if the position in the Electrochemical Series makes this look unfavourable.

The electrolysis of aqueous sodium chloride illustrates well the way in which ionic concentration can influence the products of electrolysis. The ions present in aqueous sodium chloride are $Na^+(aq)$, $Cl^-(aq)$, $H^+(aq)$ and $OH^-(aq)$. The positions of these ions in the Electrochemical Series are as follows (Table 19.2):

Table 19.2

Standard electrode potential (volts)	*Equation for reaction giving electrons*
−2.71	$Na(s) \longrightarrow Na^+(aq) + e$
0	$H_2(g) \longrightarrow 2H^+(aq) + 2e$
+0.4	$4OH^-(aq) \longrightarrow O_2(g) + 2H_2O(l) + 4e$
+1.36	$2Cl^-(aq) \longrightarrow Cl_2(aq) + 2e$

The following changes (Table 19.3) take place at the electrodes during the electrolysis of aqueous sodium chloride.

Table 19.3

At the cathode	*At the anode*
$H^+(aq)$ and $Na^+(aq)$ ions are the two positive ions present in aqueous sodium chloride. The concentration of $H^+(aq)$ ions is very small, but hydrogen is considerably lower than sodium in the ECS. Hydrogen ions are discharged in preference to sodium ions even if the concentration of $Na^+(aq)$ is high.	$OH^-(aq)$ and $Cl^-(aq)$ ions are the two negative ions present in aqueous sodium chloride. These ions are much closer in the ECS than are the ions $H^+(aq)$ and $Na^+(aq)$. If the concentration of $Cl^-(aq)$ is low, $OH^-(aq)$ ions are discharged. This is the reaction which is expected on the principle that the non-metal ion high in the ECS is discharged in preference to the non-metal ion lower in the ECS.
	If the concentration of $Cl^-(aq)$ is high, however, this ion is discharged in preference to the discharge of $OH^-(aq)$.
The reaction	$OH^-(aq)$ is discharged if the concentration of $Cl^-(aq)$ is low.
$2H^+(aq) + 2e \rightarrow H_2(g)$	$4OH^-(aq) \rightarrow O_2(g) + 2H_2O(l) + 4e$
always takes place in preference to the reaction	$Cl^-(aq)$ is discharged at high concentration of chloride ions.
$Na^+(aq) + e \rightarrow Na(s)$	$2Cl^-(aq) \rightarrow Cl_2(aq) + 2e$

Chlorine is liberated at the anode when the concentration of chloride ions is high. Use is made of this in the industrial manufacture of chlorine by electrolysing brine (saturated aqueous sodium chloride).

Active and inert electrodes

If dilute sulphuric acid is electrolysed using platinum electrodes, the cathode product is hydrogen:

$$2H^+(aq) + 2e \rightarrow H_2(g)$$

and the anode product is oxygen as $OH^-(aq)$ is discharged in preference to discharge of $SO_4^{2-}(aq)$:

$$4OH^-(aq) \rightarrow O_2(g) + 2H_2O(l) + 4e$$

But if graphite electrodes are used, the anode product is contaminated with carbon monoxide and carbon dioxide. The graphite electrode is oxidized. Platinum electrodes are inert but graphite electrodes are active in this electrolysis.

An important application of this principle is in the purification of copper. If aqueous copper(II) sulphate is electrolysed between inert electrodes, the cathode product is copper and the anode product is oxygen. If the anode is made of copper, however, it is attacked and releases $Cu^{2+}(aq)$ ions into solution.

$$Cu(s) \rightarrow Cu^{2+}(aq) + 2e$$

Very pure copper for electrical purposes is made by electrolysing aqueous copper(II) sulphate using pure copper as cathode and impure copper as anode. For each copper ion discharged at the cathode a copper ion goes into solution at the anode. The concentration of copper ions in solution remains constant and the effect is to transfer copper from anode to cathode. At the cathode only copper ions are discharged. Impurities in the copper anode sink to the bottom of the electrolysis bath.

Charge flow during electrolysis

The coulomb is the unit of electrical charge. A current of one ampere is a rate of flow of charge of one coulomb per second. The charge flow during electrolysis is calculated from a knowledge of the number of seconds for which a steady current is passed. Some examples are given below (Table 19.4).

Table 19.4

Current in circuit	*Time for which current is passed*	*Total charge flow through circuit*
1 ampere	1 second	1 coulomb
10 amperes	1 second	10 coulombs
1 ampere	20 seconds	20 coulombs
10 amperes	20 seconds	200 coulombs
A amperes	*t* seconds	*At* coulombs

The charge flow required to liberate one mole of element

Electrolysis always produces chemical change. In the reaction,

$Ag^+(aq)$	+	e	→	$Ag(s)$
1 mole of silver ions		1 mole of electrons		1 mole of silver atoms

1 mole of electrons is required to discharge 1 mole of $Ag^+(aq)$ ions. This liberates 1 mole of silver atoms. One mole of electrons represents a large charge. It is found by experiment to be 96 500 coulombs and is called one **Faraday**.

The number of Faradays (moles of electrons) required to liberate 1 mole of an element during electrolysis is found from the equation for the electrode reaction. Some examples are given below.

Sodium:

$Na^+ + e \rightarrow Na$ — 1 Faraday must be passed to liberate 1 mole of sodium atoms (23 g)

Copper:

$Cu^{2+} + 2e \rightarrow Cu$ — 2 Faradays must be passed to liberate 1 mole of copper atoms (63.5 g)

Aluminium:

$Al^{3+} + 3e \rightarrow Al$ — 3 Faradays must be passed to liberate 1 mole of aluminium atoms (27 g)

Chlorine:

$2Cl^- \rightarrow Cl_2 + 2e$ — 2 Faradays must be passed to liberate 1 mole of chlorine **molecules** (71 g)

Faraday's Laws

Problem: A steady current of 4 amperes is passed for 1800 seconds (i.e. 30 minutes) through aqueous copper(II) sulphate. What mass of copper is deposited at the cathode?

Solution:

1 ampere passed for 1 second = 1 coulomb
4 amperes passed for 1800 seconds = 4×1800 coulombs

The reaction at the cathode is

$$Cu^{2+}(aq) + 2e \rightarrow Cu(s)$$

This equation shows that 2 Faradays must be passed to liberate 1 mole of copper atoms (63.5 g).

$$2 \times 96\,500 \text{ coulombs deposit } 63.5 \text{ g of copper}$$

$$\therefore 1 \text{ coulomb deposits } \frac{63.5}{2 \times 96\,500} \text{ g of copper}$$

$$\therefore 4 \times 1800 \text{ coulombs deposit } \frac{4 \times 1800 \times 63.5}{2 \times 96\,500} \text{ g} = 4.6 \text{ g of copper}$$

$$\text{Mass of element liberated} = \frac{\text{current} \times \text{seconds} \times \text{mass of 1 mole of atoms}}{\text{charge on ion} \times 96\,500}$$

This relation summarizes the two laws of electrolysis discovered by Faraday.

1. The mass of an element liberated in electrolysis is proportional to the number of coulombs passed.
2. The masses of different elements liberated by the same number of coulombs are proportional to their molar masses (when the charges carried by the discharged ions have the same magnitude).

Summary

1. Aqueous solutions of ionic compounds / Molten ionic compounds } are electrolytes.
2. Electrolytes conduct electricity and chemical change takes place at the electrodes.
3. Electrolytes contain randomly moving ions which travel to the electrodes when current flows.
   Positive ions (cations) travel to the negative electrode (cathode).
   Negative ions (anions) travel to the positive electrode (anode).
4. Metal ions, and also $H^+(aq)$, gain electrons at the negative electrode (cathode) and the metal or hydrogen is discharged.

   $$Cu^{2+}(aq) + 2e \rightarrow Cu(s)$$

   When two types of positive ion are present in an aqueous solution, it is usually the ion lower in the ECS which is discharged. $H^+(aq)$ ions are always present in aqueous solutions and are usually discharged in preference to metal ions higher in the ECS.

   $$2H^+(aq) + 2e \rightarrow H_2(g)$$

5. Non-metal ions release electrons at the positive electrode (anode).

   $$2Cl^-(aq) \rightarrow Cl_2(aq) + 2e$$

   When two types of negative ion are present in an aqueous solution, it is usually the ion higher in the ECS which is discharged. $OH^-(aq)$ ions are always present in aqueous solutions and are usually discharged in preference to non-metal ions lower in the ECS.

   $$4OH^-(aq) \rightarrow O_2(g) + 2H_2O(l) + 4e$$

6. The reaction at the positive electrode (anode) in electrolysis involves electron release. If the anode is made of copper, these electrons can be released by dissolving of the anode.

$$Cu(s) \rightarrow Cu^{2+}(aq) + 2e$$

7. Quantitative electrolysis:

$$\text{Mass of element liberated} = \frac{\text{current} \times \text{seconds} \times \text{mass of 1 mole of atoms}}{\text{charge on ion} \times 96\,500}$$

20
Extraction of Elements
1: Electrolytic methods

The metals of ancient times such as copper, lead, zinc and tin were fairly easily smelted, or extracted from their ores, by methods described in the next chapter. Iron proved more difficult to extract, so weapons and tools of iron were not in use until long after the copper-based alloys of the Bronze Age had been discovered. Three non-metallic elements, carbon, sulphur and phosphorus, have been used for many centuries. But the extraction of such familiar elements as aluminium, magnesium, calcium, sodium, chlorine and bromine had to await the technological advances of the nineteenth century. Not until electrolysis had been discovered could these elements be obtained from their compounds in quantities of industrial importance.

Metals

Calcium, sodium, aluminium and the other metals which must be extracted by electrolytic methods are high in the Electrochemical Series. Table 20.1 gives their standard electrode potentials.

Table 20.1

Element	*Standard electrode potential (volt)*
K	−2.92
Ca	−2.87
Na	−2.71
Mg	−2.38
Al	−1.66

They are obtained by electrolysis of molten ionic compounds containing them.

Calcium

Electrolysis of molten calcium chloride gives calcium at the cathode. Since the metal reacts vigorously with air at the temperature of the melt (700 °C) it must be protected from the atmosphere. Calcium metal melts at 850 °C and can conveniently be solidified on a water-cooled steel cathode. Round the solid calcium 'carrot' which forms, calcium chloride freezes and therefore protects it. The cathode is slowly withdrawn from the melt as the 'carrot' grows.

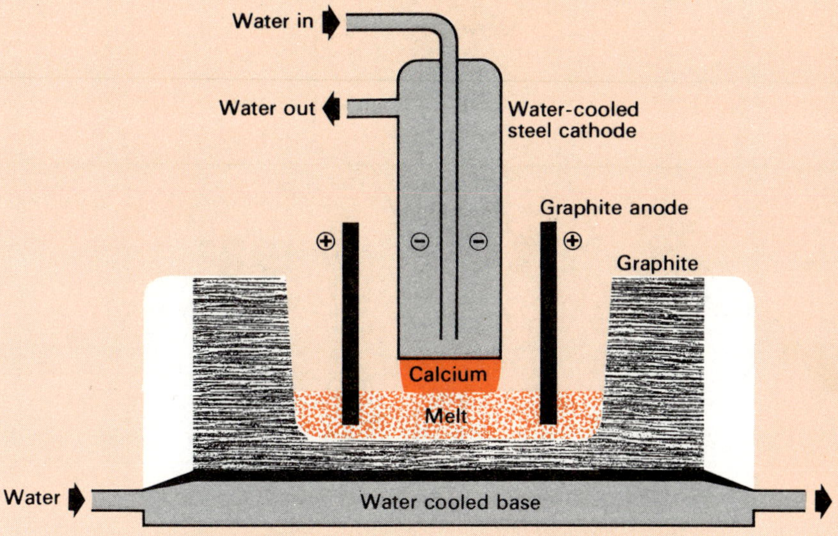

Figure 20.1 Electrolytic extraction of calcium

At the cathode

$$Ca^{2+}(l) + 2e \rightarrow Ca(s)$$

At the anode

$$2Cl^{-}(l) \rightarrow Cl_2(g) + 2e$$

Chlorine is a by-product in this electrolysis. Calcium is used in the making of high-grade steels. It reacts with any oxide in the molten steel and is termed an oxygen-getter.

Sodium

Sodium chloride melts at about 800 °C. If it is mixed with calcium chloride the mixture melts at only about 600 °C. When this melt is electrolysed the product contains calcium as well as sodium, but the calcium can be removed from the molten sodium by filtration. The melt is electrolysed between a steel cathode and a graphite anode.

At the cathode

$$Na^{+}(l) + e \rightarrow Na(l)$$

At the anode

$$2Cl^{-}(l) \rightarrow Cl_2(g) + 2e$$

Sodium metal, which melts at 97 °C, rises to the surface of the denser melt and is forced up the riser pipe into the reservoir (*see* Figure 20.2).

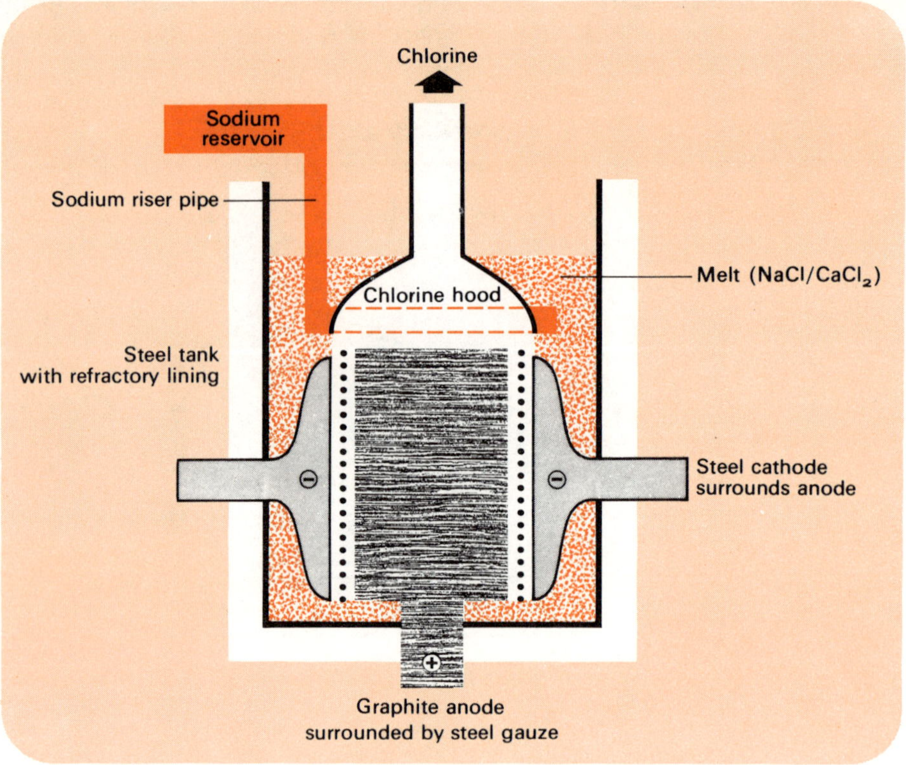

Figure 20.2 Downs cell for extraction of sodium by electrolysis of molten sodium chloride.

Sodium is an important reducing agent (electron donor) since it readily gives up electrons:

$$Na \rightarrow Na^+ + e$$

It is used in the production of the petrol anti-knock additive lead tetraethyl, although in future other anti-knock additives will be developed since lead pollution is becoming a dangerous problem.

Potassium is extracted in a similar manner to sodium.

Magnesium

This important structural metal is usually alloyed with aluminium to make high performance engines, aircraft and in other uses where lightness combined with strength is necessary. These alloys are better heat conductors than steel and several uses depend upon this property.

As in the case of sodium and calcium, the molten chloride, $MgCl_2$, is electrolysed. Whereas calcium chloride can be dehydrated by heating the hydrate, $CaCl_2,6H_2O$, heating the hydrate of magnesium chloride produces the oxide:

$$MgCl_2,6H_2O \rightarrow MgO + 2HCl + 5H_2O$$

The hydrate is therefore heated in hydrogen chloride gas.

Several rocks contain magnesium: dolomite ($MgCO_3,CaCO_3$), carnallite

($MgCl_2,KCl,6H_2O$) and magnesite ($MgCO_3$) are examples. Sea water contains 0.13 per cent magnesium and much of the metal is obtained from the sea. Magnesium hydroxide, $Mg(OH)_2$, is first precipitated from concentrated sea water by adding the more soluble calcium hydroxide, $Ca(OH)_2$. The magnesium hydroxide is then dehydrated in hydrogen chloride gas and the molten chloride electrolysed.

At the cathode

$$Mg^{2+}(l) + 2e \rightarrow Mg(s)$$

At the anode

$$2Cl^-(l) \rightarrow Cl_2(g) + 2e$$

Sodium, potassium and calcium chlorides are added to the melt to lower the melting point to about 750 °C and improve its conductivity.

Aluminium

This is the most important of the metals obtained electrolytically: over 7 000 000 tonnes are produced annually. Saucepans, wrapping foil, milk bottle tops, motor accessories and the current-carrying cables of the National Grid are only a few of the main uses.

Unlike the other metals high in the Electrochemical Series, aluminium is not produced by electrolysis of its molten chloride. A study of the melting points of the chlorides of the sodium to argon period suggests the reason:

Chloride:	(Na^+Cl^-)	$(Mg^{2+}2Cl^-)$	$AlCl_3$	$SiCl_4$	PCl_3	SCl_2	Cl_2
Melting point (°C):	808	714	180	−70	−92	−80	−101

The sharp drop from $(Mg^{2+}2Cl^-)$, 714 °C, to $AlCl_3$, 180 °C, shows aluminium chloride to be covalent. Hence it does not conduct electricity (*see* Chapter 9).

Ionic aluminium compounds such as aluminium sulphate or aluminium nitrate are not suitable for electrolysis because instead of melting when they are heated they decompose, e.g.

$$(2Al^{3+}3SO_4^{2-}) \rightarrow (2Al^{3+}3O^{2-}) + 3SO_3$$

The only compound of aluminium which is both ionic and stable enough to melt without decomposing is aluminium oxide, $(2Al^{3+}3O^{2-})$. This oxide melts at 2050 °C and until a method was found of obtaining it in liquid form at a lower temperature than this the metal could not be extracted by electrolysis. The problem was solved by Hall, working in America, and Héroult, working in France, in 1886.

Aluminium oxide is dissolved in molten cryolite, Na_3AlF_6. If the mixture is in the right proportions the melting point is about 1000 °C and at this temperature it can be electrolysed.

To the cathode are attracted Na^+ and Al^{3+} ions. The Al is lower in the Electrochemical Series and so is discharged.

$$Al^{3+}(l) + 3e \rightarrow Al(l)$$

At the anode the O^{2-} ions are discharged in preference to F^- ions since F is lower in the Electrochemical Series of the non-metals.

$$2O^{2-}(l) \rightarrow O_2(g) + 4e$$

The balanced equations for the whole process are

$$4Al^{3+} + 12e \rightarrow 4Al$$

$$6O^{2-} \rightarrow 3O_2 + 12e$$

Fresh aluminium oxide is added to the melt as the electrolysis proceeds. The electrodes are made of graphite and the anode is attacked by the hot oxygen forming a mixture of carbon monoxide and carbon dioxide.

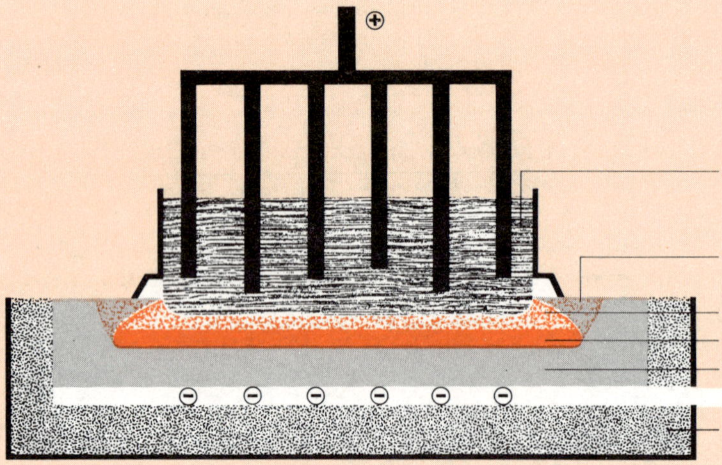

Figure 20.3 Electrolytic extraction of aluminium

Aluminium oxide for the process is required in a pure state. The bauxite, $(2Al^{3+}3O^{2-})$, from which it is obtained is found mixed with iron(III) oxide which, were it not removed, would soon render the equipment useless. The problem is solved by making use of the amphoteric nature of aluminium oxide.

If amphoteric oxides are mixed with strong alkalis such as sodium hydroxide they behave as weak acids. The oxides of lead, zinc and tin behave in this way. When powdered bauxite is heated with sodium hydroxide solution, sodium aluminate is formed:

$$(2Al^{3+}3O^{2-}) + 2(Na^+OH^-) \rightarrow 2(Na^+AlO_2^-) + H_2O$$

The iron(III) oxide and other impurities remain insoluble and can be separated by settling and filtration. Carbon dioxide is then passed into the sodium aluminate solution. Aluminium hydroxide is precipitated, leaving a solution of sodium hydrogen carbonate.

The aluminium hydroxide is then heated to form pure aluminium oxide which is electrolysed.

$$2(Al^{3+}3OH^-)(s) \rightarrow (2Al^{3+}3O^{2-})(s) + 3H_2O(g)$$

Non-metals

The standard electrode potentials of chlorine, bromine and fluorine are given in Table 20.2.

Table 20.2

Element	*Standard electrode potential (volt)*
Br	+1.07
Cl	+1.36
F	+2.85

This indicates that of the three, bromide is the most readily discharged ion and fluoride ions are least readily discharged. They can be obtained by electrolysis of molten ionic bromides, chlorides or fluorides. Chlorine and bromine do not attack water as readily as do the metals which are obtained electrolytically so they can also be extracted by electrolysing solutions of chlorides or bromides in water. Fluorine cannot be extracted in this way because its rate of attack on water is quite fast.

Chlorine

Some chlorine is made as a by-product of electrolysis of sodium chloride, magnesium chloride or calcium chloride. Most of the chlorine made in this way is used in the plant. For example, chlorine from magnesium chloride is used to make hydrogen chloride. The hydrogen chloride is passed over magnesium chloride hydrate in the dehydration process. The main source of chlorine is the electrolysis of sodium chloride solution, brine, *see* Figure 20.4 on page 184. The other product of the electrolysis is sodium hydroxide.

Several methods are in use for electrolysing brine. That illustrated is the Solvay cell. The anode is graphite and chlorine is led away through a hood.

$$2Cl^-(aq) \longrightarrow Cl_2(g) + 2e$$

The cathode is mercury. It might be expected that hydrogen would be formed at the cathode by discharge of hydrogen ions,

$$2H^+(aq) + 2e \longrightarrow H_2(g)$$

since sodium is much higher in the Electrochemical Series. However, sodium ions are in fact discharged and the sodium amalgamates with (dissolves in) the mercury cathode. The floor of the cell slopes and the amalgam slowly runs out of the cell into distilled water where it reacts to form a solution of sodium hydroxide:

$$2Na/Hg + 2H_2O \longrightarrow 2Na^+(aq) + 2OH^-(aq) + H_2(g) + 2Hg(l)$$

The mercury is then returned to the upper part of the cell. The object of this is to separate the sodium hydroxide solution from the brine.

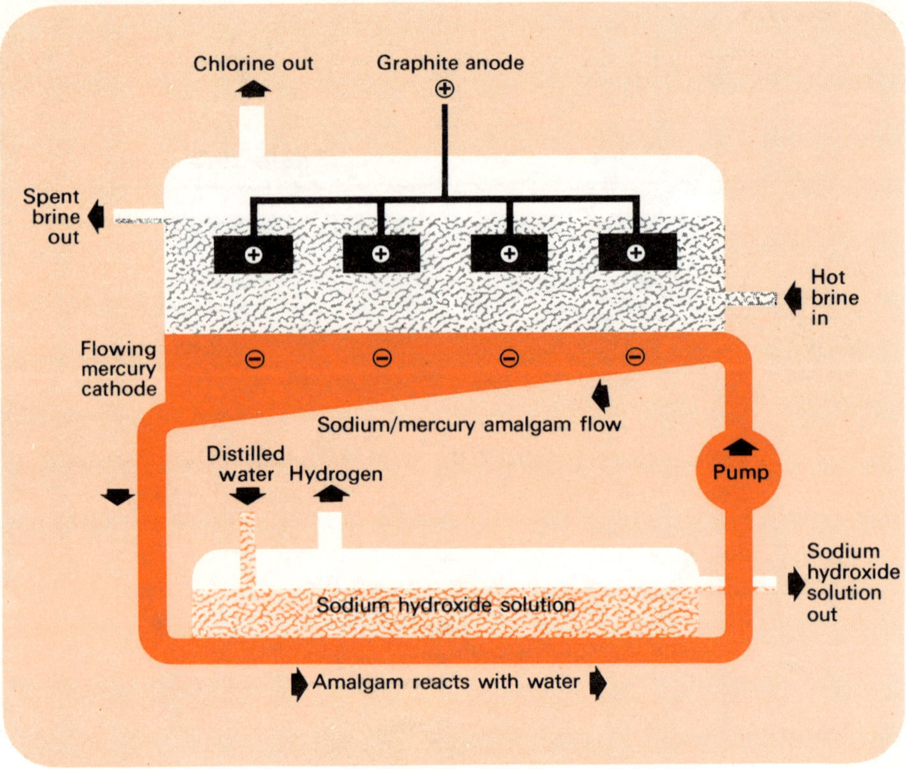

Figure 20.4 Electrolytic manufacture of chlorine and sodium hydroxide

The sodium hydroxide solution is concentrated and eventually only molten sodium hydroxide remains. It is cast into pellets, sticks or flake. This is the main industrial source of this most important alkali.

Chlorine is used in drinking water and swimming baths as a germicide. From chlorine is made a wide range of compounds which includes PVC (polyvinyl chloride) used as an electrical insulator, disinfectants such as TCP (trichlorophenol) and Dettol, insecticides such as DDT, and hydrochloric acid.

Bromine

This element is used mainly to make silver bromide for photographic emulsions. Film and enlarging paper are coated with a layer, the emulsion, of silver bromide crystals in gelatine.

If sea water is concentrated by evaporation the concentration of bromide, $Br^-(aq)$, ions is raised. If these ions are then discharged the resulting bromine can be separated from the water by blowing steam through it. The discharge is brought about by passing chlorine through the concentrated sea water.

$$Cl_2(g) + 2Br^-(aq) \rightarrow 2Cl^-(aq) + Br_2(aq)$$

This is an electron transfer reaction. The chlorine acts as the electron acceptor (oxidizing agent) and the bromide ions as the electron donor (reducing agent).

Since bromine is higher in the Electrochemical Series than chlorine, electrons are transferred from bromide ions to chlorine atoms. The chlorine is obtained electrolytically. Thus the extraction of bromine depends on an electrolytic process.

Fluorine

This is the most electronegative of all elements. Fluorine will accept electrons from all non-metallic ions or from all metallic atoms above it in the Electrochemical Series. It is of increasing importance for making the plastic Teflon (polytetrafluorethylene) and aerosol propellants.

Fluorine is made by electrolysing a mixture of molten potassium fluoride and hydrogen fluoride.

At the anode

$$2F^-(l) \rightarrow F_2(g) + 2e$$

At the cathode

$$2H^+(l) + 2e \rightarrow H_2(g)$$

The hydrogen ions are discharged in preference to potassium ions since potassium is much higher in the Electrochemical Series.

Summary

1. Metals high in the ECS must be extracted by electrolysis.
2. Lithium, Sodium, Potassium, Calcium, Magnesium } extracted by electrolysis of molten chloride
3. Aluminium: extracted by electrolysis of aluminium oxide in molten cryolite.
4. Non-metals low in the ECS are frequently extracted by electrolysis.
5. Chlorine: extracted by electrolysis of brine (saturated NaCl solution).
6. Fluorine: extracted by electrolysis of a mixture of molten potassium fluoride and hydrogen fluoride.
7. Bromine: extracted from sea water using chlorine in an electron transfer process.

21 Extraction of Elements 2: Chemical methods

When man found that tools and weapons could be made from metals instead of stone or bone he discovered the process of smelting. We can imagine that the first discoveries were accidental: stones built round fires came into contact with charcoal at high temperatures. When the fire went out there was a bead of metal in the ashes. Later, other 'stones' were tried and slowly the skills of smelting were developed. The ore of each metal, once recognized, could be collected, broken into small pieces and mixed with the substance, normally charcoal, which reduced the ore to the metal when heated. This is the origin of the word 'reduction', the reverse of oxidation. When the metal was obtained from the pile of ore there was a reduction both in the weight of the ore and in the size of the pile. Only much later was it realized that reduction and oxidation are opposites and later still that both frequently involve electron transfer processes (*see* Chapter 17).

The term smelting is usually applied to metals. When non-metals such as sulphur, phosphorus and iodine are obtained the processes are merely given the wider term extraction. In all cases of metal extraction an oxidized form of the metal is reduced to metal atoms.

Iron

More iron is produced than any other metal. Most is converted into steel. Metals had been known and used for nearly 2000 years before the method for extracting iron was discovered. This is largely because

- a very high temperature is required and
- a material other than the reducing agent has to be added.

The modern blast furnace (Figure 21.1) is usually about 50 metres (150 feet) high. It is charged with crushed iron ore (iron oxide), limestone and coke. Several reactions take place during the production of the iron. Some of these reactions are exothermic, such as the oxidation of the coke to carbon dioxide and the reduction of iron oxide by carbon monoxide. Others, such as the decomposition of the limestone and the reduction of carbon dioxide to carbon monoxide are endothermic.

The high temperature required at the bottom of the furnace is reached by using a blast of pre-heated air, often enriched with oxygen. In this blast the coke (carbon) in the lower part of the furnace burns fast

$$C(s) + O_2(g) \rightarrow CO_2(g)$$

A temperature of about 1600 °C is reached.

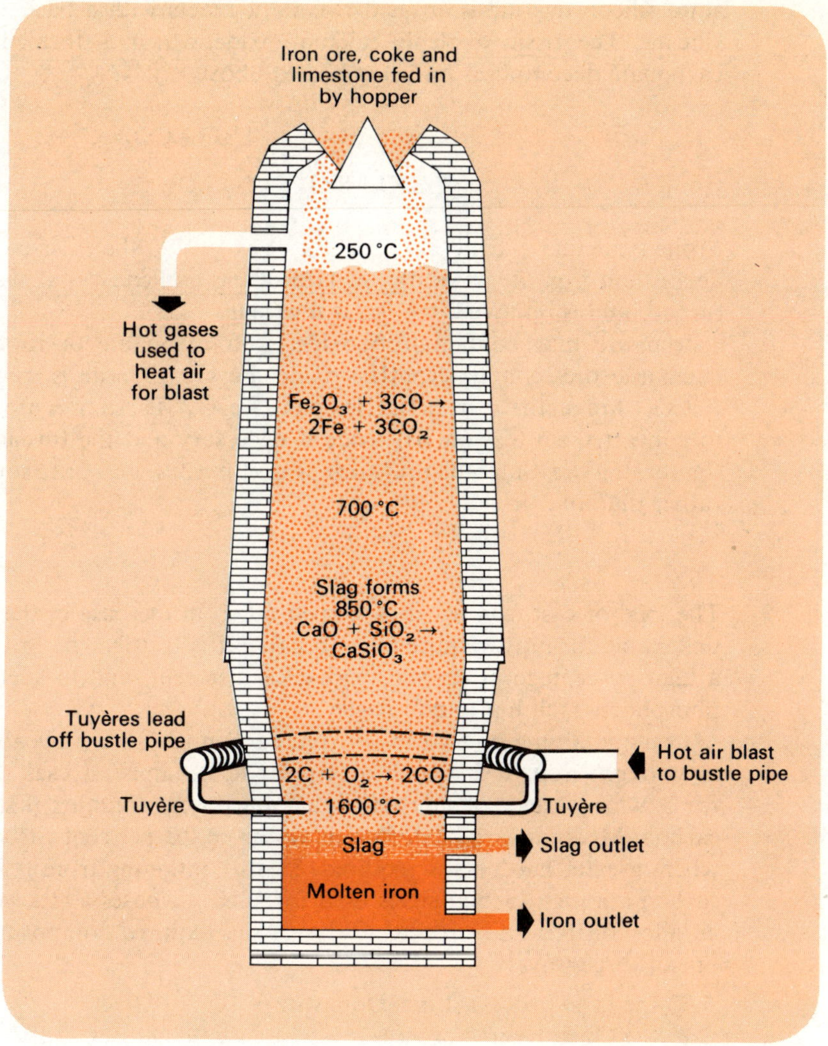

Figure 21.1 Diagram of blast furnace

As the hot carbon dioxide rises in the furnace it meets more coke. An endothermic reduction of carbon dioxide to carbon monoxide takes place

$$CO_2(g) + C(s) \rightarrow 2CO(g)$$

The carbon monoxide then reacts exothermically with the iron oxide forming iron and carbon dioxide

$$Fe_2O_3(s) + 3CO(g) \rightarrow 2Fe(l) + 3CO_2(g)$$

Molten iron trickles down inside the furnace and the carbon dioxide reacts with more coke forming carbon monoxide once again. The carbon monoxide reacts with more iron oxide.

Iron ore contains sand, silicon dioxide, as an impurity. If this were allowed to remain with the molten iron a very brittle product would result. However, since silicon dioxide is an acidic oxide it reacts with a basic oxide to form a silicate. The basic oxide is calcium oxide which is formed when calcium carbonate decomposes on being heated above 850 °C.

$$CaCO_3 \rightarrow CaO + CO_2$$
$$CaO + SiO_2 \rightarrow CaSiO_3$$

Molten calcium silicate also trickles down inside the furnace and, being less dense than iron, floats on the surface of the molten iron at the bottom of the furnace and is removed. It is known as slag.

Iron ore may contain iron oxide, iron carbonate or iron sulphide. The haematite ores contain iron(III) oxide, Fe_2O_3. Siderite is iron(II) carbonate, $FeCO_3$. Magnetite is iron(II, III) oxide, Fe_3O_4. These ores are usually roasted to convert them to iron(III) oxide if necessary and the furnace charged with the roasted ore. Sulphide ores such as iron pyrites, FeS_2, are sources of sulphur rather than of iron.

Iron to steel

The 'pig' or cast iron which is tapped off from the base of the blast furnace is unsuitable for most requirements because it is brittle. The brittleness is due to a high percentage of carbon, perhaps 4 per cent, and to such non-metals as phosphorus, sulphur and nitrogen.

Iron containing up to 0.15 per cent carbon is a soft, malleable metal known as wrought iron. It is mainly restricted to ornamental uses. Mild steel, used for a wide variety of constructional purposes, contains 0.15–0.25 per cent carbon. Medium carbon steels contain up to 0.5 per cent carbon and are used where greater hardness is required. Steels containing from 0.5 to 1.4 per cent carbon are high carbon steels used for such purposes as tool making. Steel is an alloy of iron and carbon. Alloys of the required composition are made in one of three ways:

- the L–D process (Linz–Donawitz)
- the Open Hearth process
- the Electric Furnace

Only specially high grade and special alloy steels are made in the electric furnace.

The Linz–Donawitz (L–D) process

Conversion of iron to steel involves removing a certain amount of carbon and removing such non-metals as nitrogen, sulphur and phosphorus. If the pig iron is obtained direct from the blast furnace the conversion could be done by the now obsolete Bessemer process. However, so much steel scrap is now available to the industry that methods using pig iron mixed with scrap are now of greatest importance.

In the L–D process the burning of carbon to carbon dioxide and the oxidation of other non-metals is done with a blast of oxygen ('tonnage oxygen') passed over or through the molten iron. A water-cooled copper lance is used. Powdered lime is added to form a basic slag of calcium oxide and calcium

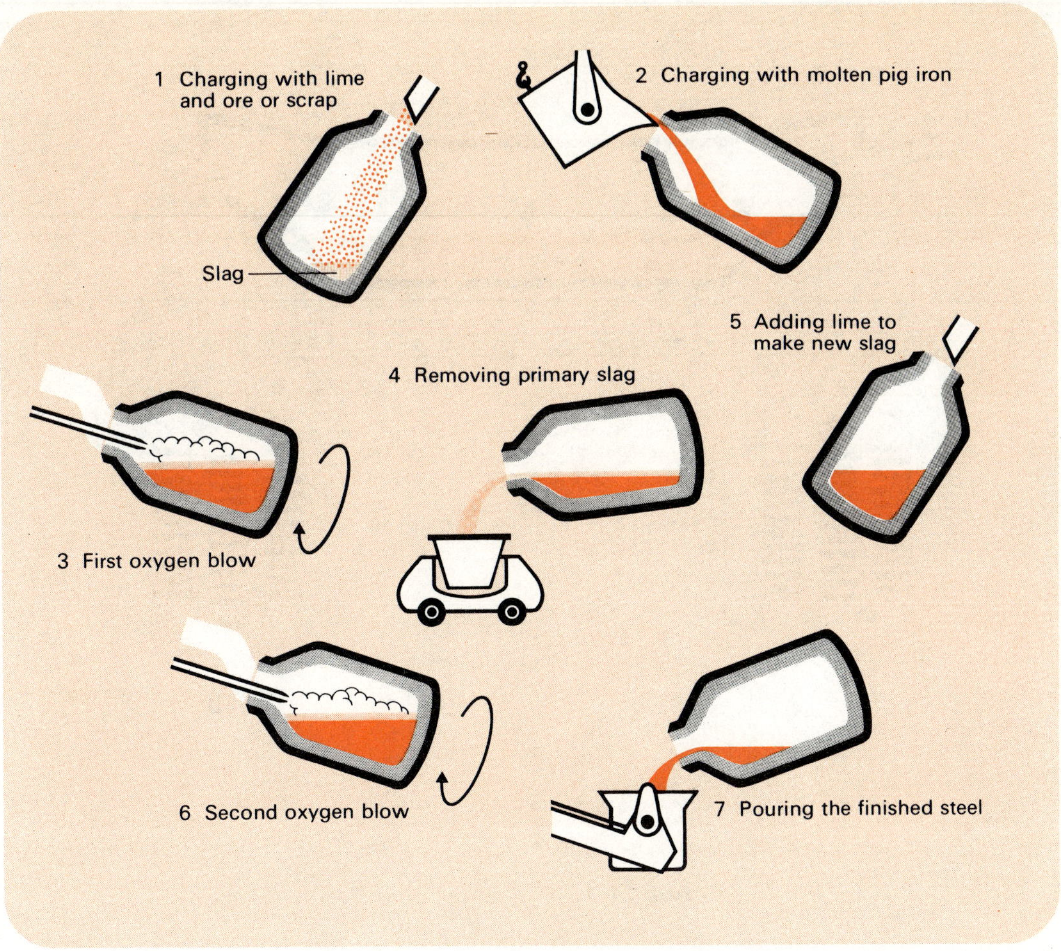

Figure 21.2 Diagrammatic representation of the Linz–Donawitz process

phosphate. The L–D process is rapid but only 15–25 per cent steel scrap can be used.

The Open Hearth process

This process can use up to 40 per cent steel scrap and although it takes longer the furnace can take a much larger charge than the L–D converter. The greater part of the world's steel is made by this process.

Cast iron and steel scrap are melted on a basic bed by an oil or gas flame. The gases, often enriched with oxygen, are passed for a time in one direction. This heats the Cowper stove. The blast is then reversed. The hot Cowper stove heats the gases to a high temperature and therefore hastens the oxidation of carbon and other non-metals. During this reversal the Cowper stove on the other side is heated.

In the later stages deoxidizers are added.

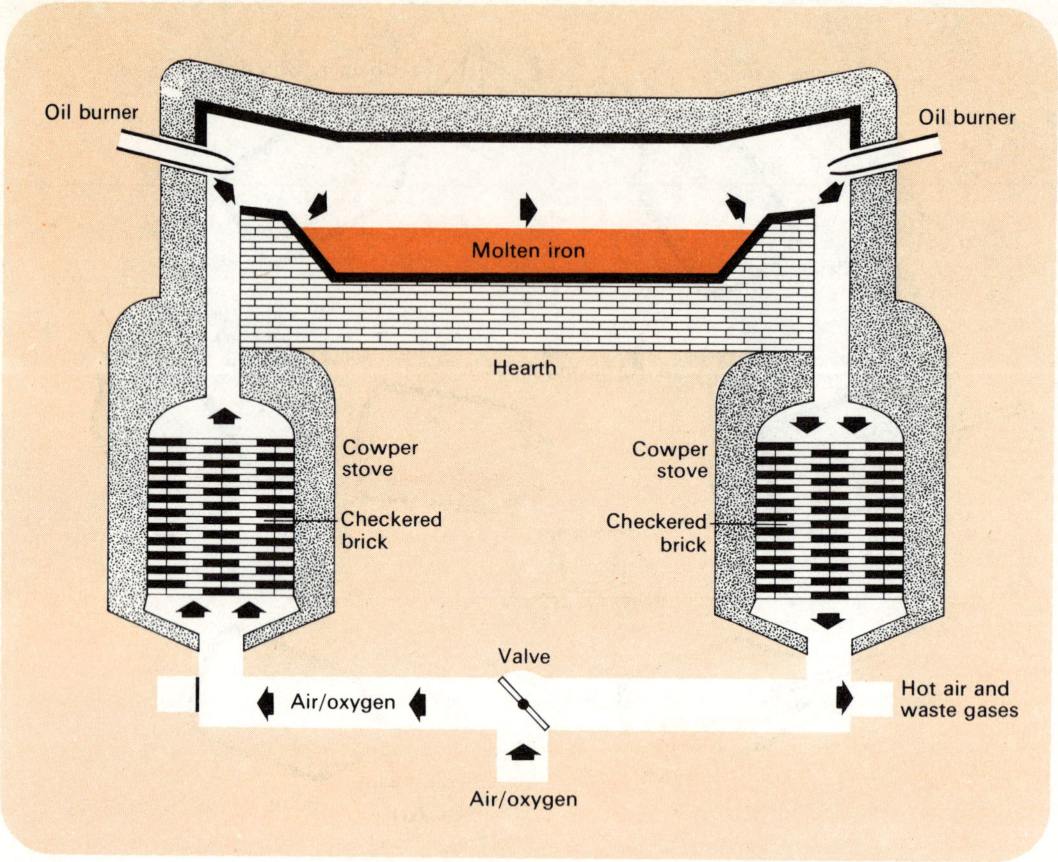

Figure 21.3 Diagram of open hearth furnace

Copper

Copper is a noble metal and resists corrosion. It does not react with water or superheated steam. It is easily bent, rolled into sheet, pulled into wire, and welded. It is a good conductor of heat. These properties make it a useful metal for boilers, car radiators, central heating installations and coinage. It is a good conductor of electricity so is used for electrical wiring.

The main ore from which copper is extracted is chalcopyrite, which has the empirical formula $CuFeS_2$. The stages of extraction are

1. The ore is roasted to burn away excess sulphur.

2. The roasted ore, mixed with silica or limestone, is heated in a reverberatory furnace. The limestone/silica forms a slag of iron silicate and other impurities.

$$2CuFeS_2 + 4O_2 \rightarrow Cu_2S + 2FeO + 3SO_2$$

$$FeO + SiO_2 \rightarrow FeSiO_3 \text{ (slag)}$$

$$2Cu_2S + 3O_2 \rightarrow 2Cu_2O + 2SO_2$$

$$2Cu_2O + Cu_2S \rightarrow 6Cu + SO_2$$

3. Impure copper is transferred to a converter of the Bessemer type and an air blast blown through it. This produces copper of up to 99 per cent purity. It is known as blister copper since the sulphur dioxide which escapes from it as it cools leaves the surface rough or 'blistered'.

4. Blister copper is then refined thermally in another converter. Air or oxygen is blown through molten copper and the last traces of oxide are removed by a process known as poling. Freshly cut branches of beech, oak and other hardwoods are pushed into the molten copper. Gases which have a reducing action are released from the green wood. Thermally refined copper is suitable for all but electrical purposes.

5. Copper for electrical uses needs to be further refined since small traces of impurity reduce its conductivity. Thermally refined copper is made the anode in a bath of copper sulphate solution. The cathode is of pure copper. As electrolysis proceeds, pure copper is deposited on the cathode:

$$Cu^{2+}(aq) + 2e \rightarrow Cu(s)$$

Copper ionizes from the impure anode:

$$Cu(s) - 2e \rightarrow Cu^{2+}(aq)$$

These copper ions move to the cathode where they are discharged and deposited as **pure** copper.

Lead

Among the main uses of lead are for making solder, for the plates of car batteries and for making the petrol additive, lead tetraethyl. This pattern may change in the near future as new designs of battery are introduced and as opposition to lead pollution increases.

The chief ore from which lead is extracted is the sulphide galena, PbS. Galena is often found with the main ore of zinc, zinc blende, ZnS. The two metals are extracted together in one design of modern furnace.

Lead is extracted in a form of blast furnace smaller than that used for iron.

1. The sulphide, PbS, is first roasted to convert it to lead oxide:

$$2PbS + 3O_2 \rightarrow 2PbO + 2SO_2$$

The sulphur dioxide is used to make sulphuric acid.

2. Lead oxide is mixed with coke and limestone. A little scrap iron is added to remove unchanged lead sulphide and the mixture is heated in the blast furnace. The following reactions take place:

$$PbO + C \rightarrow Pb + CO$$

$$PbS + Fe \rightarrow Pb + FeS$$

Molten lead trickles down to the bottom of the furnace and can be tapped off at intervals.

3. Lead from the blast furnace contains such impurities as tin, copper, silver, arsenic, antimony and sulphur. These are removed by further processes.

4. Lead of high purity is refined electrolytically. The cathode is of pure lead and the anode of impure lead.

Zinc

Zinc boils at 907 °C. This comparatively low boiling point makes it possible for zinc to be distilled out of the furnace, thereby separating it from lead which boils at about 1750 °C. Both are produced in the same furnace.

1. Zinc blende, ZnS, is roasted to produce zinc oxide and sulphur dioxide. The sulphur dioxide is used to make sulphuric acid.

$$2ZnS + 3O_2 \rightarrow 2ZnO + 2SO_2$$

2. The zinc oxide is then heated with coke. The zinc distils from the furnace. Its oxidation back to zinc oxide is prevented in older methods by ensuring that it only comes in contact with carbon monoxide. Hence excess carbon is used.

$$ZnO + C \rightarrow Zn + CO$$

$$ZnO + CO \rightarrow Zn + CO_2$$

$$CO_2 + C \rightarrow 2CO$$

In a more modern method the oxidation of zinc back to zinc oxide is prevented by condensing it on molten lead.

A method of extracting zinc which is of increasing importance is electrolysis of zinc sulphate solution. The cathode is of aluminium and the anode of lead.

Zinc is used in galvanizing iron, a process in which the iron is coated with a thin layer of zinc which then acts as a sacrificial metal to prevent the iron from rusting. Galvanized iron is used for roofing, nails, wire and vessels such as buckets. It is also used in several important alloys. Brass and gunmetal are alloys of zinc with copper. Another zinc alloy is used for die-casting, where intricate detail and precision are required: petrol pumps, model cars, carburettors and vacuum cleaner parts are examples. The casing of dry cell batteries is of zinc.

Non-metals

Nitrogen, oxygen and sulphur are extracted by physical rather than chemical means.

Nitrogen and oxygen are obtained from air as follows.

1. Air from which water vapour and carbon dioxide have been removed is compressed. It is then cooled by expansion until it liquefies.

2. The liquid air is then fractionally distilled. Nitrogen, which boils at −196 °C, distils first. Oxygen distils 13 degrees higher at −183 °C; it contains about 5 per cent argon, boiling point −186 °C.

Oxygen is conveniently transported in liquid form and used largely in steel production ('tonnage oxygen'). Oxygen gas is stored in cylinders for use in welding and cutting metals (oxyacetylene blowlamps) and for medical purposes. Liquid nitrogen is used as a low-temperature coolant. The gas is used to make ammonia, NH_3.

Sulphur is found native in surface rocks in some parts of the world, e.g. Sicily. It is separated by heating the sulphur-bearing rock to melt the sulphur, m.p. 96 °C. Other deposits of sulphur were discovered in Texas and other places while drilling for oil. The deposits are some 150 metres (about 500 feet) below the surface. The Frasch process was devised to extract it (Figure 21.4).

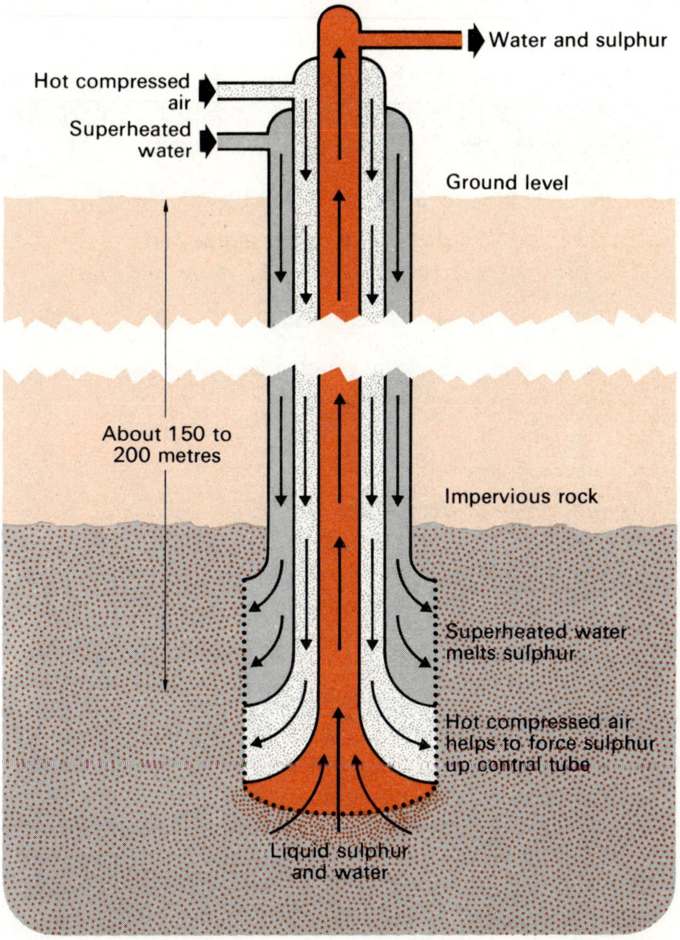

Figure 21.4 Diagram of Frasch pump

Compressed air and superheated water at 160 °C are forced down the triple concentric pipe. The sulphur melts and the compressed air forces it to the surface up the middle section.

Sulphur is also extracted from petroleum during the process of making motor spirit (gasoline), and from natural gas.

Much of the world's sulphur is used to make sulphuric acid. The rest is used for drugs, fireworks, gunpowder, matches, fungicides and to harden (vulcanize) rubber. With the introduction of artificial rubbers vulcanizing is becoming less important.

Hydrogen

Although some hydrogen is made as a by-product of electrolytic processes, this is not an important source of the element. Most hydrogen is obtained either from water or from petroleum.

Hydrogen from water

If steam is passed over white-hot coke at a temperature of well over 1000 °C, a mixture of carbon monoxide and hydrogen is produced. The mixture is called water gas.

$$H_2O + C \rightarrow CO + H_2$$

The water gas is then mixed with steam and passed over an iron oxide catalyst. The product is carbon dioxide and more hydrogen.

$$CO + H_2 + H_2O \rightarrow CO_2 + 2H_2$$

Since carbon dioxide is much more soluble in water than hydrogen the two gases are easily separated.

Hydrogen from petroleum

Petroleum contains hydrocarbons of high molecular weight. Since the main demand is for compounds of lower molecular weight, these compounds have to be 'cracked'. Hydrogen is often a by-product in this process.

The main uses of hydrogen are

1. for making ammonia
2. for use in the high temperature oxy-hydrogen flame
3. for making margarine: unsaturated hydrocarbons are converted to saturated compounds with higher melting points (*see* page 228)
4. in 'hydroforming' motor spirit
5. in the extraction of germanium and some other metals from their oxides:

$$GeO_2 + 2H_2 \rightarrow Ge + 2H_2O$$

6. for filling balloons

Phosphorus

A very large use for phosphorus is in the manufacture of matches. In strike-anywhere matches the head contains a compound of phosphorus and sulphur mixed with potassium nitrate and glue. Safety matchheads contain sulphur and potassium chlorate. The side of the box contains phosphorus, antimony sulphide and iron(III) oxide. Compounds of phosphorus are extensively used as fertilizers since the element is essential to life. Phosphates are used in some water-softening processes.

Phosphorus is extracted in a high-temperature electric furnace. Phosphate ore contains fluorapatite, $3Ca_3(PO_4)_2,CaF_2$ or chlorapatite, $3Ca_3(PO_4)_2,CaCl_2$. It is heated to about 2000 °C with silica and coke. Probably phosphorus(V) oxide, P_4O_{10}, is produced first and is then reduced to phosphorus. The overall reaction is

$$\underset{\text{Apatite}}{2Ca_3(PO_4)_2} + \underset{\text{Silica}}{6SiO_2} + \underset{\text{Coke}}{10C} \rightarrow \underset{\substack{\text{Calcium}\\\text{silicate}}}{6CaSiO_3} + P_4 + 10CO$$

Iodine

The most important source of iodine is the sodium iodate which is mixed with sodium nitrate in South American caliche. Less important sources are natural brines from oil wells, and seaweed.

After the sodium nitrate has been crystallized, sodium iodate, $NaIO_3$, remains in solution. To the solution is added a calculated amount of sodium hydrogen sulphite, $NaHSO_3$, to reduce part of the iodate to iodide:

$$3HSO_3^- + IO_3^- \rightarrow I^- + 3HSO_4^-$$

The iodide produced reacts with the remaining iodate to produce iodine

$$IO_3^- + 5I^- + 6HSO_4^- \rightarrow 3I_2 + 3H_2O + 6SO_4^{2-}$$

The solid iodine is washed with water and sublimed to purify it.

Iodine is used in medical products, in the preparation of potassium iodide, for making organic iodides such as iodobenzene and alkyl iodides, for manufacturing polaroid, and in the manufacture of photographic film.

Summary

1. Metals and non-metals below aluminium and above bromine in the ECS are extracted by chemical methods.
2. Iron: extracted in blast furnace using limestone for flux and coke for reducing agent.
3. Zinc: extracted in blast furnace using coke.
4. Lead: extracted in blast furnace using coke.
5. Copper: extracted in reverberatory furnace by autoreductive process:

$$Cu_2S + 2Cu_2O \rightarrow 6Cu + SO_2$$

6. Phosphorus: extracted in electric furnace using coke and silica.
7. Hydrogen: extracted either from water ($CO + H_2$ = water gas) or by cracking petroleum.
8. Iodine: extracted from the sodium iodate contained in caliche (Chile saltpetre).
9. Oxygen: extracted from air by fractional distillation of liquid air.
10. Nitrogen: extracted from air by fractional distillation of liquid air.
11. Sulphur: found native (uncombined) and is separated from rock by melting.
12. Noble gases: all present in air from which they are separated by physical means, e.g. fractional distillation and diffusion.

22 Group Properties of Normal Elements

Similarities and trends in properties for the elements within each group are pointed out in this chapter. Discussion is limited to a small number of important properties. When you are familiar with the general patterns of the Periodic Table, you should be able to predict these properties. There is no need to remember the properties of each separate element, but it is necessary to be familiar with the general patterns.

Group 1, the sodium group

						H	He
Li	Be	B	C	N	O	F	Ne
Na	Mg	Al	Si	P	S	Cl	Ar
K	Ca	Ga	Ge	As	Se	Br	Kr
Rb	Sr	In	Sn	Sb	Te	I	Xe

These are metallic elements with fairly high boiling points, but rather low melting points. The elements all react vigorously with water to form hydrogen and an alkaline solution of the very soluble metal hydroxide, ElOH. The symbol El stands for any element.

$$2El(s) + 2H_2O(l) \longrightarrow 2ElOH(aq) + H_2(g)$$

The elements react very vigorously with the hydrogen ions present in aqueous solutions of acids. This reaction can be explosive with the elements below lithium.

$$2El(s) + 2H^+(aq) \longrightarrow 2El^+(aq) + H_2(g)$$

The elements tarnish immediately on exposure to air. They burn with a brilliant flame when heated in air, forming the oxide El_2O. The peroxide El_2O_2 is usually formed as well.

$$4El(s) + O_2(g) \longrightarrow 2El_2O(s)$$

The oxides are high melting point ionic solids, $(2El^+O^{2-})$. They react with water to give alkaline solutions of the metal hydroxides.

$$El_2O(s) + H_2O(l) \longrightarrow 2ElOH(aq)$$

The oxides and hydroxides of all the elements are basic. They react with acids to form a salt and water.

$$El_2O(s) + 2HCl(aq) \rightarrow 2ElCl(aq) + H_2O(l)$$

$$ElOH(aq) + HNO_3(aq) \rightarrow ElNO_3(aq) + H_2O(l)$$

The high reactivity of the elements in Group 1 shows also in their ready and exothermic reaction with chlorine. They form solid chlorides which have high melting points and high boiling points. The chlorides contain El^+ and Cl^- ions.

$$2El(s) + Cl_2(g) \rightarrow 2ElCl(s)$$

The colourless El^+ ion is present in all compounds of the Group 1 elements. All compounds are colourless unless the negative ion (e.g. chromate) is coloured. The ionic nature of the compounds makes them solids with high melting points and high boiling points. Almost all compounds are soluble in water. Aqueous solutions of all compounds, and molten compounds, are good conductors of electricity.

The elements of Group 1 all have very similar properties.

Group 2, the magnesium group

						H	He
Li	Be	B	C	N	O	F	Ne
Na	Mg	Al	Si	P	S	Cl	Ar
K	Ca	Ga	Ge	As	Se	Br	Kr
Rb	Sr	In	Sn	Sb	Te	I	Xe

Apart from beryllium these are reactive metals, but not as reactive as the elements in the sodium group. The lower reactivity is partly due to the metal being covered by an oxide layer. The melting and boiling points of the metals are higher than in the sodium group, indicating stronger metallic bonding.

The elements calcium and strontium (the lower members of the group) react vigorously with water to form hydrogen and an alkaline solution of the sparingly soluble hydroxide, $El(OH)_2$. Magnesium reacts only slowly with water because magnesium hydroxide is almost insoluble. Beryllium has no reaction with water.

$$El(s) + 2H_2O(l) \rightarrow El(OH)_2(aq) + H_2(g)$$

The metals react readily with dilute acids to produce hydrogen. Formation of a layer of insoluble sulphate reduces the rate of reaction when dilute sulphuric acid reacts with calcium and strontium. Salts of these metals are frequently less soluble than similar salts of metals in the sodium group.

$$El(s) + 2H^+(aq) \rightarrow El^{2+}(aq) + H_2(g)$$

The elements become covered with an oxide layer on exposure to air. This corrosion is, however, much slower than is found for elements in the sodium group. The elements burn in air with a brilliant flame when heated. The oxide ElO is formed.

$$2El(s) + O_2(g) \rightarrow 2ElO(s)$$

The oxides are high melting point solids. BeO is composed of a giant structure of atoms, the others are ionic ($El^{2+}O^{2-}$). The ionic oxides react with water to form the hydroxides. The hydroxides are slightly soluble in water, solubility increasing on going down the group. These solutions are alkaline.

$$ElO(s) + H_2O(l) \rightarrow El(OH)_2(s)$$

The oxides and hydroxides of all the elements are basic. They react with acids to form a salt and water.

$$ElO(s) + 2HCl(aq) \rightarrow ElCl_2(aq) + H_2O(l)$$

$$El(OH)_2(s) + 2HNO_3(aq) \rightarrow El(NO_3)_2(aq) + 2H_2O(l)$$

All the elements react vigorously with chlorine to form the chloride $ElCl_2$. The chlorides of the elements below beryllium are solids which have high melting and boiling points. The chlorides contain El^{2+} and Cl^- ions.

$$El(s) + Cl_2(g) \rightarrow ElCl_2(s)$$

The colourless El^{2+} ion is present in compounds of the Group 2 elements. (Beryllium is on the diagonal separating metals from non-metals, and some of its compounds are covalent). All compounds are colourless unless the negative ion is coloured. The ionic nature of the compounds makes them solids with high melting points and high boiling points. Many compounds are soluble in water, but certain important compounds are insoluble or only slightly soluble. Carbonates are insoluble. Sulphates are soluble for beryllium and magnesium, becoming insoluble on going further down the group. Hydroxides are insoluble for beryllium, almost insoluble for magnesium, becoming more soluble on going further down the group.

The elements of Group 2 have many similar chemical properties. Differences in reactivity, solubility and ionic nature of compounds are, however, noticeable.

Group 3, the aluminium group

						H	He
Li	Be	B	C	N	O	F	Ne
Na	Mg	Al	Si	P	S	Cl	Ar
K	Ca	Ga	Ge	As	Se	Br	Kr
Rb	Sr	In	Sn	Sb	Te	I	Xe

Elements in the aluminium group are closer to the diagonal and therefore less metallic in properties than the elements of the sodium and magnesium groups. Boron is above the diagonal and has all the properties of a non-metal. Boron does not react with acids to produce hydrogen, it forms an acidic oxide, B_2O_3, and its chloride, BCl_3, is covalent and gaseous.

The elements below boron react with acids to produce hydrogen, but do not react with water as they are covered with a waterproof layer of insoluble oxide.

$$2El(s) + 6H^+(aq) \rightarrow 2El^{3+}(aq) + 3H_2(g)$$

The elements are covered with a thin layer of oxide which makes them resistant to further corrosion (compare with the elements in the sodium and magnesium groups). When heated to a high temperature in air, they burn to form the oxide El_2O_3.

$$4El(s) + 3O_2(g) \rightarrow 2El_2O_3(s)$$

The oxides are high melting point solids. They are insoluble in water. They do not react with water to form the hydroxides (compare with oxides of the sodium and magnesium groups). The oxides and hydroxides react with acids and bases.

$$El_2O_3(s) + 6HCl(aq) \rightarrow 2ElCl_3(aq) + 3H_2O(l)$$

$$El_2O_3(s) + 2NaOH(aq) \rightarrow 2NaElO_2(aq) + H_2O(l)$$

Oxides and hydroxides which react with acids and bases are called **amphoteric.**

The elements react readily when heated in chlorine to form the chloride $ElCl_3$.

$$2El(s) + 3Cl_2(g) \rightarrow 2ElCl_3(s)$$

The chlorides are solids, but their melting and boiling points are far lower than for chlorides in the sodium and magnesium groups (*see* Figures 9.12 and 9.13). Many compounds contain the El^{3+} ion, but the low melting and boiling points indicate that certain compounds have a structure which is on the border-line between ionic and covalent.

Group 4, the silicon group

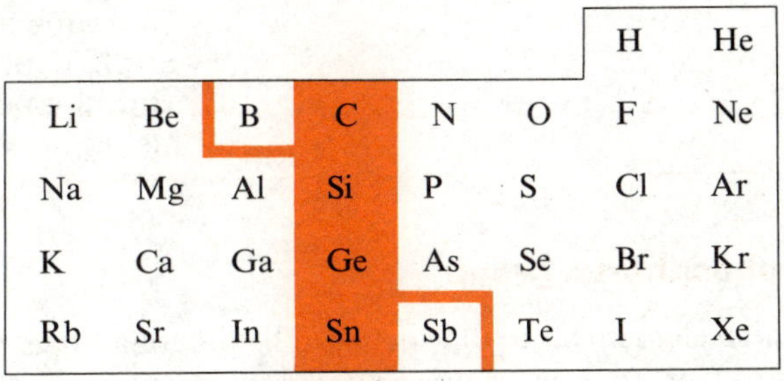

These elements resemble each other in the formulae of their oxides, ElO_2, and chlorides, $ElCl_4$, and in the gaseous nature and formulae of their compounds with hydrogen, ElH_4. The diagonal passes through the middle of the group. Trends from non-metallic properties with carbon and silicon to metallic/non-metallic properties with germanium and tin are as important as the group similarities (Table 22.1).

Table 22.1

Carbon and silicon	*Germanium and tin*
Carbon has two polymorphic forms, both with a giant structure of atoms. Diamond is a non-conductor of electricity, while graphite does conduct along the plane of the sheets of atoms. Silicon has the diamond structure but does conduct to some extent.	Germanium has the appearance of a metal. It has the diamond structure and is a better conductor than silicon. Tin has two polymorphic forms. One form, with the diamond structure, is non-metallic. The other form is completely metallic.
These elements do not react with acids to form hydrogen.	Tin is sufficiently metallic to react slowly with acids to produce hydrogen. $Sn(s)+2H^+(aq) \rightarrow Sn^{2+}(aq)+H_2(g)$
The oxides CO_2 and SiO_2 are acidic. Carbon forms a second oxide, CO, carbon monoxide, which is neutral under most conditions. Acidic oxides react with bases to form salts. $CO_2(g)+Ca(OH)_2(aq)$ $\rightarrow CaCO_3(s) + H_2O(l)$	The oxides ElO and ElO_2 are formed by both elements. These oxides are amphoteric—they show acidic and basic properties.
The chlorides CCl_4 and $SiCl_4$ are covalent compounds. They have low melting points and low boiling points.	The chlorides $GeCl_4$ and $SnCl_4$ are covalent compounds. They have low melting points and low boiling points. Tin forms an ionic chloride $SnCl_2$ containing the Sn^{2+} ion. This chloride has a fairly high melting and boiling point.

Group 5, the phosphorus group

These elements are non-metals. The diagonal does pass close to antimony and this element shows some metallic properties. All elements form acidic oxides of empirical formulae El_2O_3 and El_2O_5. All elements form a gaseous compound with hydrogen, ElH_3. All elements form a chloride $ElCl_3$. The chloride $ElCl_3$ is a covalent liquid for nitrogen, phosphorus and arsenic. $SbCl_3$ is a solid with

						H	He
Li	Be	B	C	**N**	O	F	Ne
Na	Mg	Al	Si	**P**	S	Cl	Ar
K	Ca	Ga	Ge	**As**	Se	Br	Kr
Rb	Sr	In	Sn	**Sb**	Te	I	Xe

boiling point 220 °C, this fairly high boiling point indicating that the compound may be on the borderline between ionic and covalent. Antimony does form the ion Sb^{3+} in its sulphate. Phosphorus, arsenic and antimony all show polymorphism. One polymorphic form of each element has some metallic properties. The other polymorphs are non-metallic.

Group 6, the sulphur group

						H	He
Li	Be	B	C	N	**O**	F	Ne
Na	Mg	Al	Si	P	**S**	Cl	Ar
K	Ca	Ga	Ge	As	**Se**	Br	Kr
Rb	Sr	In	Sn	Sb	**Te**	I	Xe

These elements are non-metals. All elements form acidic oxides ElO_2 and ElO_3. All elements form a low boiling point compound with hydrogen, H_2El. Each element forms more than one chloride. The chlorides of oxygen, sulphur and selenium are covalent. Tellurium forms solid chlorides $TeCl_2$ and $TeCl_4$. These chlorides have fairly high boiling points, indicating that they may be on the borderline between ionic and covalent. One polymorphic form of selenium and the only form of tellurium do show some metallic properties.

Group 7, the chlorine group

						H	He
Li	Be	B	C	N	O	**F**	Ne
Na	Mg	Al	Si	P	S	**Cl**	Ar
K	Ca	Ga	Ge	As	Se	**Br**	Kr
Rb	Sr	In	Sn	Sb	Te	**I**	Xe

These elements are frequently called the halogens. They are all non-metallic and very similar in properties. An important feature of the chemistry of the elements in Groups 3–6 is the trend from non-metallic to metallic properties on going down the group. This trend is unimportant or absent in the sodium, magnesium, chlorine and argon groups. Group similarity is the important feature of the two groups on the far left and the two groups on the far right of the Periodic Table.

Some ways in which the halogens resemble each other

The halogens resemble each other in the following ways.

1. They are all composed of covalent molecules containing two atoms, El_2. This accounts for their low melting and boiling points, and for their solubility in covalent liquids such as tetrachloromethane, CCl_4.

2. All halogens combine readily with most metals to form ionic compounds containing the halide ion El^-.

$$2Na(s) + El_2 \rightarrow 2NaEl(s)$$

3. All halogens combine with most non-metals to form covalent halides.

$$2P + 3El_2 \rightarrow 2PEl_3$$

4. All halogens combine with hydrogen to form gaseous covalent compounds. These compounds are very soluble in water. Aqueous solutions of the compounds are strongly acidic.

$$H_2(g) + El_2 \rightarrow 2HEl(g)$$

5. The halogens form compounds with silver of formula AgEl. Silver chloride, silver bromide and silver iodide are all insoluble in water. They are precipitated from solution when aqueous silver nitrate is added to an aqueous solution of halide ions. This can be used as a test for halides (*see* page 241).

$Ag^+(aq) + Cl^-(aq) \rightarrow AgCl(s)$	White precipitate
$Ag^+(aq) + Br^-(aq) \rightarrow AgBr(s)$	Pale yellow precipitate
$Ag^+(aq) + I^-(aq) \rightarrow AgI(s)$	Bright yellow precipitate

Trends in properties within the halogen group

Fluorine is the most reactive of the halogens. Reactivity decreases on going down the group, iodine being the least reactive of the halogens (*see* Chapter 18, pages 167 and 168).

1. Fluorine reacts immediately and completely with water to form oxygen and hydrogen fluoride. Hydrogen fluoride dissolves in any remaining water to form hydrofluoric acid.

$$2F_2(g) + 2H_2O(l) \rightarrow 4HF(aq) + O_2(g)$$

The other halogens dissolve in water to form coloured solutions of halogen molecules. Iodine has no further reaction with water. Chlorine and bromine very slowly decompose water. The final products after days (with chlorine) or weeks (with bromine) are the halogen acid and oxygen.

2. Halogens higher in the group (more reactive halogens) react with the halide ions of less reactive halogens in aqueous solution. The lower halogen is displaced from solution.

Chlorine displaces bromine from bromides and iodine from iodides.

$$Cl_2(aq) + 2Br^-(aq) \rightarrow 2Cl^-(aq) + Br_2(aq)$$

$$Cl_2(aq) + 2I^-(aq) \rightarrow 2Cl^-(aq) + I_2(aq)$$

Bromine displaces iodine from iodides.

$$Br_2(aq) + 2I^-(aq) \rightarrow 2Br^-(aq) + I_2(aq)$$

The position of hydrogen in Group 7

Hydrogen resembles the halogens in several ways.

1. Hydrogen forms a covalent molecule containing two atoms, H_2.

2. Hydrogen combines with reactive metals to form ionic compounds containing the hydride ion, H^-.

$$2Na(s) + H_2(g) \rightarrow 2NaH(s)$$

3. Hydrogen forms covalent compounds with many non-metals. Examples are methane, CH_4, ammonia, NH_3, and water, H_2O. This is a property of a non-metal.

For these reasons hydrogen can be placed in Group 7. Hydrogen does, however, form the ion $H^+(aq)$ in aqueous solution. This is a property of an element in Group 1, and hydrogen is frequently included in this group of elements.

Group 0, the argon group

						H	He
Li	Be	B	C	N	O	F	Ne
Na	Mg	Al	Si	P	S	Cl	Ar
K	Ca	Ga	Ge	As	Se	Br	Kr
Rb	Sr	In	Sn	Sb	Te	I	Xe

In a negative way these elements show group similarities as strongly as do the elements in the sodium, magnesium and chlorine groups. The elements resemble each other in their chemical inertness. The elements of higher atomic number have been forced to form compounds (*see* page 95), but this only happens when they are subjected to vigorous chemical attack. This family of elements is frequently called the noble gases because of the lack of chemical activity.

Summary

Some of the periodic features found in chemical properties are summarized in the tables which follow.

						H	He
Li	**Be**	B	C	N	O	F	Ne
Na	**Mg**	**Al**	Si	P	S	Cl	Ar
K	**Ca**	**Ga**	Ge	As	Se	Br	Kr
Rb	**Sr**	**In**	**Sn**	Sb	Te	I	Xe

Metallic elements.
Combine with non-metallic elements only.
Usually form positive ions in compounds.
Oxides are basic.
Chlorides are solids, usually composed of ions.

						H	He
Li	Be	B	C	N	O	F	Ne
Na	**Mg**	Al	Si	P	S	Cl	Ar
K	**Ca**	Ga	Ge	As	Se	Br	Kr
Rb	**Sr**	In	Sn	Sb	Te	I	Xe

Some metallic elements react with cold water to produce hydrogen.
Hydroxides form alkaline aqueous solutions.
Chlorides have boiling points above 1000 °C and contain ions.

						H	He
Li	Be	**B**	**C**	**N**	**O**	**F**	Ne
Na	Mg	Al	**Si**	**P**	**S**	**Cl**	Ar
K	Ca	Ga	Ge	**As**	**Se**	**Br**	Kr
Rb	Sr	In	Sn	Sb	**Te**	**I**	Xe

Non-metallic elements.
Combine with both metals and other non-metals.
Oxides are frequently acidic.
Chlorides are usually liquids or gases.

23 Elements in the Period Sodium to Argon

General data

Physical and chemical properties of the elements in the period sodium to argon are given below, together with information about some important compounds. *See* Table 23.1 for page references.

Table 23.1

Property	*Page*
Symbol	206
Atomic number	206
Relative atomic mass (Atomic weight)	206
Main isotopes	206
Atomic volume, cm^3	206
Approximate volume of gramme atom at r.t.p., cm^3	207
Electron configuration	207
Standard electrode potential, volt	207
First ionization energy, kJ/mole	208
Important ions	208
Density of solid, g/cm^3	208
Melting point °C	208
Boiling point °C	208
Reactions of element with oxygen	209
Reaction of element with chlorine	209
Reactions of element with water	209
Formula of oxide(s)	210
Heat of formation of oxide(s), kJ/mole	210
Nature of oxide(s)	210
Reaction of oxide(s) with water	210
Bonding in oxide(s)	211
Formula of chloride(s)	211
Heat of formation of chloride(s), kJ/mole	211
Bonding in chloride(s)	211
Reaction of chloride(s) with water	212
Main ores	212
Method of extraction	212

Element	*Symbol*	*Atomic number*	*Relative atomic mass*	*Main isotopes (per cent)*	*Atomic volume (cm^3)*
Sodium	Na	11	22.99	^{23}Na 100	23.8
Magnesium	Mg	12	24.30	^{24}Mg 78.6 ^{25}Mg 10.1 ^{26}Mg 11.3	14.0
Aluminium	Al	13	26.98	^{27}Al 100	10.0
Silicon	Si	14	28.01	^{28}Si 92.2 ^{29}Si 4.7 ^{30}Si 3.1	13.1
Phosphorus	P	15	30.97	^{31}P 100	17.0 (white) 13.5 (red)
Sulphur	S	16	32.06	^{32}S 95.0 ^{33}S 0.76 ^{34}S 4.2 ^{36}S 0.02	15.5 (rhombic) 16.4 (monoclinic)
Chlorine	Cl	17	35.45	^{35}Cl 75.5 ^{37}Cl 24.5	22.8
Argon	Ar	18	39.95	^{36}Ar 0.34 ^{38}Ar 0.06 ^{40}Ar 99.6	35.5

Element	*Approximate volume of gramme atom at r.t.p. (cm^3)*	*Electron configuration*	*Standard electrode potential (volt)*
Sodium	24	2.8.1	−2.71 $Na \rightarrow Na^{+} + e$
Magnesium	14	2.8.2	−2.38 $Mg \rightarrow Mg^{2+} + 2e$
Aluminium	10	2.8.3	−1.66 $Al \rightarrow Al^{3+} + 3e$
Silicon	13	2.8.4	—
Phosphorus	17 (white) 14 (red)	2.8.5	—
Sulphur	15 (rhombic) 16 (monoclinic)	2.8.6	−0.51 $S^{2-} \rightarrow S + 2e$
Chlorine	12 000	2.8.7	+1.36 $2Cl^{-} \rightarrow Cl_2 + 2e$
Argon	24 000	2.8.8	—

Element	*First ionization energy (kJ/mole)*	*Important ions*	*Density of solid (g/cm³)*	*Melting point (°C)*	*Boiling point (°C)*
Sodium	494	Na^+	0.97	98	890
Magnesium	736	Mg^{2+}	1.74	650	1110
Aluminium	577	Al^{3+}	2.70	660	2470
Silicon	786	SiO_3^{2-}	2.33	1410	2360
Phosphorus	1060	PO_4^{3-} HPO_3^{2-}	1.82 (white) 2.34 (red)	44 (white) 590 (red)	280 (white)
Sulphur	1000	SO_4^{2-} SO_3^{2-} S^{2-}	2.07 (rhombic) 1.96 (monoclinic)	113 (rhombic) 119 (monoclinic)	445
Chlorine	1260	ClO_4^- ClO_3^- ClO^- Cl^-	1.56	−101	−34.7
Argon	1520	—	1.40	−189	−186

Element	Reactions of element: with oxygen	with chlorine	with water
Sodium	$4Na+O_2\rightarrow 2Na_2O$ $2Na+O_2\rightarrow Na_2O_2$	$2Na+Cl_2\rightarrow 2NaCl$	$2Na+2H_2O$ $\rightarrow 2NaOH+H_2$
Magnesium	$2Mg+O_2\rightarrow 2MgO$	$Mg+Cl_2\rightarrow MgCl_2$	$Mg+2H_2O$ $\rightarrow Mg(OH)_2+H_2$ (slowly in cold) $Mg+H_2O$ $\rightarrow MgO+H_2$ (superheated steam)
Aluminium	$4Al+3O_2\rightarrow 2Al_2O_3$	$2Al+3Cl_2\rightarrow 2AlCl_3$	Hydroxide water-proof: reaction stops unless hydroxide film removed
Silicon	$Si+O_2\rightarrow SiO_2$	$Si+2Cl_2\rightarrow SiCl_4$	No reaction
Phosphorus	$4P+3O_2\rightarrow P_4O_6$ $4P+5O_2\rightarrow P_4O_{10}$	$2P+3Cl_2\rightarrow 2PCl_3$ $2P+5Cl_2\rightarrow 2PCl_5$	No reaction
Sulphur	$S+O_2\rightarrow SO_2$	$2S+Cl_2\rightarrow S_2Cl_2$ $S+Cl_2\rightarrow SCl_2$	No reaction
Chlorine	No reaction	—	Cl_2+H_2O $\rightarrow HCl+HClO$ $2HClO\rightarrow O_2$ $+2HCl$ (both reactions slow)
Argon	No reaction	No reaction	No reaction

Element	*Formula of oxide(s)*	*Heat of formation of oxide(s) (kJ/mole)*	*Nature of oxide(s)*	*Reaction of oxide(s) with water*
Sodium	Na_2O	−416	Basic	$Na_2O + H_2O \rightarrow 2NaOH$ (strong alkali)
	Na_2O_2	−505	Peroxide	$2Na_2O_2 + 2H_2O \rightarrow 4NaOH + O_2$
Magnesium	MgO	−602	Basic	$MgO + H_2O \rightarrow Mg(OH)_2$ (solution alkaline)
Aluminium	Al_2O_3	−1680	Amphoteric	Insoluble
Silicon	SiO_2	−910	Acidic	Insoluble
Phosphorus	P_4O_6	−1640	P_4O_6 (acidic)	$P_4O_6 + 6H_2O \rightarrow 4H_3PO_3$ (phosphorous acid)
	P_4O_{10}	−2980	P_4O_{10} (acidic)	$P_4O_{10} + 6H_2O \rightarrow 4H_3PO_4$ (phosphoric acid)
Sulphur	SO_2	−297	SO_2 (acidic)	$SO_2 + H_2O \rightarrow H_2SO_3$ (sulphurous acid)
	SO_3	−395	SO_3 (acidic)	$SO_3 + H_2O \rightarrow H_2SO_4$ (sulphuric acid)
Chlorine	Cl_2O	+80	Cl_2O (acidic)	$Cl_2O + H_2O \rightarrow 2HClO$
	ClO_2	+103	ClO_2 (acidic)	(soluble)
	Cl_2O_7	+265	Cl_2O_7 (acidic)	$Cl_2O_7 + H_2O \rightarrow 2HClO_4$ (perchloric acid)
Argon	None	—	—	—

Element	*Bonding in oxides*	*Formula of chloride(s)*	*Heat of formation of chloride(s) (kJ/mole)*	*Bonding in chloride(s)* —
Sodium	Na_2O (giant lattice of ions) Na_2O_2 (giant lattice of ions)	NaCl	−411	Giant lattice of ions
Magnesium	MgO (giant lattice of ions)	$MgCl_2$	−642	Giant lattice of ions
Aluminium	Al_2O_3 (giant lattice of atoms)	$AlCl_3$	−704	Covalent
Silicon	SiO_2 (giant lattice of atoms)	$SiCl_4$	−640	Covalent
Phosphorus	P_4O_6 (covalent) P_4O_{14} (covalent)	PCl_3 PCl_5	−320 −463	PCl_3 (covalent) PCl_5 (covalent)
Sulphur	SO_2 (covalent) SO_3 (covalent)	S_2Cl_2 SCl_2	−60	S_2Cl_2 (covalent) SCl_2 (covalent)
Chlorine	Cl_2O (covalent) ClO_2 (covalent) Cl_2O_7 (covalent)	Cl_2	0	Cl_2 (covalent)
Argon	No oxide formed	No chloride	—	—

Element	*Reaction of chloride(s) with water*	*Main ores*	*Method of extraction*
Sodium	NaCl soluble to give $Na^+(aq)$ and $Cl^-(aq)$ ions	Rock salt, NaCl	Electrolysis of molten NaCl
Magnesium	$MgCl_2$ soluble to give $Mg^{2+}(aq)$ and $Cl^-(aq)$ ions	Dolomite, $CaCO_3, MgCO_3$; sea water; magnesite, $MgCO_3$	Electrolysis of molten $MgCl_2$
Aluminium	$AlCl_3 + 3H_2O \rightarrow Al(OH)_3 + 3HCl$	Bauxite, $Al_2O_3, 2H_2O$	Electrolysis of Al_2O_3 in molten cryolite
Silicon	$SiCl_4 + 2H_2O \rightarrow SiO_2 + 4HCl$	Sand, quartz, SiO_2	Reduction of SiO_2 with magnesium or coke
Phosphorus	$PCl_3 + 3H_2O \rightarrow H_3PO_3 + 3HCl$ $PCl_5 + H_2O \rightarrow POCl_3 + 2HCl$	Apatite, $Ca_3(PO_4)_2$	Coke reduction of apatite
Sulphur	$S_2Cl_2 + 3H_2O \rightarrow 4HCl + H_2SO_3 + 3S$ $SCl_2 + H_2O \rightarrow$ Products include HCl and S	Pyrites, FeS_2; native S	Roasting to SO_2 Melting by Sicilian or Frasch process
Chlorine	$Cl_2 + H_2O \rightarrow HCl + HClO$ then $2HClO \rightarrow O_2 + 2HCl$	Sodium chloride, NaCl	Electrolysis of brine (saturated NaCl(aq))
Argon	No chloride formed	Obtained from air	Fractional distillation of liquid air

24 Transition Metals

Comparison of normal and transition metals

21 Sc Scandium 45	22 Ti Titanium 48	23 V Vanadium 51	24 Cr Chromium 52	25 Mn Manganese 55	26 Fe Iron 56	27 Co Cobalt 59	28 Ni Nickel 59	29 Cu Copper 64	30 Zn Zinc 65
39 Y Yttrium 89	40 Zr Zirconium 91	41 Nb Niobium 93	42 Mo Molybdenum 96	43 Tc Technetium 99	44 Ru Ruthenium 101	45 Rh Rhodium 103	46 Pd Palladium 106	47 Ag Silver 108	48 Cd Cadmium 112
57 La Lanthanum 139	72 Hf Hafnium 178	73 Ta Tantalum 181	74 W Tungsten 184	75 Re Rhenium 186	76 Os Osmium 190	77 Ir Iridium 192	78 Pt Platinum 195	79 Au Gold 197	80 Hg Mercury 201

Figure 24.1 Transition metals

The properties of the normal elements are discussed in Chapters 22 and 23. In this chapter the properties of two typical normal metals, sodium and magnesium, are compared with the properties of two typical transition metals, iron and copper. Table 24.1 summarizes this comparison.

Table 24.1
Comparison of Normal Metals and Transition Metals *

	Sodium	*Magnesium*	*Iron*	*Copper*
Electronic configuration	2.8.1.	2.8.2.	2.8.14.2.	2.8.17.2.
Density in g/cm³	0.97	1.7	7.9	8.9
Volume of one mole of atoms (atomic volume) in cm³	24	14	7	7
Melting point in °C	98	650	1540	1080

* Table 24.1 continues on page 214.

	Sodium	*Magnesium*	*Iron*	*Copper*
Boiling point in °C	890	1110	3000	2600
Ions formed by element	Na^+	Mg^{2+}	Fe^{2+}, Fe^{3+}	Cu^+, Cu^{2+}
Formulae of chlorides	NaCl	$MgCl_2$	$FeCl_2$, $FeCl_3$	CuCl, $CuCl_2$
Position in ECS	Very high	High	Middle	Low
Reaction with water	Vigorous, H_2 evolved	Very slow, H_2 evolved	No reaction, rusts very slowly if air is present	No reaction
Reaction with H^+(aq) ions in dilute acids	Explosive, H_2 evolved	Vigorous, H_2 evolved	Slow, H_2 evolved	No reaction
Reaction when heated in HCl gas	Vigorous, NaCl formed	Vigorous, $MgCl_2$ formed	Vigorous, $FeCl_2$ formed	No reaction
Reaction when heated in Cl_2 gas	Vigorous NaCl formed	Vigorous $MgCl_2$ formed	Vigorous $FeCl_3$ formed	Vigorous $CuCl_2$ formed
Solubility of hydroxides in water	Very soluble, alkaline solution formed	Slightly soluble, alkaline solution formed	Insoluble	Insoluble
Colour of ions in aqueous solution	Na^+ colourless	Mg^{2+} colourless	Both Fe^{2+} and Fe^{3+} are frequently coloured	Cu^+ colourless, Cu^{2+} coloured
Catalytic activity of element and compounds	Poor	Poor	Good	Good

Electronic configuration

Atoms of normal metals have a small number of electrons in the outer shell. **All the inner shells are full.**

Atoms of transition metals have a small number of electrons in the outer shell. **The shell before the outer one is not full.** The difference between the properties of normal and transition metals results from this difference in electronic configuration.

Bonding in metals

The atoms of metallic elements are closely packed in regular giant structures. The electrons in the incomplete shells of each atom are used to hold the structure together. These electrons act as an 'electron cement' (*see* page 103).

Transition metal atoms have more electrons in their two incomplete shells than normal metal atoms have in their one incomplete shell. This gives rise to stronger bonding between transition metal atoms than between atoms of normal metals. The strong bonding in transition metals shows in their high density, small volume of one mole of atoms, high melting point and high boiling point.

Formulae of compounds

Atoms of normal metals lose their outer electrons when they form ionic compounds. All the outer electrons are usually lost and normal metals form ions with a constant number of positive charges. Sodium always forms Na^+ ions, magnesium always forms Mg^{2+} ions. Sodium chloride always has the formula NaCl, magnesium chloride always has the formula $MgCl_2$. (Thallium is one example of an exception to the general rule, *see* page 81.)

Atoms of transition metals are usually able to lose more than one number of outer electrons when they form ionic compounds. Iron atoms can lose two or three electrons. This gives rise to Fe^{2+} and Fe^{3+} ions, and chlorides of formula $FeCl_2$ and $FeCl_3$. Copper atoms can lose one or two electrons. This gives rise to Cu^+ and Cu^{2+} ions, and chlorides of formula CuCl and $CuCl_2$.

Reactivity of metals

The order of some important metals in the ECS is shown below. (*See* Table 18.3 for more information.)

Lithium
Potassium
Calcium
Sodium
Magnesium
Aluminium
Zinc ←
Iron ←
Lead
HYDROGEN   Transition metals
Copper ←
Silver ←

Normal metals are all above hydrogen in the ECS, and are frequently very high. Transition metals are far lower in the ECS. Many of them come below hydrogen.

Reaction of metals with water

The normal metals of Groups 1 and 2 (except beryllium) react with water to give hydrogen and an alkaline solution of the metal hydroxide. The reaction of magnesium with water is slow.

$$2Na(s) + 2H_2O(l) \rightarrow 2NaOH(aq) + H_2(g)$$

$$Mg(s) + 2H_2O(l) \rightarrow Mg(OH)_2(s) + H_2(g)$$

Transition metals do not react with water. The transition metals higher in the ECS do react with steam when strongly heated. Red-hot iron, for example, reacts reversibly with steam.

$$3Fe(s) + 4H_2O(g) \rightleftharpoons Fe_3O_4(s) + 4H_2(g)$$

Reaction of metals with acids in aqueous solution

Normal metals high in the ECS react explosively with dilute acids. Magnesium is sufficiently low to react safely, but still at a rapid rate.

$$Mg(s) + 2H^+(aq) \rightarrow Mg^{2+}(aq) + H_2(g)$$

Iron is above hydrogen in the ECS and reacts steadily with dilute acids to give hydrogen and an aqueous solution of iron(II) ions. Transition metals such as copper which are below hydrogen in the ECS do not react with the $H^+(aq)$ ions present in dilute acids.

$$Fe(s) + 2H^+(aq) \rightarrow Fe^{2+}(aq) + H_2(g)$$

All aqueous acids contain $H^+(aq)$ ions. These ions accept electrons from metals above hydrogen in the ECS. The metal atoms form positive ions by electron loss.

$$\begin{array}{l} Mg(s) \rightarrow Mg^{2+}(aq) + 2e \\ \qquad\qquad\qquad\qquad\quad \downarrow \\ \qquad\qquad\qquad\qquad\quad 2e + 2H^+(aq) \rightarrow H_2(g) \end{array}$$

Dilute nitric acid contains nitrate ions, $NO_3^-(aq)$, as well as $H^+(aq)$ ions. These two types of ion together are able to accept electrons more readily than $H^+(aq)$ alone accepts electrons (*see* Table 18.3). The half-reaction is

$$3e + 4H^+(aq) + NO_3^-(aq) \rightarrow NO(g) + 2H_2O(l)$$

Dilute nitric acid reacts readily with most metals, including those below hydrogen in the ECS. Copper, for example, dissolves in dilute nitric acid to form an aqueous solution of copper(II) ions. Nitrogen monoxide gas is evolved.

$$\begin{array}{l} 3Cu(s) \rightarrow 3Cu^{2+}(aq) + 6e \\ \qquad\qquad\qquad\qquad\quad \downarrow \\ \qquad\qquad\qquad\qquad\quad 6e + 8H^+(aq) + 2NO_3^-(aq) \rightarrow 2NO(g) + 4H_2O(l) \end{array}$$

Note that dilute nitric acid converts copper to copper(II), not to copper(I). In the same way dilute nitric acid converts iron to iron(III), not to iron(II). Dilute nitric acid removes electrons so readily from metal atoms that the metal ion with the higher number of positive charges is formed.

Reaction of metals with hydrogen chloride

Chlorides are also formed when chlorine gas is passed over many heated metals. Reaction is vigorous and exothermic.

$$2Na(s) + 2HCl(g) \rightarrow 2NaCl(s) + H_2(g)$$

$$Mg(s) + 2HCl(g) \rightarrow MgCl_2(s) + H_2(g)$$

$$Fe(s) + 2HCl(g) \rightarrow FeCl_2(s) + H_2(g)$$

- Iron forms iron(II) chloride, not iron(III) chloride.
- Metals low in the ECS, such as copper, do not react with hydrogen chloride.

Reaction of metals with chlorine

Chlorides are also formed when chlorine gas is passed over heated metals. Reaction is once again vigorous and exothermic.

$$2Na(s) + Cl_2(g) \rightarrow 2NaCl(s)$$

$$Mg(s) + Cl_2(g) \rightarrow MgCl_2(s)$$

$$2Fe(s) + 3Cl_2(g) \rightarrow 2FeCl_3(s)$$

$$Cu(s) + Cl_2(g) \rightarrow CuCl_2(s)$$

- Chlorine removes electrons very readily from metals. The metal ion with the highest possible number of charges is formed. Iron forms iron(III) chloride, not iron(II) chloride. Copper forms copper(II) chloride, not copper(I) chloride.
- Copper reacts with chlorine, but not with hydrogen chloride.

Preparation of copper (I) compounds

Copper does not readily form ions. Vigorous oxidizing agents such as dilute nitric acid or chlorine are necessary to make copper ionize. These oxidizing agents are so powerful in their action that copper(II) is formed, not copper(I). Compounds of copper(I) can be made by first converting copper to a copper(II) compound. The copper(II) compound is then reduced.

Preparation of copper(I) chloride

A copper(II) salt such as the chloride is dissolved in concentrated hydrochloric acid and is boiled with excess copper. This produces copper(I) ions in solution. White insoluble copper(I) chloride is precipitated when this solution is diluted with water. The precipitate is filtered, washed with water and dried.

$$Cu + Cu^{2+} \rightarrow 2Cu^+$$

on dilution

$$Cu^+ + Cl^- \rightarrow CuCl$$

Preparation of copper(I) oxide

Copper(II) salts can be reduced to copper(I) oxide when alkali is present. A mild reducing agent such as glucose is boiled with an aqueous mixture of a copper(II) salt, alkali and a stabilizing agent. The stabilizing agent, which is a citrate or tartrate, prevents copper(II) hydroxide decomposing into copper(II) oxide. After boiling the mixture for some time, copper(I) oxide is precipitated as a red solid. The precipitate is filtered, washed with water and dried.

$$2Cu(OH)_2 \xrightarrow[\text{glucose}]{\text{reduction by}} [2CuOH] \xrightarrow[\text{hydroxide decomposes}]{\text{unstable copper(I)}} Cu_2O + H_2O$$

Solubility of metal hydroxides

Many normal metals are high in the ECS. The hydroxides of these metals are soluble in water to give alkaline solutions.

- NaOH is very soluble in water and forms a strongly alkaline solution.
- $Mg(OH)_2$ is slightly soluble in water and forms a moderately alkaline solution.

Transition metals are far lower in the ECS than the most reactive normal metals. The hydroxides, and oxides, of transition metals are all insoluble in water. The hydroxides, and oxides, of both normal and transition metals are, however, basic. They dissolve in acids to give aqueous solutions of the metal ions.

$$NaOH(s) + HCl(aq) \rightarrow NaCl(aq) + H_2O(l)$$

$$Mg(OH)_2(s) + 2HCl(aq) \rightarrow MgCl_2(aq) + 2H_2O(l)$$

$$Fe(OH)_3(s) + 3HCl(aq) \rightarrow FeCl_3(aq) + 3H_2O(l)$$

$$Cu(OH)_2(s) + 2HCl(aq) \rightarrow CuCl_2(aq) + 2H_2O(l)$$

Colour of metal ions

The ions of normal metals are colourless. Compounds containing normal metals are colourless unless the negative ion is coloured.

The ions of transition metals are frequently coloured, particularly if the compound containing the ion is hydrated or in aqueous solution.

Hydrated iron(II) sulphate is pale green.
Iron(II) hydroxide is green.
Hydrated iron(III) chloride is yellow.
Iron(III) hydroxide is brown.
Hydrated copper(II) sulphate is blue.
Copper(II) hydroxide is blue.
Copper(I) oxide is red, but most copper(I) salts are colourless.

Compounds containing transition metals as positive ions can often be readily recognized in analysis. Aqueous sodium hydroxide is added to an aqueous solution of the compound. The insoluble metal hydroxide is precipitated. The transition metal can be identified by the colour of this precipitate.

Iron(II) salts give a green precipitate	$Fe^{2+}(aq) + 2OH^-(aq) \rightarrow Fe(OH)_2(s)$
Iron(III) salts give a brown precipitate	$Fe^{3+}(aq) + 3OH^-(aq) \rightarrow Fe(OH)_3(s)$
Copper(II) salts give a blue precipitate	$Cu^{2+}(aq) + 2OH^-(aq) \rightarrow Cu(OH)_2(s)$

Catalytic activity

Catalysts are substances which, although usually present in only small amounts, can increase considerably the rate of certain reactions. Catalysts are not consumed during the reaction. The rate at which hydrogen peroxide decomposes into water and oxygen is catalysed by many transition metals and their oxides (*see* page 142).

$$2H_2O_2(aq) \rightarrow 2H_2O(l) + O_2(g)$$

Catalytic activity in this and other reactions is not confined to transition metals and their compounds. The decomposition of hydrogen peroxide is also catalysed by the normal metal lead and by its oxides. $H^+(aq)$ ions and $OH^-(aq)$ ions catalyse many reactions. Iodine, phosphorus, nitrogen monoxide, silicon dioxide and tin(IV) bromide are other examples of normal elements and their compounds which can act as catalysts in certain reactions.

Important industrial processes where catalysts are used include the following:

the Haber process for producing ammonia (*see* Figure 24.2)

$$N_2(g) + 3H_2(g) \xrightarrow{\text{iron catalyst}} 2NH_3(g)$$

the contact process for producing sulphuric acid via SO_3 (*see* Figure 24.3)

$$2SO_2(g) + O_2(g) \xrightarrow[\text{oxide catalyst}]{\text{vanadium(V)}} 2SO_3(g)$$

the oxidation of ammonia for producing nitric acid via NO (*see* Figure 24.4)

$$4NH_3(g) + 5O_2(g) \xrightarrow[\text{catalyst}]{\text{platinum}} 6H_2O(g) + 4NO(g)$$

Air (a mixture of N_2 and O_2) is the source of nitrogen.

1. Air can be used for the partial oxidation of hydrocarbons to H_2. This consumes the O_2 and leaves a gaseous mixture containing H_2 and N_2.
2. Air can be separated into N_2 and O_2 by fractional distillation of liquid air. The O_2 can then be used for partial oxidation of hydrocarbon to H_2.

Hydrocarbons (petroleum or natural gas) are the main sources of hydrogen. Two processes used on a large scale are steam reforming and partial oxidation of natural gas, methane.

Steam reforming: Steam and methane at high temperature and high pressure are passed over a nickel catalyst.

$$CH_4(g) + H_2O(g) \rightarrow CO(g) + 3H_2(g)$$

Partial oxidation: This is achieved using methane and air (or oxygen) at high temperature and high pressure. A catalyst and steam is also present in some processes.

$$2CH_4(g) + O_2(g) \rightarrow 2CO(g) + 4H_2(g)$$

Both steam reforming and partial oxidation produce a mixture of CO and H_2. This is mixed with steam and passed over an iron catalyst. The CO_2 formed is then removed.

$$CO(g) + H_2O(g) \rightarrow CO_2(g) + H_2(g)$$

An **iron catalyst** is used to increase the rate at which equilibrium is reached. The iron is very finely divided. This gives a large surface area of catalyst which increases the rate of reaction.

Ammonia must be separated from uncombined N_2 and H_2.

1. Ammonia is dissolved in water under pressure.
2. The gases are cooled until NH_3 separates as a liquid.

Unused N_2 and H_2 are then recycled to make more ammonia.

$$N_2(g) + 3H_2(g) \overset{Fe}{\rightleftharpoons} 2NH_3(g) \quad \Delta H = -92 \text{ kJ/mole}$$

The reaction is exothermic. The catalyst is kept hot by the heat of reaction.

exothermic direction

Temperature →

The equilibrium position shifts to the right with **decrease** in temperature (*see* page 150). The catalyst at present in use, however, needs a fairly high temperature (about 400 °C) before it becomes effective. This high temperature reduces the yield of NH_3 to about 15 per cent. The uncombined N_2 and H_2 are recycled so that eventually they are completely converted to NH_3.

4 moles of gaseous molecules form 2 moles of gaseous molecules

Pressure →

The equilibrium position shifts to the right with **increase** in pressure (*see* page 152). A pressure of about 200 atmospheres is used. The high pressure favours the formation of NH_3 in the equilibrium mixture. High pressure also increases the rate of reaction because the molecules are closer together. The molecules collide more frequently at high pressure and this increases the rate of reaction.

Figure 24.2 The Haber process for the production of ammonia

Sulphur dioxide is obtained by:

1. burning sulphur or sulphur compounds in air

$$S + O_2 \rightarrow SO_2$$

2. heating calcium sulphate with coke (carbon)

$$2CaSO_4 + C \rightarrow 2CaO + 2SO_2 + CO_2$$

Air is used as a source of **oxygen**.

A **catalyst** containing vanadium(V) oxide, V_2O_5, is used. The catalyst is spread out on an inert material to give greater surface area. The large surface area increases the rate of reaction.

SO_3 is dissolved in concentrated sulphuric acid. Water is then added.

$$SO_3 + H_2O \rightarrow H_2SO_4$$

In some plants, concentrated sulphuric acid is sprayed into the equilibrium mixture of SO_2 and SO_3. This dissolves the SO_3 and removes it from the equilibrium mixture. The equilibrium position then shifts to form more SO_3 (*see* page 153). This gives an increased yield of SO_3.

$$2SO_2(g) + O_2(g) \overset{V_2O_5}{\rightleftharpoons} 2SO_3(g) \quad \Delta H = -192 \text{ kJ/mole}$$

The reaction is exothermic. The catalyst is kept hot by the heat of reaction.

exothermic direction

Temperature ⟶

The equilibrium position shifts to the right with **decrease** in temperature (*see* page 150).

The catalyst, however, requires a fairly high temperature (about 450 °C) to give a fast rate of reaction. The yield of SO_3 is about 95 per cent at this temperature. Compare this with the low yield of NH_3 in the Haber process.

3 moles of gaseous molecules form 2 moles of gaseous molecules

Pressure ⟶

The equilibrium position shifts to the right with **increase** in pressure (*see* page 152).

High pressure favours the formation of SO_3, and also increases the rate of reaction. At atmospheric pressure the yield of SO_3 is high (about 95 per cent) and reaction is rapid. Extra pressure is not needed and is not used.

Figure 24.3 The contact process for producing sulphuric acid via sulphur trioxide

Ammonia is produced from N_2 and H_2 by the Haber process.

Air (a mixture of N_2 and O_2) is the source of **oxygen**.

A **catalyst** of platinum and rhodium is used. The catalyst is in the form of gauze to give a large surface area and rapid rate of reaction.

The nitrogen monoxide is cooled and mixed with air to give nitrogen dioxide. Nitric acid is formed when nitrogen dioxide dissolves in water with air present.

$$2NO(g) + O_2(g) \rightarrow 2NO_2(g)$$

$$4NO_2(g) + 2H_2O(l) + O_2(g) \rightarrow 4HNO_3(aq)$$

$$4NH_3(g) + 5O_2(g) \overset{Pt/Rh}{\rightleftharpoons} 6H_2O(g) + 4NO(g) \quad \Delta H = -907 \text{ kJ/mole}$$

The reaction is exothermic. The catalyst is kept hot by the heat of reaction.

Temperature —— exothermic direction ——→

The equilibrium position shifts to the right with **decrease** in temperature (*see* page 150). The catalyst, however, requires a high temperature (about 900 °C) to give a fast rate of reaction. At this temperature the equilibrium is still very favourable for the formation of NO.

Pressure —— 9 moles of gaseous molecules form 10 moles of gaseous molecules ——→

The equilibrium position shifts to the right with **decrease** in pressure (*see* page 152). Decreasing the pressure would, however, decrease the rate of reaction as the molecules would be further apart and collide less frequently. The equilibrium position is very favourable for the formation of NO, and in practice the pressure is often increased to a few atmospheres to give an increased rate of reaction.

Figure 24.4 The oxidation of ammonia for producing nitric acid via nitrogen monoxide

Summary of properties of transition metals

1. Atoms of transition metals have an electronic configuration with two incomplete shells.
2. Bonding between transition metal atoms is strong. The metals have high density, small volume for one mole of atoms, high melting point and high boiling point.
3. Transition metals can usually form positive ions with more than one value for the charge, such as Fe^{2+} and Fe^{3+}. They form more than one chloride, such as $FeCl_2$ and $FeCl_3$.
4. Transition metals are not found at the top of the ECS.
5. Transition metals do not react with cold water.
6. Only the most reactive transition metals release hydrogen from acids, but most transition metals dissolve in dilute nitric acid.
7. Only the most reactive transition metals are attacked by hydrogen chloride gas. Chlorine, which is a stronger oxidizing agent, converts most transition metals to their chlorides.
8. The hydroxides of all transition metals are insoluble in water.
9. Many compounds containing transition metals are coloured.
10. Transition metals and their compounds are frequently good catalysts.

25 Carbon Chemistry

The compounds of carbon are more numerous than those of any other element. This is due to the ability of carbon atoms to join together in **chains** or, to a lesser extent, in **rings**. In Chapter 10 the type of bonding in carbon compounds is discussed.

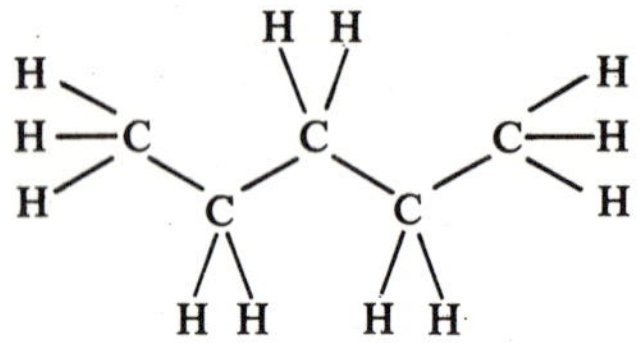

Pentane: a compound of carbon and hydrogen containing a *chain* of 5 carbon atoms

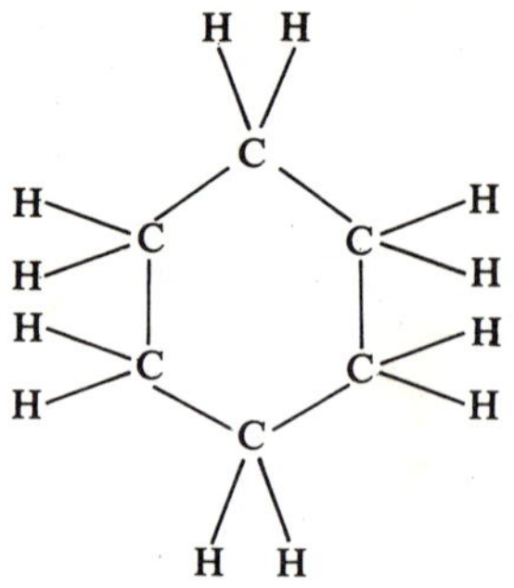

Cyclohexane: a *ring* compound of carbon containing 6 carbon atoms

Although the property of chain formation is not unique to carbon, it is so much more extensive than in any other element that compounds of carbon form a special branch of the study of chemistry: **organic chemistry**. This name was given in the belief that only living organisms were able to synthesize all but the simplest compounds of carbon. This belief is no longer held, but the name organic chemistry has remained.

Petroleum

The raw material for over 80 per cent by weight of manufactured carbon compounds is crude petroleum. Petroleum is a mixture of hydrocarbons (compounds of carbon and hydrogen only) of fairly high molar mass. **Fractional distillation** is used to separate the mixture into fractions containing fewer compounds:

- methane, propane and butane are gases
- compounds containing five or more carbon atoms in their molecules are liquids or solids of low melting point

A process known as **cracking** is then used to break large molecules into smaller, simpler ones.

Dodecane, $C_{12}H_{26}$ might be cracked to butene, C_4H_8, and butane, C_4H_{10}.

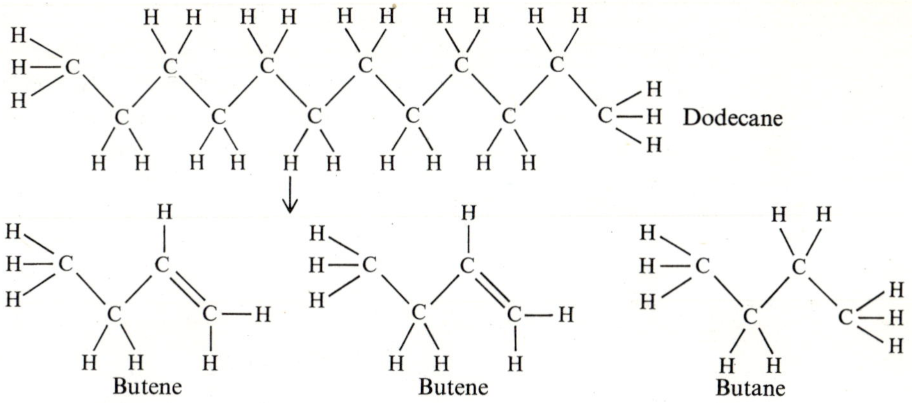

Dodecane might be cracked even further:

$$C_{12}H_{26} \rightarrow 5C_2H_4 + C_2H_6$$

or further still to carbon and hydrogen. This is an important source of hydrogen gas.

The simpler compounds such as C_2H_4, ethene (ethylene) are starting materials for synthesis of a vast number of substances. They range from drugs, dyestuffs and photographic chemicals to plastics, artificial fibres, insecticides and explosives. Petrol, diesel oil, paraffin and other fuels are also obtained from petroleum, together with lubricating oils and the bitumen used for surfacing roads with asphalt.

The compounds obtained from distillation or cracking of petroleum are all hydrocarbons. They have a number of properties in common.

1. They burn to carbon dioxide and water:

$$\underset{\text{Propane}}{C_3H_8} + 5O_2 \rightarrow 3CO_2 + 4H_2O$$

$$\underset{\text{Ethene}}{C_2H_4} + 3O_2 \rightarrow 2CO_2 + 2H_2O$$

2. They can be made to react with chlorine to give chloro-substituted hydrocarbons:

$$C_3H_8 + Cl_2 \rightarrow \underset{\text{Chloropropane}}{C_3H_7Cl} + HCl$$

$$C_3H_8 + 2Cl_2 \rightarrow \underset{\text{Dichloropropane}}{C_3H_6Cl_2} + 2HCl$$

3. They are all insoluble in water.
4. They are less dense than water and float on its surface.

Alkanes

All hydrocarbons

- share a number of properties with other hydrocarbons
- differ from each other in the number of carbon atoms in their molecules
- differ in physical properties rather than general chemical properties

Hydrocarbons in which all carbon atoms are tetrahedrally bonded to the other atoms in the molecule (*see* Chapter 10) and are therefore compounds with only single covalent bonds, form a group or series called the alkanes.

Each alkane differs from the previous member of the series by CH_2. The chemical properties of alkanes are very similar. Series such as the alkanes are frequent in organic chemistry. They are termed **homologous series** on account of the similarity in properties of all members or homologues. Their physical properties show a progression. For example, melting and boiling points increase with molar mass. Table 25.1 gives two physical properties for a number of alkanes.

Table 25.1

Name of alkane	*Formula*	*Molar mass (g)*	*Melting point (°C)*	*Boiling point (°C)*
Methane	CH_4	16	−182	−162
Ethane	C_2H_6	30	−183	−89
Propane	C_3H_8	44	−188	−42
Butane	C_4H_{10}	58	−138	−0.5
Pentane	C_5H_{12}	72	−130	36
Hexane	C_6H_{14}	86	−95	70

Inspection of Table 25.1 shows that

1. each hydrocarbon contains one carbon atom and two hydrogen atoms more than the preceding one
2. as the molar mass increases the values of both melting and boiling points increase

Characteristics such as 1. and 2. above are typical for all homologous series. The alkanes have the general formula $C_nH_{(2n+2)}$ where n is a whole number. Other homologous series of hydrocarbons are

alkenes: C_nH_{2n} alkynes: $C_nH_{(2n-2)}$

Alkanes react with chlorine under suitable conditions. When reaction takes place atoms of chlorine replace atoms of hydrogen in a substitution reaction. Chloroalkanes are the product.

$$\underset{\text{Methane}}{CH_4} + Cl_2 \rightarrow \underset{\text{Chloromethane}}{CH_3Cl} + HCl$$

$$\underset{\text{Propane}}{C_3H_8} + Cl_2 \rightarrow \underset{\text{Chloropropane}}{C_3H_7Cl} + HCl$$

Further substitution is possible if more chlorine is available:

$$\underset{\text{Chloromethane}}{CH_3Cl} + Cl_2 \rightarrow \underset{\text{Dichloromethane}}{CH_2Cl_2} + HCl$$

$$\underset{\text{Dichloromethane}}{CH_2Cl_2} + Cl_2 \rightarrow \underset{\text{Trichloromethane}}{CHCl_3} + HCl$$

$$\underset{\text{Trichloromethane}}{CHCl_3} + Cl_2 \rightarrow \underset{\text{Tetrachloromethane}}{CCl_4} + HCl$$

Compounds such as alkanes and chloroalkanes in which there are single covalent bonds only are called **saturated** compounds.

Alkenes

This homologous series has the general formula C_nH_{2n}. Its members contain a double bond and are **unsaturated**. Compounds which are saturated can only react by **substitution**; the unsaturated alkenes react by **addition**.

$$\underset{\text{Ethene}}{C_2H_4} + Cl_2 \rightarrow \underset{\text{Dichloroethane}}{C_2H_4Cl_2}$$

Table 25.2 gives the structural formula and physical properties of the first five members of the series.

Table 25.2

Name of alkene	*Structural formula*	*Molar mass (g)*	*Melting point (°C)*	*Boiling point (°C)*
Ethene	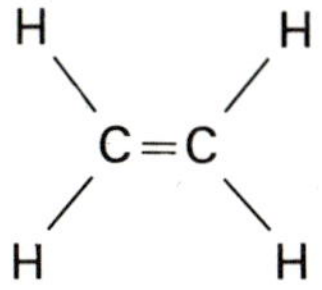	28	−169	−104
Propene	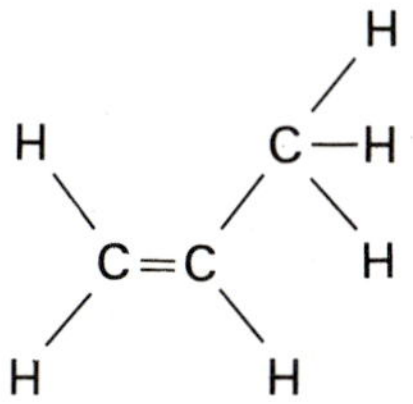	42	−185	−48
Butene	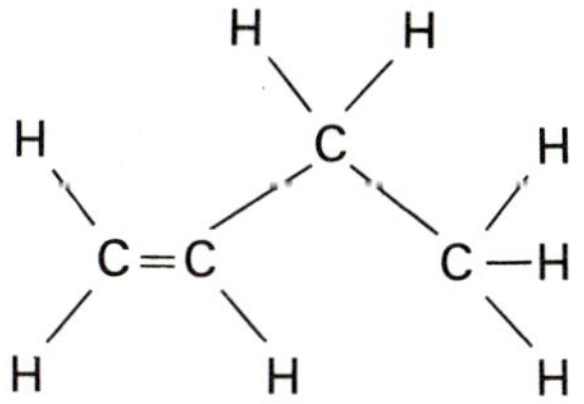	56	−185	−6
Pentene	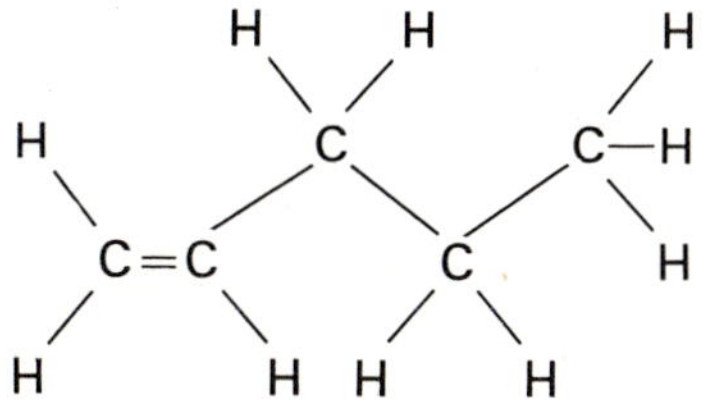	70	−139	30
Hexene	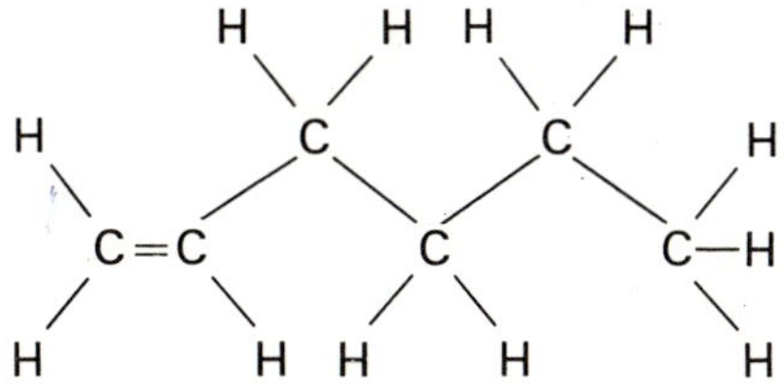	84	−98	64

Unlike alkanes, which are sluggish in most of their reactions, alkenes react readily with a number of substances. Table 25.3 lists some of these reactions.

Table 25.3

Reagent	*Equation of reaction*	*Name of product*	*Type of compound*
Bromine	$C_2H_4 + Br_2 \rightarrow C_2H_4Br_2$	Dibromoethane	Dibromoalkane
Alkaline permanganate solution	$C_2H_4 + H_2O + \hat{O} \rightarrow C_2H_4(OH)_2$	Ethanediol (glycol)	Dihydric alcohol
Hydrogen chloride	$C_2H_4 + HCl \rightarrow C_2H_5Cl$	Chloroethane	Chloroalkane

Other alkenes react in a similar way with these reagents.

Alkynes

Another homologous series of hydrocarbons is the alkynes, $C_nH_{(2n-2)}$, which contain a triple bond in the molecule. The simplest member is ethyne, usually called acetylene, C_2H_2.

Table 25.4 gives structures for several alkynes.

Table 25.4

Name of alkyne	*Molecular formula*	*Structural formula*
Ethyne (acetylene)	C_2H_2	H—C≡C—H
Propyne	C_3H_4	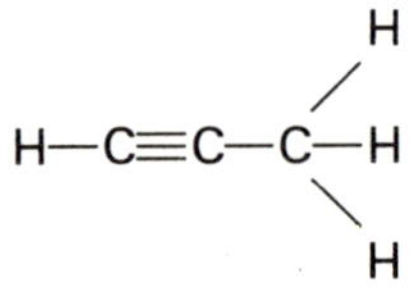
Butyne	C_4H_6	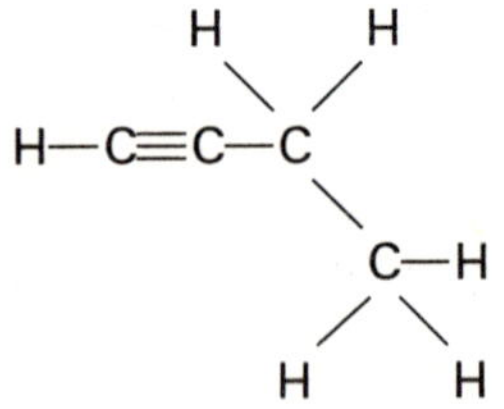
Pentyne	C_5H_8	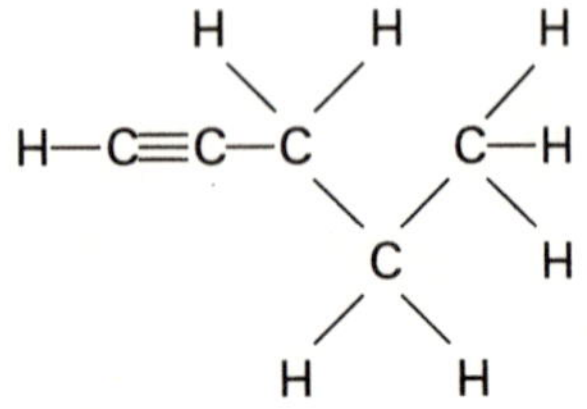

Isomerism

Compounds with a given molecular formula may have different structural formulae. This is termed isomerism (*see* also Chapter 10). Isomerism is very common in organic chemistry. It may arise

- from chain branching or
- from variations in the position of attached groups

Figure 25.1 shows three structures for the alkane C_5H_{12}. Figure 25.2 shows eight possible structures for the chloroalkane $C_5H_{11}Cl$.

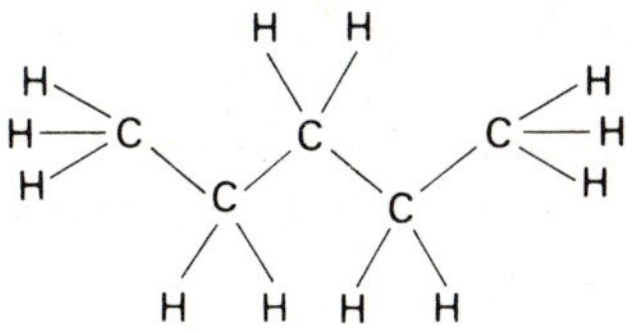

Pentane

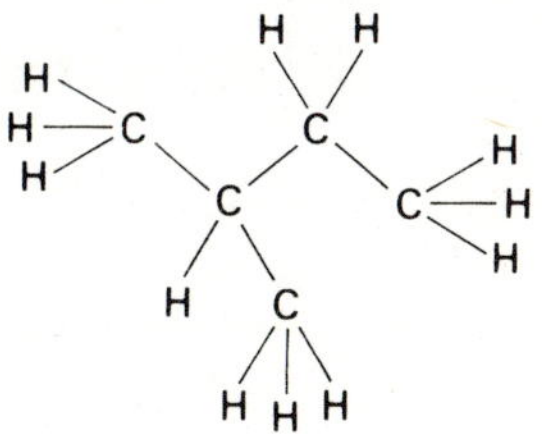

2-Methylbutane

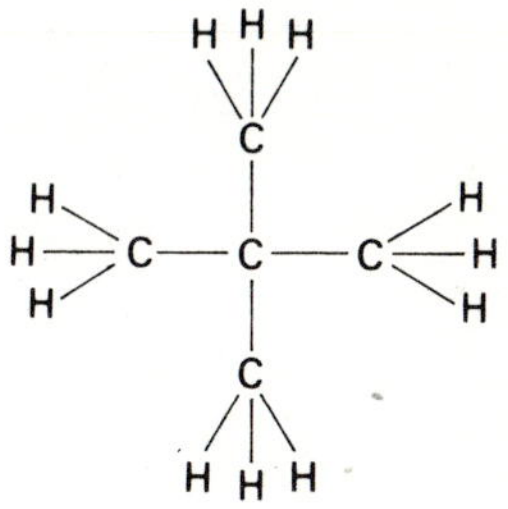

2-2-Dimethylpropane

Figure 25.1

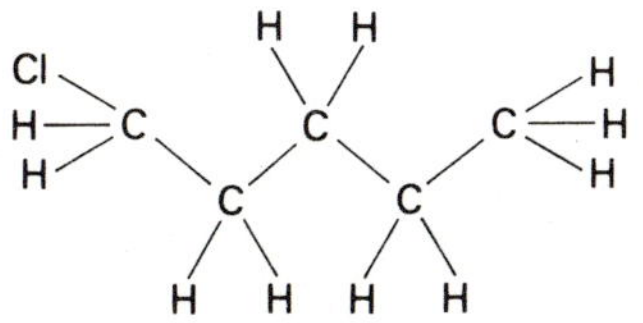

1-Chloropentane

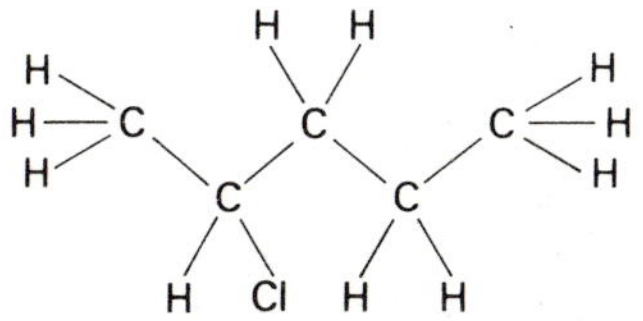

2-Chloropentane

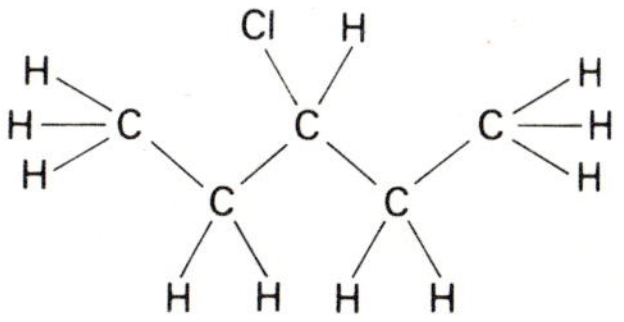

3-Chloropentane

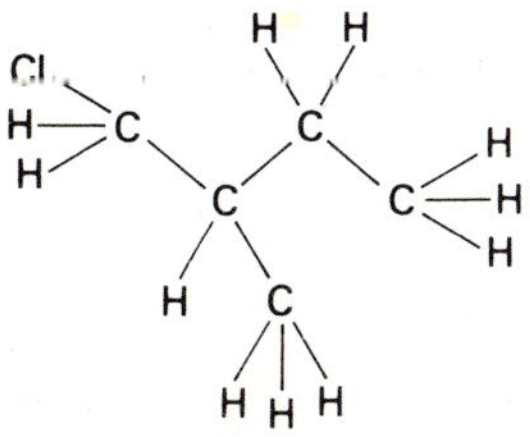

1-Chloro-2-methylbutane

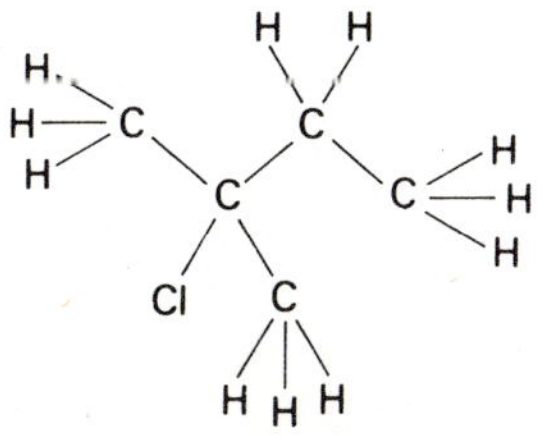

2-Chloro-2-methylbutane

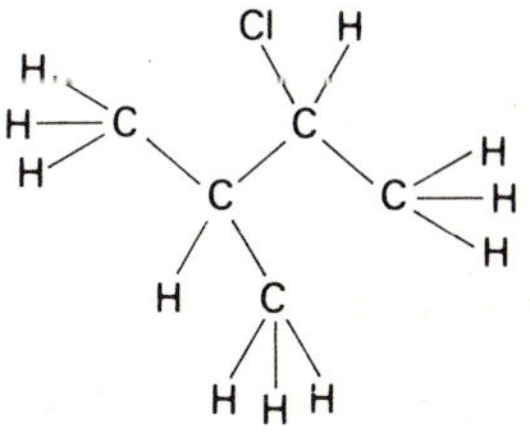

3-Chloro-2-methylbutane

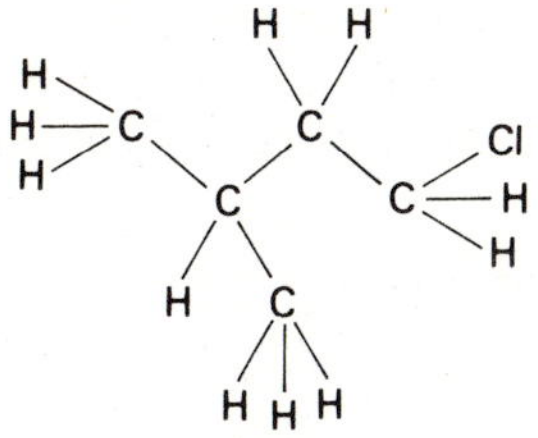

1-Chloro-3-methylbutane

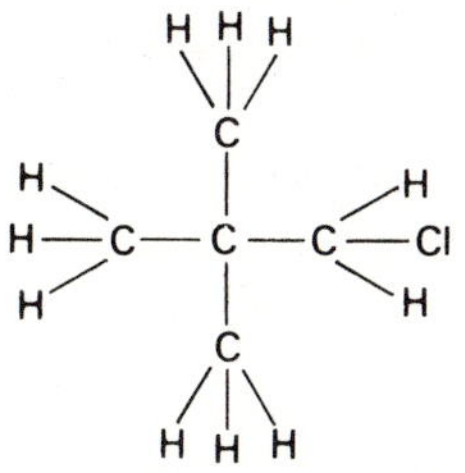

1-Chloro -2-2-dimethylpropane

Figure 25.2

In the case of alkenes and alkynes isomerism can arise from the position of the double or triple bond. Two isomers of butene, but-1-ene and but-2-ene are possible (Figure 25.3).

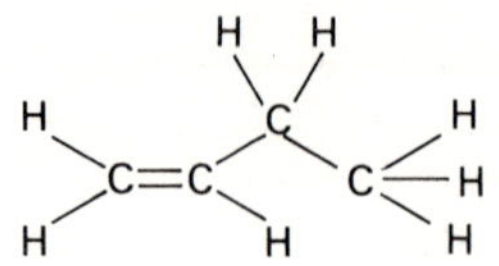

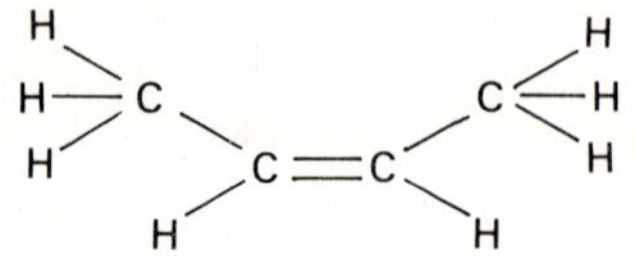

Figure 25.3

The number of possible isomers rises rapidly with the number of carbon atoms in the molecule. Table 25.5 gives the number of possible isomers of alkanes with various numbers of carbon atoms.

Table 25.5

Carbon atoms in alkane	*Number of possible isomers*
1	1
2	1
3	1
4	2
5	3
6	5
7	9
8	18
⋮	⋮
40	62 491 178 805 831

Alcohols

If yeast is mixed with a solution of sugar or glucose it feeds on the sugar and produces alcohol and carbon dioxide. The process is called **fermentation** and is used in the making of beer, wines, spirits and alcohol and for other purposes. Beer contains only 2 or 3 per cent alcohol, wine up to 14 per cent and spirits a much higher percentage. For making spirits or pure alcohol, fractional distillation must be used since above 14 per cent of alcohol the yeast ceases to function and dies. By fractionally distilling the mixture, alcohol mixed with 4 per cent of water is obtained.

The name alcohol is given to another homologous series. The alcohol in alcoholic drinks is one member, ethanol, of this series. Alcohols contain an —OH group attached to carbon. They may be regarded as alkanes in which a hydrogen atom has been replaced by an —OH group. Alcohols have the general formula $C_nH_{(2n+1)}OH$. The ethanol in alcoholic drinks is made by fermentation but much of the alcohol required for other purposes is made from ethene, a product of cracking petroleum. Ethene and steam are reacted under suitable conditions:

$$\underset{\text{Ethene}}{C_2H_4} + H_2O \rightarrow \underset{\text{Ethanol}}{C_2H_5OH}$$

Properties and reactions of ethanol

Ethanol is a colourless liquid of density about 0.8 g/cm^3. It burns with oxygen to carbon dioxide and water:

$$C_2H_5OH + 3O_2 \rightarrow 2CO_2 + 3H_2O$$

Sodium reacts with ethanol, replacing the hydrogen of the —OH group, but not any other, to form an alkoxide:

$$2Na + 2C_2H_5OH \rightarrow \underset{\text{Sodium ethoxide}}{2C_2H_5ONa} + H_2$$

Phosphorus tri- or pentachloride react with ethanol, replacing the —OH group with chlorine to form a chloroalkane:

$$PCl_3 + 3C_2H_5OH \rightarrow \underset{\text{Chloroethane}}{3C_2H_5Cl} + H_3PO_3$$

$$PCl_5 + C_2H_5OH \rightarrow C_2H_5Cl + POCl_3 + HCl$$

A mixture of sodium dichromate and sulphuric acid oxidizes ethanol to the organic acid, acetic acid. Vinegar contains this acid owing to atmospheric oxidation of the ethanol in wine. Organic acids contain the group —CO_2H:

$$C_2H_5OH + \hat{O} \rightarrow CH_3CO_2H$$

With organic acids, ethanol reacts to form esters. The reaction is slow and reversible. Some concentrated sulphuric acid is added to increase the speed at which the ester is formed:

$$C_2H_5OH + \underset{\text{Acetic acid}}{HO_2CCH_3} \rightleftharpoons \underset{\text{Ethyl acetate}}{C_2H_5O_2CCH_3} + H_2O$$

Amino acids and proteins

Acetic acid has the structure

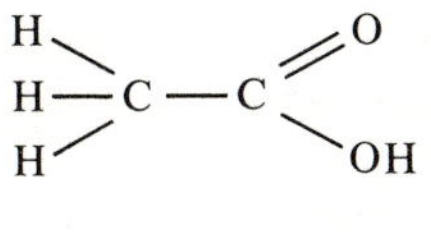

If one of the hydrogens attached to carbon is replaced by an amino group, —NH_2, the compound aminoacetic acid or glycine is made

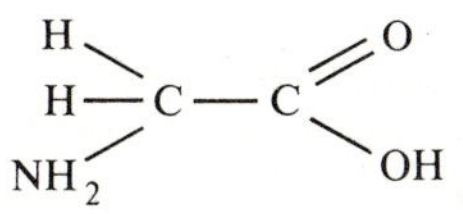

Amino acids have the ability to polymerize, *see* below. Two molecules join in a **peptide bond** between the amino group of one and the carboxylate group of the other

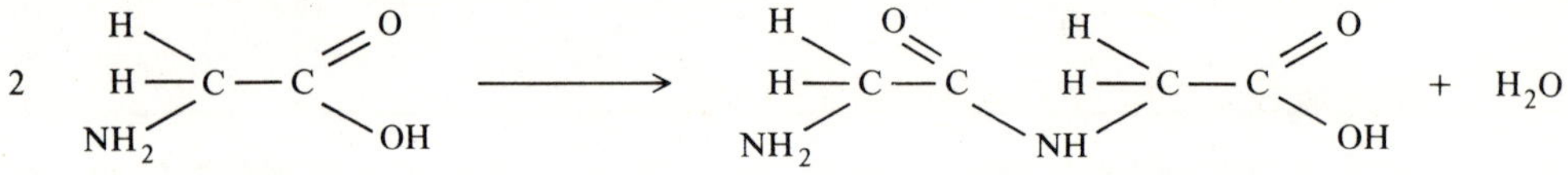

The peptide glycylglycine

This linkage leaves the resulting molecule with a carboxylate and an amino group not involved in bonding. They are thus able to link with further amino acids to form a long chain. In this way a polypeptide can be built up.

In the naturally occurring polypeptides the amino acids are not all glycine but may be selected from twenty or more amino acids. The sequence of acids along the polymer molecule determines the identity of the polypeptide. Further polymerization leads to proteins when polypeptide chains link. Hydrolysis of proteins leads to amino acids.

Polymers

Under certain conditions ethene molecules can combine into long chains. The resulting substance, called polythene, resembles an alkane with a very high molar mass. Ethene is an unsaturated compound, containing a double bond, but polythene is a saturated compound containing only single bonds.

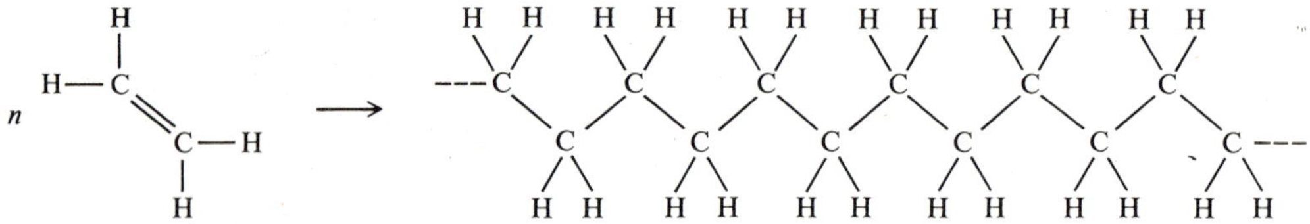

The formula of polythene can be written $(CH_2)_n$ where n is a very large number. Polythene is a **polymer** and the process of making ethene molecules combine into polythene is called **polymerization**. Most plastics are polymers.

In the polymerization of ethene to polythene there is only one product. The ethene units are added together and polythene is called an **addition polymer**. Another example of a polymer plastic is Perspex. It is made by polymerizing methyl methacrylate:

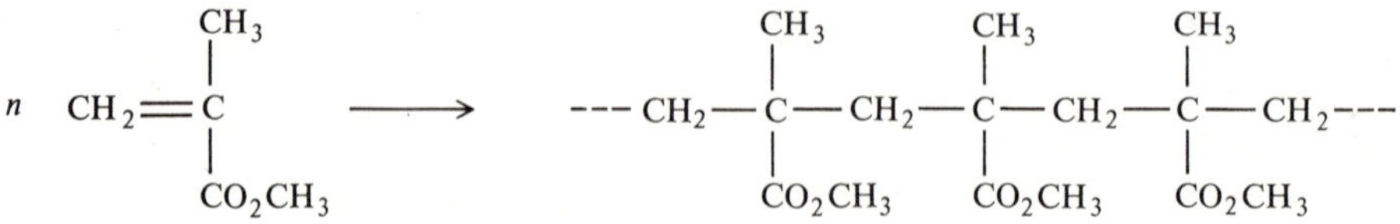

In another type of polymerization a small molecule such as water (H_2O) or hydrogen chloride (HCl) is produced as well as the polymer. The small molecule

is said to be **eliminated** and the process called an **elimination** or **condensation reaction**. Polymers made in this way are termed **condensation polymers**. Examples are 'Terylene', nylon, starch, cellulose and proteins.

'Terylene' is made by polymerizing the alcohol ethanediol (glycol) and the dimethyl ester of terephthalic acid. Methanol is eliminated:

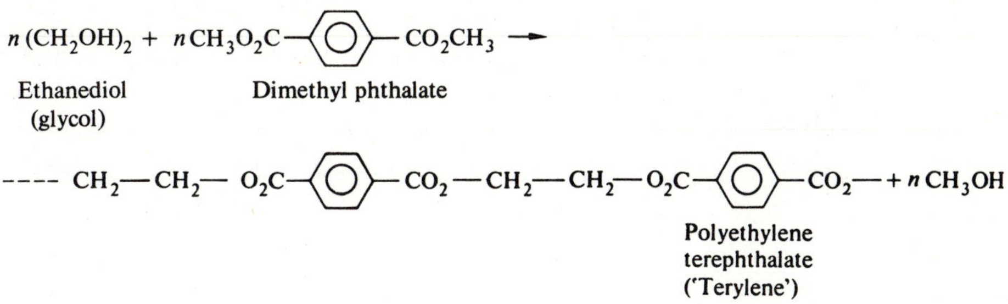

Nylon is a polypeptide (*see* page 234). It is formed by condensing the chloride of adipic acid with hexamethylene diamine. Hydrogen chloride is eliminated.

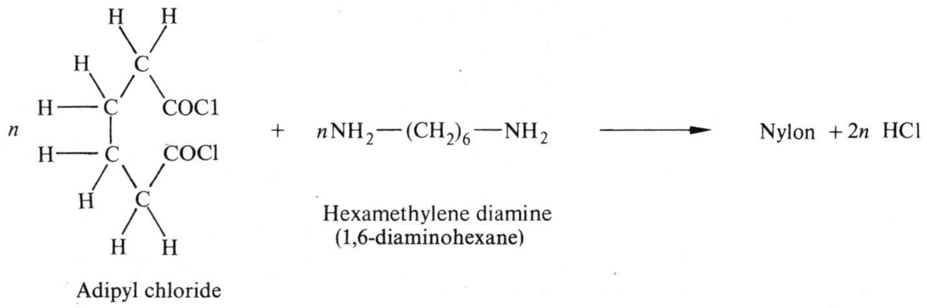

Nylon is a polypeptide (*see* page 234). It is formed by condensing the chloride
the peptide bonds clearly

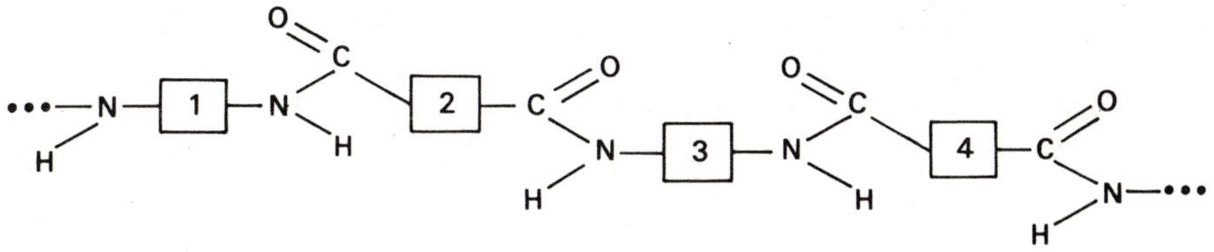

Figure 25.4

The odd-numbered rectangles stand for the group —$(CH_2)_6$— and the even-numbered rectangles for the group —$(CH_2)_4$—.

Starch and cellulose

Glucose has the formula $C_6H_{12}O_6$. It is able to form condensation polymers of several hundred units as **starch** and about 3000 to 4000 units as **cellulose**. All are examples of **carbohydrates**, so-called because the hydrogen and oxygen are in the same ratio as in water. Polymerization of glucose to starch or cellulose is by elimination of water. Under suitable conditions the starch or cellulose can be **hydrolysed** back to glucose.

$$\underset{\text{Starch}}{(C_6H_{10}O_5)_n} + n\,H_2O \rightarrow n\,\underset{\text{Glucose}}{C_6H_{12}O_6}$$

Soap

Natural fats are esters of the alcohol propanetriol (glycerol) and an acid high in the homologous series of acids. An example is propan-1,2,3-trioctadecanoate (glyceryl tristearate):

$$\begin{array}{l} H_2C\text{—}O_2C\text{—}C_{17}H_{35} \\ \;\;| \\ HC\text{—}O_2C\text{—}C_{17}H_{35} \\ \;\;| \\ H_2C\text{—}O_2C\text{—}C_{17}H_{35} \end{array}$$

If this ester is hydrolysed using sodium hydroxide, propanetriol (glycerol) and sodium octadecanoate (sodium stearate) are formed. The sodium octadecanoate is an example of a soap.

$$\begin{array}{l} H_2C\text{—}O_2C\text{—}C_{17}H_{35} \\ \;\;| \\ HC\text{—}O_2C\text{—}C_{17}H_{35} + 3NaOH(aq) \rightarrow 3C_{17}H_{35}CO_2Na + C_3H_5(OH)_3 \\ \;\;| \\ H_2C\text{—}O_2C\text{—}C_{17}H_{35} \end{array}$$

Sodium octadecanoate (stearate)   Propanetriol

If sodium octadecanoate encounters calcium or magnesium ions a precipitate of calcium or magnesium octadecanoate is formed. This is the scum which forms when soap is put into hard water.

Detergents

Sodium alkyl sulphates and alkyl benzene sulphonates do not suffer the disadvantages of soaps. They are more soluble in water and the calcium or magnesium ions in hard water do not produce precipitation since the calcium and magnesium salts are also soluble. Sodium dodecanyl sulphate, $C_{12}H_{25}SO_4Na$ and sodium tetradecanyl sulphate, $C_{14}H_{29}SO_4Na$ are examples. Detergents also lather in salt water.

Summary

1. Compounds of carbon are very numerous owing to its ability to form chains or rings of carbon atoms.
2. Isomerism is very common in organic compounds.
3. Most organic substances are manufactured from petroleum.
4. Saturated compounds have only single bonds between atoms.
5. Alkanes are saturated compounds of the general formula $C_nH_{(2n+2)}$. Alkanes are one of many homologous series of compounds.
6. Alkenes are unsaturated hydrocarbons containing a double bond. They have the general formula C_nH_{2n}.

7. Alkynes are also unsaturated and contain a triple bond. Their general formula is $C_nH_{(2n-2)}$.
8. Alkenes and alkynes react by addition. Alkanes react by substitution.
9. Carbon forms many groups of compounds called homologous series. The members of each series have several chemical properties in common. Their physical properties show a progression through the series. Each member differs from the next by $—CH_2—$.
10. Alcohols have the general formula $C_nH_{(2n+1)}OH$.
11. Some carbon compounds can polymerize. Addition polymers include polythene and Perspex. Condensation polymers include nylon, starch, cellulose and proteins.
12. Soaps are sodium salts of long-chain carboxylic acids. They are made by hydrolysing esters. The calcium and magnesium salts are insoluble in water, and precipitate as scum when soap is added to hard water.
13. Detergents are sodium salts of alkylsulphuric or alkyl benzenesulphonic acids. The calcium and magnesium salts of these acids are soluble in water so they lather in hard or salt water.

26 Analysis

Cations

Flame tests

These ions give a colour to the hot flame when they are evaporated from nichrome wire (Table 26.1):

$$Li^+ \quad Na^+ \quad K^+ \quad Ca^{2+} \quad Ba^{2+} \quad Pb^{2+} \quad Cu^{2+}$$

The wire is first heated to evaporate ions which might have contaminated the surface. It is then dipped into the substance under investigation and returned to the flame. Some colours appear only briefly since the ions evaporate rapidly. For K^+, the flame should be viewed through a blue glass to cut out the colour for Na^+.

Table 26.1

Colour of flame	*Ion present*
Scarlet	Li^+
Yellow/orange	Na^+
Lilac or pink	K^+
Brick red	Ca^{2+}
Apple green	Ba^{2+}
Blue/white	Pb^{2+}
Blue/green	Cu^{2+}

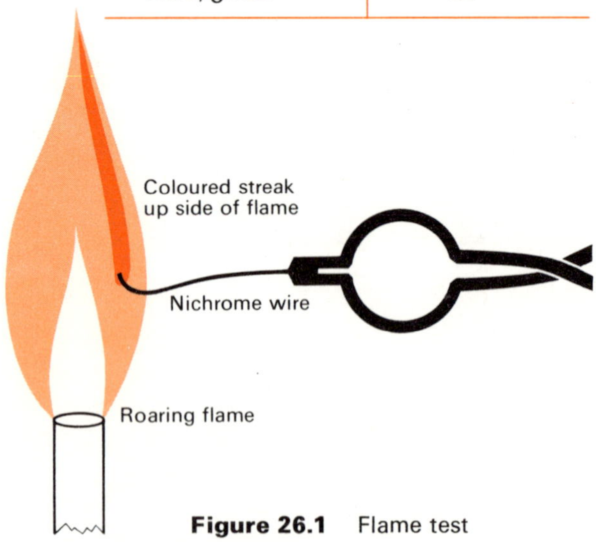

Figure 26.1 Flame test

For the ions Li^+ Na^+ K^+ Ba^{2+} and Ca^{2+} the flame test is sufficient confirmation at this level.

Cations in solution

To 1 cm^3 portions of a concentrated solution of the substance under investigation the reagents shown in Table 26.2 are added dropwise, stirring after each addition.

Table 26.2
Add Each of the Following Dropwise

Cation	*Ammonia solution*	*Sodium hydroxide solution*
NH_4^+	Solutions mix; no further change	Ammonia given off on warming (*see* Table 26.5); can be detected by characteristic smell
Cu^{2+}	Blue/green precipitate; dissolves to give royal blue solution	Jelly-like blue precipitate; goes black on warming
Fe^{2+}	Pale green precipitate; very dark if contaminated with Fe^{3+} ions	Pale green precipitate; very dark if contaminated with Fe^{3+} ions
Fe^{3+}	Reddish-brown precipitate	Reddish-brown precipitate
Pb^{2+}	White precipitate	White precipitate; dissolves in excess NaOH(aq)
Mg^{2+}	White precipitate	White precipitate
Zn^{2+}	White precipitate; dissolves in excess	White precipitate; dissolves in excess
Li^+, Na^+, K^+, Ca^{2+}, Ba^{2+}	These ions give no precipitate with either ammonia or sodium hydroxide solution. The flame test is sufficient confirmation of their presence	

Specific tests

Magnesium: Magneson reagent gives a blue precipitate with Mg^{2+} ions.

Lead: Pb^{2+} ions give a white precipitate with chloride ions. The lead chloride, $PbCl_2$, formed is soluble in hot water and reappears on cooling.

Zinc: Some compounds of zinc give zinc oxide, ZnO, on heating. Zinc oxide is yellow when hot, but turns white on cooling.

Anions

Table 26.3
Tests on Solid Substance*

Anion	*Heat in hard glass tube*	*Warm with a few drops concentrated sulphuric acid*	*Add dilute nitric acid to acid*
Bicarbonate, (hydrogen-carbonate) HCO_3^-	CO_2(g) given off; steam given off	CO_2(g) given off	CO_2(g) given off
Bromide, Br^-	—	Br_2(g), brown gas, given off	—
Carbonate, CO_3^{2-}	CO_2(g) may be given off; steam may be given off	CO_2(g) given off	CO_2(g) given off
Chloride, Cl^-	—	HCl(g) given off	—
Iodide, I^-	—	I_2 (dark solid or purple vapour)	—
Nitrate, NO_3^-	O_2(g) given off; NO_2(g), brown gas, may be given off also	HNO_3(g) and NO_2(g), brown gas, given off	—
Nitrite, NO_2^-	—	NO_2(g), brown gas, given off	NO_2(g), brown gas, given off
Sulphate, SO_4^{2-}	White SO_3 fumes sometimes given off	—	—
Sulphite, SO_3^{2-}	—	SO_2(g) given off	SO_2(g) given off
Sulphide, S^{2-}	—	H_2S(g) given off	H_2S(g) given off

*Tests for gases are given in Table 26.5

Table 26.4
Tests on Solution

Anion	*Add barium nitrate solution dropwise followed by dilute nitric acid*	*Add silver nitrate solution dropwise followed by dilute nitric acid*
Bicarbonate (hydrogen-carbonate) HCO_3^-	White precipitate $BaCO_3$; dissolves in nitric acid	White precipitate Ag_2CO_3; dissolves in nitric acid
Bromide, Br^-	—	Pale yellow precipitate AgBr; insoluble in nitric acid
Carbonate, CO_3^{2-}	White precipitate $BaCO_3$; dissolves in nitric acid	White precipitate Ag_2CO_3; dissolves in nitric acid
Chloride, Cl^-	—	White precipitate AgCl; insoluble in nitric acid
Iodide, I^-	—	Bright yellow precipitate AgI; insoluble in nitric acid
Sulphate, SO_4^{2-}	White precipitate $BaSO_4$; insoluble in nitric acid	—
Sulphite, SO_3^{2-}	White precipitate $BaSO_3$; dissolves in nitric acid	—
Sulphide, S^{2-}	—	Black precipitate Ag_2S; insoluble in nitric acid
Nitrate, NO_3^- or nitrite, NO_2^-	To a strongly alkaline solution of a nitrate or nitrite add Devarda's alloy and warm. Ammonia is given off (*see* Table 26.5) mixed with hydrogen. This test is only valid if ammonium ions are absent from the substance being investigated	

Table 26.5
Tests for Gases and Iodine

Gas	*Colour and smell*	*Method and observation*
Ammonia, NH_3	Colourless; pungent smell	Hold wet red litmus in gas. Litmus turns blue. With hydrogen chloride gas a blue/white smoke forms
Bromine, Br_2	Reddish-brown; choking smell IRRITANT POISON	Hold wet blue litmus in gas. Litmus turns red then rapidly bleached
Carbon dioxide, CO_2	Colourless; odourless	Bubble through lime water; cloudiness produced
Hydrogen chloride, HCl	Colourless; pungent smell IRRITANT	Hold wet blue litmus in gas. Litmus turns red. With NH_3(g) a blue/white smoke forms
Hydrogen sulphide, H_2S	Colourless; bad eggs smell POISON	Soak filter paper in lead acetate solution. Turns black when held in gas
Nitrogen dioxide, NO_2	Reddish-brown; pungent smell POISON	Hold wet blue litmus in gas. Litmus turns red but is not bleached
Sulphur dioxide, SO_2	Colourless; choking smell IRRITANT	Soak filter paper in purple potassium permanganate solution. Turns from purple to colourless. OR, soak paper in orange potassium dichromate solution. Turns from orange to green
Oxygen, O_2	Colourless; odourless	Glowing splint held in gas glows brighter or bursts into flame
Iodine, I_2	Dark solid, purple vapour IRRITANT	Shake a tiny portion of iodine with organic solvent such as trichloroethane. Iodine dissolves giving violet solution

Questions

Most of these questions ask you to interpret information and to make predictions. Do not be put off if a problem is set on a situation with which you are not familiar. The purpose is to test your understanding of chemistry rather than whether you can simply remember facts.

To answer many of these questions you will need to look at the tables, diagrams and text in the chapter to which the questions refer. If you require relative atomic masses (atomic weights) these will be found on the Periodic Table inside the front cover of this book. These relative atomic masses are given to the nearest whole number except in the case of chlorine. If you require the volume of a mole of gaseous molecules at room temperature and pressure you can take this to be 24 000 cm^3.

Chapter 1

1. Use the material in this chapter to find the answers to the following:
   (a) Name a substance which turns blue when water is added.
   (b) Which white solid gives two gases and a yellow solid on heating?
   (c) Which substance could be heated to obtain black copper oxide?
   (d) Name two substances which give lead monoxide when heated.
   (e) Name three substances which melt between 450 °C and 600 °C.
   (f) How many substances mentioned in this chapter are denser than water, as far as you can tell from the information given?
   (g) Name ten substances which contain oxygen.
   (h) Name five substances which decompose on heating and five which do not.
   (i) Describe, briefly, how you would show that carbon dioxide is different from oxygen.
   (j) What mass of magnesium would combine with 2.67 g of oxygen?

2. Substances A to G in the accompanying table can be assumed not to react together even on quite strong heating. Look at the properties of each and answer the questions below the table.

Substance	*Solubility in water*	*Solubility in petrol*	*Density g/cm³*	*M.p., °C*	*B.p., °C*
A	insoluble	soluble	0.5	−30	250
B	insoluble	insoluble	0.6	does not melt but burns on heating in air	
C	insoluble	insoluble	7.0	1500	3000
D	soluble	insoluble	3.5	650	2800
E	soluble	soluble	0.8	−120	80
F	insoluble	soluble	2.4	114	444
G	insoluble	soluble	0.5	75	320

(a) From a mixture of all substances how would you separate a sample of E?
(b) How would you obtain pure, dry samples of A, B and C from a mixture of these three alone?
(c) Which of the substances A to G are liquids at room temperature?
(d) Which of the substances A to G would melt at the temperature of steam?
(e) The substances A to G are, not in order, wax, wood, iron, oil, magnesium, sulphur, alcohol. Use your general knowledge to assign the correct letter to each substance. Use a data book if you really need to.

3. During part of the process of oil refining, crude petroleum is heated. This is called thermal cracking. Draw up a chart showing as many uses as possible for each product obtained from thermal cracking. Information might be obtained from the library or from such firms as Shell Petroleum.

Chapter 2

1. Study the following table and answer the questions below.

Substance	*Melting point, °C*
Mercury	−40
Tin	230
Sulphur	119
Sodium chloride	850
Magnesium chloride	1050
Potassium nitrate	750
Wax	120
Ice	0
Iron	1530
Aluminium	660

(a) List the substances which conduct electricity at room temperature.
(b) If the temperature of each substance was raised until it melted, which would then conduct electricity?

2. Write 'True or 'False' for each of the statements (a) to (e).
(a) No solid non-metal is a good conductor of electricity.
(b) Some molten non-metals are poor conductors of electricity.
(c) All molten metals conduct electricity well.
(d) Some solid compounds conduct electricity well.
(e) Some molten compounds conduct electricity well.

3. Which of the following substances are good conductors of electricity in the solid state?

iron magnesium wood copper
sodium sulphate oil

4. Liquid water is a very poor conductor of electricity, but solutions of some substances in water are good conductors. Use the information about solubility in Chapter 1 to help answer this question.
Which of the following substances dissolve in water to give a solution which conducts electricity well?

potassium permanganate lead nitrate
alcohol iron magnesium oxide sugar
copper oxide

Chapter 3

1. Which element is present in all the compounds of Table 1.3?

2. Find the names of four compounds which decompose on heating to give oxygen as one product.

3. (a) Which is the most abundant element in the Earth's crust?
(b) What percentage does sea water contain of this element?
(c) What percentage does the whole Earth contain of this element?

4. What are the main uses of oxygen?

5. Write an account in note and table form of the compound water. Headings you could use are: occurrence, short list of substances soluble in water, short list of insoluble substances, substances which react with water, uses for water.

6. Make lists of basic oxides, acidic oxides and salts mentioned in Chapters 1, 2, and 3.

7. (a) Name the salts which could be made using the following substances only:

nitric acid	sodium hydroxide
sulphuric acid	magnesium hydroxide
hydrochloric acid	copper oxide
acetic acid	potassium

(b) Describe how you would make one of these salts.
(c) When an acid reacts with a base, what are the products?
(d) When an acid reacts with a metal, what are the products?
(e) Acids react with some metals but not with others. What names are given to the two kinds of metal? Name one example of each.

8. A pupil has been given a number of aqueous solutions and has determined the pH of each using indicator paper. Here are the results.

Solution	A	B	C	D	E	F	G
pH	3	4	9	6	8	5	7

(a) Which solution is the most alkaline?
(b) Which solution is the most acidic?
(c) Which solution is closest to neutral?
(d) Which of the following will increase the pH of B? There may be more than one correct answer.
(i) addition of water
(ii) addition of sodium hydroxide
(iii) addition of A
(iv) addition of E

Chapter 4

1. You should use the information in Table 4.2 to answer these questions.

(a) What mass of calcium is combined with 1000 mg of oxygen in calcium oxide?
(b) If one atom of calcium combines with every one atom of oxygen in calcium oxide, which element has the atoms with the greater mass?
(c) What mass of lead is combined with 1000 mg of oxygen in lead(II) oxide?
(d) If one atom of lead combines with every one atom of oxygen in lead(II) oxide, which element has the atoms with the greater mass?
(e) Arrange the elements lead, oxygen and calcium in order of the mass of their atoms with the lowest mass first.

2. Write down the names and symbols of:

(a) two liquid elements
(b) four solid metallic elements
(c) four gaseous elements
(d) four solid non-metallic elements

3. Refer to Figure 4.2 to answer the following questions.

(a) Why is it necessary to burn the excess hydrogen in this experiment?
(b) The hydrogen should not be ignited until all the air has been displaced from the apparatus. Explain why this is so.
(c) A solid is heated during this experiment. Give the name and colour of this solid at the beginning of the experiment.
(d) A solid remains at the end of the experiment. Give the name and colour of this solid.

4. The relative atomic mass of sodium is 23. Write down the mass of

(a) 1 mole of sodium atoms
(b) 10 moles of sodium atoms
(c) 0.05 mole of sodium atoms

5. The relative atomic mass of lead is 207. Write down the mass of

(a) 1 mole of lead atoms
(b) 100 moles of lead atoms
(c) 0.1 mole of lead atoms

6. Use the information in Table 4.4 to write down the mass of 1 mole of atoms of hydrogen, carbon, nitrogen, oxygen, magnesium, aluminium, sulphur, chlorine and copper.

7. Use the information in Table 4.4 to write down the mass of 10 moles of atoms of lithium, fluorine, phosphorus, argon, calcium and silver.

8. Use the information in Table 4.4 to write down the mass of 0.1 mole of helium, beryllium, boron, neon, silicon and potassium.

9. How many moles of atoms are there in the following masses? You can use the information in Table 4.4 to answer this question.

(a) 46 g of sodium
(b) 76 g of fluorine
(c) 120.3 g of calcium
(d) 10.1 g of neon
(e) 4 g of oxygen

10. Carbon and oxygen form a compound containing 24 g of carbon to every 64 g of oxygen. Calculate the numbers of moles of atoms in these masses of elements and suggest a formula for the compound. Relative atomic masses are given in Table 4.4.

11. Suggest formulae for the following chlorides using the information given about their compositions by mass. Relative atomic masses are given in Table 4.4.

(a) 460 g of sodium and 710 g of chlorine
(b) 2.43 g of magnesium and 7.1 g of chlorine
(cf 270 g of aluminium and 1065 g of chlorine
(d) 2.84 g of silicon and 14.2 g of chlorine
(e) 62 g of phosphorus and 213 g of chlorine
(f) 9.63 g of sulphur and 2.13 g of chlorine

12. Using the values given in Table 4.4, plot a graph of atomic number (horizontal axis) against relative atomic mass. Use this graph to predict the relative atomic masses of the following elements.

Element	*Atomic number*
Rubidium	37
Caesium	55
Bromine	35
Iodine	53
Tungsten	74
Lawrencium	103

13. Write down the names and formulae of

(a) three oxides
(b) two alkalies
(c) two acids
(d) two salts

14. Write down the name and the formula for a molecule of each of the following substances:

(a) two gaseous elements
(b) two gaseous compounds
(c) one noble gas
(d) two liquid compounds

Chapter 5

1. You may have used the 'drop of a drop' method to obtain information about the size of molecules. This question is concerned with an investigation of the size of oleic acid molecules.

A student filled a burette with oleic acid and slowly ran out 3 cm^3, counting the drops produced. They numbered 60. He then ran out one more drop onto a microscope slide. He had been provided with a glass rod drawn out to a fine solid point, and he touched this against the drop of oleic acid and withdrew a 'drop of a drop'. He wiped the point on a cloth and then withdrew another drop of a drop. He managed to do this 100 times before the drop had disappeared.

The student lightly dusted the surface of a large tray of water with fine white powder. He took a 'drop of a drop' of oleic acid and touched the tip of the glass rod on the water surface. The white powder rushed away from the point of contact leaving a clear area of 250 cm^2.

(a) What is the volume of one drop from the burette?
(b) What is the volume of a 'drop of a drop'?
(c) If the clear area of 250 cm^2 on the water surface is caused by a layer of oleic acid molecules, what is the thickness of the oleic acid film?
(d) What information does this experiment give about the size of oleic acid molecules? Word your answer carefully.

2. This question is concerned with a second investigation of the size of an oleic acid molecule.

A solution containing 0.4 cm^3 of oleic acid in a litre of pentane was put in a burette and 50 drops were run out. The burette reading changed by 2 cm^3.

A drop of this solution was allowed to fall on a water surface which had been lightly dusted with white powder. The white powder rushed away from the point where the drop had fallen on the water. The area of clear surface formed was 80 cm^2.

(a) What is the volume of one drop of the solution?
(b) What volume of oleic acid is contained in one drop of solution?
(c) If the clear area of 80 cm^2 on the water surface is caused by a layer of oleic acid molecules, what is the thickness of the oleic acid film?
(d) What information does this experiment give about the size of oleic acid molecules? Word your answer carefully.
(e) Make a list of the properties that you think pentane should have if it is to be used in this experiment.

3. Compare the values obtained for the thickness of the oleic acid layer in questions 2 and 3. If one of the results gives a value for the size of an oleic acid molecule in one dimension, which of the two values will do this?

4. Some properties of lithium, calcium and rubidium are given in the following table.

Element	*Volume of 1 mole of atoms*	*Diameter of atom*
Lithium	13 cm^3	3×10^{-8} cm
Calcium	26 cm^3	4×10^{-8} cm
Rubidium	56 cm^3	5×10^{-8} cm

(a) If lithium atoms are regarded as cubes of side 3×10^{-8} cm, what is the volume of one lithium atom?
(b) Use the volume of one lithium atom from (a) to calculate the number of lithium atoms in 13 cm^3 of lithium (1 mole of lithium atoms).
(c) Calculate in a similar way the number of calcium atoms in 26 cm^3 of calcium and the number of rubidium atoms in 56 cm^3 of rubidium.
(d) Your answers in parts (b) and (c) should be similar and very large numbers. What is the number that is being calculated and what is its correct value?

5. (a) A lorry load of sand was found to have the dimensions:

500 cm × 200 cm × 100 cm

What is the volume of sand in one lorry load?
(b) One cm^3 of sand was taken from the lorry and the number of grains of sand in this volume was counted. There were found to be 10 000 grains of sand. How many grains of sand were there in one lorry load?
(c) A chemist wants to make a model of 1 mole of carbon atoms using a grain of sand to represent one carbon atom. How much sand will it be necessary to order for delivery by lorry if this model is to be constructed?

6. The following table gives information about five elements.

Element	*Physical state at room temperature*	*Packing of atoms at room temperature*	*Electrical conductivity of element at room temperature*
A	liquid	fairly close	poor
B	liquid	fairly close	good
C	solid	close	poor
D	solid	close	good
E	gas	far apart	poor

Which element, or elements, in this table are metals? Explain briefly your reasons for making this choice.

Questions 7–15 are concerned with certain elements for which information is given in the table below.

Element	*M.p., °C*	*B.p., °C*	*Volume of 1 mole of atoms at room temperature and pressure, cm^3*
A	−189	−186	24 000
B	−100	−35	12 000
C	−7	58	26
D	30	2240	12
E	44	280	17
F	660	2450	10
G	3410	5930	10

Use this information to answer the questions which follow.

7. Which elements have a structure at room temperature which is mostly empty space? Explain how you have made your selection.

8. Which elements at room temperature will have a structure in which the atoms are packed in a close and regular way? Explain how you have made your selection.

9. Which element at room temperature will have a structure in which the particles are fairly closely packed but in which the packing is not completely regular? Explain how you have made your choice.

10. Which elements are composed of particles which only diffuse very slowly at room temperature? Explain how you have made your selection.

11. Which elements will completely fill any container into which they are put? Explain how you have made your selection.

12. Which two substances will be far more compressible than any of the others? Explain why you have selected these two.

13. Which elements will be gaseous at 100 °C?

14. Which elements will be solid at 0 °C?

15. Which two elements will diffuse readily at room temperature? Explain how you have made your selection.

Chapters 6 and 7

1. Modern chemistry syllabuses tend to stress the importance of principles and theory rather than factual information. Facts are interesting, useful and important as well and it is easy to acquire facts by the activity of data gathering.

   Prepare a number of pieces of paper or thin card about half the size of a postcard. Each will be a data card for one element and in time a large collection can be built up. If possible use a different colour ink for each group in the Periodic Table. Since hydrogen does not clearly belong to any group, use a separate colour.

   Each card should be set out in the same format with the elemental symbol and name written boldly in the middle.

   On the backs of the cards further information can be added later in the course, e.g. formula of chloride, formula of oxide, whether the chloride is covalent or ionic, etc.

   It is useful to compile the cards in the following order:

   (a) the first twenty elements
   (b) rubidium and caesium
   (c) bromine and iodine
   (d) strontium and barium
   (e) iron, copper, zinc, silver
   (f) any other elements of interest

> Relative atomic mass
> Melting point
> Boiling point
>
> **SODIUM**
> **Na**
>
> Density
> Price per kg
> Three uses
> Occurrence
> Atomic number

2. Write out several blank periodic tables as shown below with spaces for the first eighteen elements (hydrogen to argon).

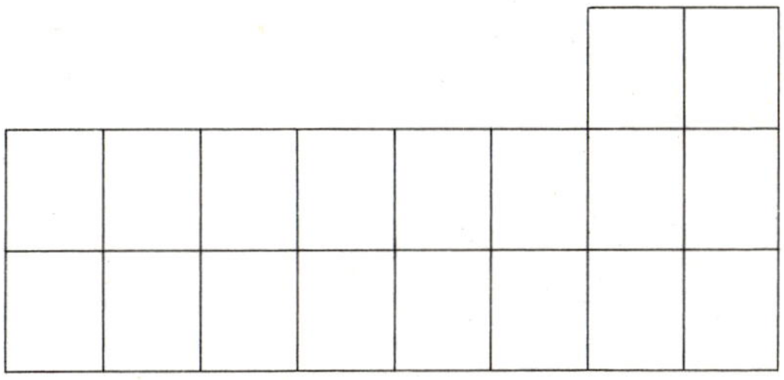

Use these blank periodic tables to answer the questions which follow.

(a) Show the positions of metals, non-metals and the diagonal.
(b) Show the elements which react with cold water to produce hydrogen.
(c) Show the elements which have oxides and hydroxides which dissolve in water to form alkaline solutions.
(d) Show the elements which form acidic oxides.
(e) Show the elements which can form ions with charge +1.
(f) Show the elements which can form ions with charge +2.
(g) Show the elements which can form ions with charge +3.
(h) Show the elements which form chlorides with very high boiling points (above 1000 °C).
(i) Show the elements which form chlorides which are liquid or gaseous at room temperature.
(j) Show the elements which are called noble gases.

3. Write out a Periodic Table showing at least the first eighteen elements (hydrogen to argon) and use this to explain the meaning of the following terms:

   (a) group
   (b) period
   (c) diagonal

4. Decide from the descriptions given below on which side of the diagonal each of the following elements lies. In each case answer either 'left' or 'right'.

   (a) Element A is a non-metal.
   (b) Element B has a low boiling point.
   (c) Element C is a good conductor of heat and electricity.
   (d) Element D forms positive ions in its compounds.
   (e) Element E forms an oxide which dissolves in water to give an acidic solution.
   (f) Element F forms compounds with metallic elements and also forms compounds with non-metallic elements.
   (g) Element G forms a basic oxide.
   (h) Element H combines with elements from the right of the diagonal to form compounds with high boiling points which contain ions.

5. Each of the following questions is in the form of a brief description of an element followed by three suggested elements that this might be. Choose the element from these suggestions which best fits the description.

(a) Element J is metallic and forms a chloride with formula JCl_2.	sodium, magnesium, sulphur.
(b) Element K forms an acidic oxide.	lithium, calcium, phosphorus.
(c) Element L reacts with cold water to produce hydrogen.	potassium, silicon, chlorine.
(d) Element M forms a liquid chloride at room temperature.	sodium, magnesium, sulphur.
(e) Element N forms a chloride with a high boiling point.	magnesium, silicon, phosphorus.
(f) Element O is an unreactive gas.	phosphorus, chlorine, argon.
(g) Element P forms a liquid compound with hydrogen of formula H_2P.	sodium, oxygen, chlorine.

6. The following table gives information about six elements.

Element	*Physical state at room temperature*	*Nature of oxide*	*Does element form a compound with magnesium*	*Does element form a compound with chlorine*
A	gas	no oxide	no	no
B	liquid	solid, basic	no	yes
C	solid	solid, acidic	yes	yes
D	solid	solid, basic	no	yes
E	liquid	solid, acidic	yes	yes
F	solid	gaseous, acidic	yes	yes

Which element, or elements, in this table are metals? Explain briefly your reasons for making this choice.

7. You probably know that many chemical and physical properties of elements vary in a periodic way with the relative atomic mass of the element. The following table gives the approximate date at which each of the first nineteen elements in the Periodic Table was isolated as the free element. Plot a graph of date of isolation (vertical axis) against atomic number and then answer the questions which follow.

Element	*Date of isolation*	*Element*	*Date of isolation*
H	1766	Na	1807
He	1895	Mg	1808
Li	1817	Al	1824
Be	1828	Si	1809
B	1808	P	1669
C	Known to ancients	S	Known to ancients
N	1775	Cl	1774
O	1775	Ar	1894
F	1886	K	1808
Ne	1900		

   (a) Discuss the periodic trends, if any, which show in this graph.
   (b) Several very metallic elements (Li, Na, Mg, Al, K) were isolated in the free state shortly after 1800. Can you find any reason for this?
   (c) B and Si were also isolated shortly after 1800. Can you suggest any reason for this?
   (d) Predict the date of isolation of calcium (atomic number 20).
   (e) Can you suggest why FOUR of the elements were not isolated until the end of the nineteenth century?
   (f) Predict the date of the following elements and explain briefly how you have made your prediction:

   Element A. This element is metallic. Its abundance in the earth's crust is low, although it occurs in isolated deposits at high concentration. It does not react with water or oxygen – in fact it is a very unreactive element.

   Element B. This element is metallic. It is found in many deposits in high concentration. It reacts with water and with oxygen. Its oxide does not react with carbon.

Element C. This element is non-metallic and is one of the most reactive of all elements. It is fairly abundant and is found at high concentration in certain deposits.

Element D. This element is metallic. It is abundant and is found in high concentration in certain deposits. It reacts with water containing dissolved oxygen. Its oxide reacts with carbon.

8. The melting and boiling points of the chlorides of lithium to potassium are given in the table below. Use this information to answer the questions which follow.

Atomic number	*Formula*	*M.p., °C*	*B.p., °C*
3	LiCl	610	1350
4	$BeCl_2$	405	487
5	BCl_3	−107	12
6	CCl_4	−23	77
7	NCl_3	−27	71
8	OCl_2	−20	2
9	FCl	−154	−101
10	chloride of neon not yet isolated		
11	NaCl	801	1465
12	$MgCl_2$	712	1418
13	$AlCl_3$	190	180
14	$SiCl_4$	−68	57
15	PCl_3		
16	SCl_2	−80	59
17	ClCl	−101	−35
18	chloride of argon not yet isolated		
19	KCl		

(a) On ONE sheet of graph paper, plot melting point and boiling point (vertical axis) against atomic number.
(b) Do you consider that the melting and boiling points of chlorides are 'periodic' properties?
(c) What is unexpected about the information given for $AlCl_3$? Can you suggest a reason for this?
(d) Is there anything 'periodic' about the number of atoms of chlorine that combine with one atom of each element?
(e) Which chlorides will be liquid at normal temperature?
(f) Which chlorides will be gaseous at normal temperature?
(g) Which chloride has the longest temperature range over which it is liquid?
(h) Predict the melting and boiling points of PCl_3.
(i) Predict the melting and boiling points of KCl.
(j) Is it possible to predict what the melting and boiling points of the chlorides of neon and argon will be if they are ever isolated? Explain your answer briefly.
(k) Suggest the chloride in which there is probably the weakest attraction between individual molecules.

9. The following table gives information about the approximate abundance and price of the first twenty elements in the Periodic Table. Use this information to answer the questions which follow.

Element	*Abundance in Earth's crust in parts per million by mass*	*Price in pence/kg*
H	1000	350
He	1	4000
Li	50	3000
Be	4	13 000
B	3	5500
C	400	65
N	47	40
O	470 000	50
F	500	130
Ne	1	17 000
Na	30 000	170
Mg	20 000	140
Al	85 000	110
Si	280 000	115
P	1000	200
S	500	30
Cl	300	40
Ar	1	65
K	26 000	850
Ca	36 000	750

The Earth's crust includes the atmosphere.

(a) Plot a graph of log abundance (vertical axis) against atomic number. On the same sheet of graph paper plot log price (vertical axis) against atomic number. (The abundance of the rare gases is less than 1 ppm in each case, but the value is put at 1 ppm to assist in plotting the graph.)

Comment on any periodic trends that you can see in the two graphs.

(b) Arrange the seven most abundant elements in the table in order of abundance and for these elements calculate their percentage contribution to the mass of the Earth's crust.

What percentage of the Earth's crust is made up of elements other than the seven most abundant? What contribution to this percentage is made by an element not in the table, iron, whose abundance is 50 000 ppm?

10. Can you account for the following observations that could be made about the table in question 9?

(a) Oxygen is 10 000 times more abundant than nitrogen but both elements are almost the same price.
(b) Sodium and potassium are of almost equal abundance but potassium is much more expensive than sodium.
(c) The abundance of sulphur and chlorine is low but they are both among the cheapest of elements.
(d) Argon is very rare but is relatively inexpensive.
(e) The price of no element is less than 30 p/kg.

11. The following table gives information about the specific heat capacity of several solid elements – the specific heat capacity is the number of joules required to raise the temperature of 1 g of the element by 1 °C.

Element	*Relative atomic mass*	*Specific heat capacity* (joules/g/°C)
Li	6.9	3.30
Be	9	1.90
B	10	1.30
C	12	0.71
Na	23	1.25
Mg	24	1.05
Al	27	0.92
Si	28	0.67
P	31	0.76
S	32	0.76
K	39	0.76
Ca	40	0.63
Sc	45	0.55
Fe	56	0.46
Cu	64	0.38
Rb	85	0.34

(a) Plot a graph of specific heat capacity (vertical axis) against relative atomic mass. Is the curve 'periodic'?
(b) Plot a graph of

specific heat capacity (vertical axis) against

$$\frac{1}{\text{relative atomic mass}}$$

Suggest an equation which connects specific heat capacity and relative atomic mass for most elements.
(c) An element of specific heat capacity 0.50 joules/g/°C forms a chloride in which 17 g of the element combines with 1 mole of chlorine atoms. Suggest an accurate value for the relative atomic mass of the element.

Chapter 8

1. The following table gives information about the atoms of certain elements. Some numbers in the table have been replaced by letters. For what number does each of these letters stand?

Number of protons in nucleus of atom	*Number of neutrons in nucleus of atom*	*Total number of electrons around nucleus of atom*	*Relative atomic mass*
22	26	A	B
C	74	D	127
34	E	F	79
G	H	82	207
J	30	26	K
L	20	M	39

2. The following table gives information about various atoms. Select from this table the pair of atoms which are isotopes of the same element and explain briefly how you have made your choice.

Atom	*Number of protons*	*Number of neutrons*	*Total number of electrons*	*Relative atomic mass*
A	45	35	45	80
B	35	45	35	80
C	80	80	80	160
D	45	35	45	80
E	35	44	35	79
F	44	35	44	79

3. Table 8.2 gives a way of showing the numbers of protons, neutrons and electrons in an atom. Draw similar diagrams for atoms of the following elements.

Element	*Atomic number*	*Relative atomic mass*
Carbon	6	12
Neon	10	20
Neon	10	22
Magnesium	12	24
Argon	18	40
Iron	26	56
Strontium	38	88

4. The following table gives information about the electronic configuration of various elements. Use Use this information to complete the final two columns in the table.

Element	Electronic configuration	Atomic number of element	Element's group number in Periodic Table
A	2.1		
B	2.3		
C	2.8.1		
D	2.8.4		
E	2.8.6		
F	2.8.8.2		

5. Use the information in the following table to select two pairs of elements which come from the same group in the Periodic Table.

Element	Atomic number	Relative atomic mass	Electronic configuration
A	9	19	2.7
B	11	23	2.8.1
C	12	24	2.8.2
D	15	28	2.8.5
E	17	35	2.8.7
F	18	40	2.8.8
G	19	39	2.8.8.1

6. Most of the information is missing from the following table. Make a copy of this table and include in it all the missing information.

Total number of electrons around nucleus of atom	Number of outer electrons	Atomic number of element	Electronic configuration of element	Group number of element in Periodic Table
			2.8.3	
		9		
19				

7. Look at Fig. 8.2 and explain what information this gives about the electronic configuration of potassium.

8. Look at Fig. 8.3 and explain what information this gives about the electronic configurations of the elements from hydrogen to magnesium.

Chapter 9

1. The following table gives information about some chlorides of some elements in the period potassium–krypton.

	KCl	$CaCl_2$	$GaCl_3$	$GeCl_4$	$AsCl_3$	Se_2Cl_2	$BrCl$
Group number of element combined with chlorine	1	2	3	4	5	6	7
Number of outer electrons in atom of element combined with chlorine	1	2	3	4	5	6	7
Melting point of chloride, °C	780	770	78	−50	−16	−85	−50
Boiling point of chloride, °C	1400	1600	200	87	122	130	5

Select chlorides from this table to answer the questions which follow.

(a) Which chlorides do you expect to be liquid at room temperature?
(b) Which chloride do you expect to be gaseous at room temperature?
(c) The formulae of all the chlorides except one fit a definite pattern. Which chloride does not fit this pattern in formula? Explain briefly your reason for making this choice.
(d) Which chlorides are likely to be composed of ions?
(e) Which chlorides are likely to be poor conductors of electricity when in the liquid state?
(f) Which chlorides are likely to have high molar heats of vaporization?
(g) Select two chlorides from the table and draw diagrams to show what happens to each atom's outer electrons when bonding takes place.

2. This question is concerned with the information about various chlorides given in Figs. 9.11 to 9.14.

(a) List the eight chlorides which have the highest electrical conductivity when molten.
(b) List the eight chlorides with the highest melting points.
(c) List the eight chlorides with the highest boiling points.
(d) List the eight chlorides with the highest molar heats of vaporization.
(e) Compare your answers in parts (a) to (d) and decide which eight chlorides are most likely to be composed of ions.
(f) Write the formulae of the chlorides you have chosen in part (e) in ionic form (e.g. Cs^+Cl^-, $Ba^{2+}2Cl^-$).
(g) Select one ionic chloride in which the atom of the element combined with chlorine has one electron in the outer shell. Draw a diagram to show what happens to outer electrons when this ionic chloride is formed from atoms of its elements.
(h) Select one ionic chloride in which the atom of the element combined with chlorine has two electrons in the outer shell. Draw a diagram to show what happens to outer electrons when this ionic chloride is formed from atoms of its elements.

(i) Select one ionic chloride in which the atom of the element combined with chlorine has three electrons in the outer shell. Draw a diagram to show what happens to outer electrons when this ionic chloride is formed from atoms of its elements.
(j) Select one ionic chloride and sketch the arrangement of the ions in the crystal.

3. The following table gives information about some fluorides of some elements in the period lithium–neon. Use this information to answer the questions which follow.

	LiF	*BeF*$_2$	*BF*$_3$	*CF*$_4$	*NF*$_3$	*OF*$_2$	*FF*
Number of outer electrons in atom of element combined with fluorine	1	2	3	4	5	6	7
Melting point of fluoride, °C	870	800	−128	−184	−217	−224	−220
Boiling point of fluoride, °C	1670	1330	−102	−128	−120	−145	−188
Electrical conductivity of fluoride when liquid	good	poor	poor	poor	poor	poor	poor

(a) Which fluorides are gaseous at room temperature?
(b) Which fluoride or fluorides are likely to be composed of ions? Explain briefly how you have made your choice.
(c) Select a fluoride in which the bonding is ionic and draw a diagram to show what happens to each atom's outer electrons when ionic bonding takes place.
(d) Select a fluoride in which the bonding is non-ionic (covalent). Draw a diagram to show what happens to each atom's outer electrons when bonding takes place. It is not necessary to answer this part of the question unless you have already studied non-ionic bonding.

4. The following table gives information about some oxides of some elements in the period lithium–neon. Use this information to answer the questions which follow.

	Li$_2$*O*	*BeO*	*B*$_2$*O*$_3$	*CO*$_2$	*N*$_2$*O*$_4$	*OO*	*F*$_2$*O*
Number of outer electrons in atom of element combined with oxygen	1	2	3	4	5	6	7
Melting point of oxide, °C	1700	2500	450	−57	−9	−219	−224
Boiling point of oxide, °C	2300	3900	1500	−78	21	−183	−145
Electrical conductivity of oxide when liquid	good	poor	poor	poor	poor	poor	poor

(a) Which oxides are gaseous at room temperature?
(b) Which oxide is most likely to be liquid at room temperature?
(c) Which oxide will sublime on heating? (Sublimation takes place when a solid turns to vapour before the melting point has been reached.)
(d) Which oxide or oxides are likely to be composed of ions? Explain briefly how you have made your choice.
(e) Select an oxide in which the bonding is ionic and draw a diagram to show what happens to each atom's outer electrons when ionic bonding takes place.
(f) Select an oxide in which the bonding is non-ionic (covalent). Draw a diagram to show what happens to each atom's outer electrons when bonding takes place. It is not necessary to answer this part of the question unless you have already studied non-ionic bonding.

5. Use the information in Table 9.2 to predict the formulae of the following compounds:

(a) sodium carbonate
(b) potassium hydroxide
(c) lithium nitrate
(d) calcium bicarbonate
(e) aluminium oxide
(f) barium sulphate
(g) lead(II) bromide
(h) copper(II) sulphide
(i) ammonium phosphate
(j) silver chloride
(k) Magnesium acetate
(l) iron(II) fluoride
(m) iron(III) fluoride
(n) copper(I) oxide
(o) potassium permanganate
(p) silver iodide
(q) potassium dichromate
(r) sodium sulphite

Chapter 10

1. Question 1 on Chapter 9 has a table giving information about chlorides of elements in the period potassium–krypton. Use this table to answer the following questions.

(a) In which chlorides is the bonding covalent? Explain briefly why you have chosen these chlorides.
(b) Draw diagrams for the covalent chlorides in the table to show what happens to each atom's outer electrons when covalent bonding takes place.
(c) Predict the shape of the molecule for each of these covalent chlorides and explain briefly why you have chosen each shape.

2. Question 3 on Chapter 9 has a table giving information about fluorides of elements in the period lithium–neon. Use this table to answer the following questions.

(a) In which fluorides is the bonding covalent? Explain briefly why you have chosen these fluorides.
(b) One of the covalent fluorides in this table has a giant molecular structure. Which fluoride is

this? Explain briefly your reason for making this choice.
(c) Select four fluorides from the table in which the bonding is covalent. Draw diagrams to show what happens to each atom's outer electrons when covalent bonding takes place.
(d) Predict the shape of the molecule for each of the fluorides you have chosen in part (c) and explain briefly how you have made these predictions.

3. Question 4 on Chapter 9 has a table giving information about oxides of elements in the perion lithium–neon. Use this information to answer the following questions.

   (a) In which oxides is the bonding covalent? Explain briefly why you have chosen these oxides.
   (b) Two of the covalent oxides in this table have giant molecular structures. Which are these two oxides? Explain briefly your reason for making this choice.
   (c) Select three oxides from the table in which the bonding is covalent. Draw diagrams to show what happens to each atom's outer electrons when covalent bonding takes place.
   (d) Predict the shape of the molecule for each of the oxides you have chosen in part (c) and explain briefly how you have made these predictions.

4. This question is concerned with Figure 10.4. Choose one compound from each small square in the figure. For each of the compounds you have chosen:

   (a) write down the formula and say whether the bonding is ionic or covalent.
   (b) draw a diagram to show what happens to each atom's outer electrons when the compound is formed from atoms of its elements.
   (c) if the compound is covalent, predict the shape of its molecules and explain briefly why you have chosen the shape.

5. This question is concerned with the carbon compounds shown on page 97. Choose five compounds from this figure and in each case:

   (a) draw a diagram to show what happens to each atom's outer electrons when the compound is formed from its elements.
   (b) predict and sketch the shape of a molecule of the compound.

6. This question is concerned with the carbon compounds shown in Figure 10.6. Choose five compounds from this figure and in each case:

   (a) draw a diagram to show what happens to each atom's outer electrons when the compound is formed from its elements.
   (b) predict and sketch the shape of a molecule of the compound.

7. This question is concerned with the compounds shown in Figure 10.7. Choose five compounds from this figure and in each case:

   (a) draw a diagram to show what happens to each atom's outer electrons when the compound is formed from its elements.
   (b) predict and sketch the shape of a molecule of the compound.

8. This question is concerned with the compounds shown in Figure 10.8. Choose two compounds from this figure and in each case:

   (a) draw a diagram to show what happens to each atom's outer electrons when the compound is formed from its elements.
   (b) predict and sketch the shape of a molecule of the compound.

9. Make a copy of three of the structures shown in Figure 10.10 and on these show clearly how each atom's outer electrons are used in covalent bonding.

Questions 10–23 are concerned only with those elements whose symbols appear in the incomplete Periodic Table given below.

						H	He
			C		O		
	Mg		Si	P	S	Cl	

Choose from these elements the ones which fit the following descriptions:

10. Three elements, solid at room temperature, which can exist in two polymorphic (allotropic) forms.

11. The non-metal whose oxide is a high melting solid in which one mole of atoms of the element is combined with two moles of oxygen atoms.

12. Four elements whose molecules are far apart at room temperature.

13. The element whose atoms are packed most closely at room temperature.

14. Two elements which burn in oxygen to form oxides in which the molecules are far apart at room temperature.

15. Two non-metals with a giant molecular structure at room temperature.

16. An element whose chloride is composed of a giant structure of ions at room temperature. What ions are present in this giant structure?

17. Four elements which will fill any container into which they are put.

18. The element with the greatest volume for one mole of atoms at room temperature.

19. The element which will combine with oxygen to form an oxide with a giant structure of ions at room temperature.

20. The two elements which will form molecular crystals at room temperature.

21. The element which could be readily obtained in the form of a strong wire. How would you attempt to convert a lump of this element into a thin wire?

22. The solid element in which certain electrons are able to move freely in all directions through the structure.

23. The element which will form the chloride with the highest boiling point.

Questions 24–33 are concerned with the substances for which information is given in the table below.

Substance	*M.p., °C*	*B.p., °C*	*Solubility in water*	*Electrical conductivity of aqueous solution*	*Electrical conductivity of solid*	*Electrical conductivity of molten substance*
A	1700	2500	insoluble	insoluble	poor	poor
B	800	1470	soluble without reaction	good	poor	good
C	500	950	insoluble	insoluble	poor	good
D	100	890	reacts to give solution, H_2 evolved	good	good	good
E	−23	77	insoluble	insoluble	poor	poor

Choose substances from this table to answer the questions which follow.

24. Which two substances are most likely to be composed of a giant structure of ions at room temperature? Explain why you have made this choice.

25. Which substance is most likely to be a metallic element? Explain why you have chosen this substance.

26. Which substance is most likely to be composed of simple molecules at room temperature? Explain why you have chosen this substance.

27. Which substances are solids at room temperature and pressure?

28. Two substances are compounds formed between a metal and a non-metal. Which two substances are these? What properties have you used to make this choice?

29. Which substance is most likely to be silicon dioxide? Explain why you have chosen this substance.

30. Which substances will be composed of a regular arrangement of particles at room temperature? Explain why you have chosen these substances.

31. Which substances will be liquid at 1000 °C?

32. Which substance will probably have the lowest molar heat of vaporization? Explain why you have chosen this substance.

33. Which will probably be the most easily compressible substance? How will the compressibility of this substance compare with that of oxygen? Explain why you have given these answers.

Questions 34–42 are concerned with the following substances.

A graphite
B sulphur
C chlorine
D diamond
E phosphorus

Choose from the above list the ONE which in each case most closely fits the descriptions given below.

34. Each atom is bonded to four others in a 3-dimensional network. All bond lengths and bond angles are identical.

35. Atoms bond strongly to form molecules containing two atoms. Forces between molecules are weak.

36. Atoms are arranged in flat layers with strong bonding between atoms within each layer. The forces between layers are weak and the distance between layers is relatively large.

37. Each atom bonds strongly to two other atoms to form a ring of eight atoms. Forces between different rings are relatively weak.

38. Each atom bonds strongly to three other atoms to form a molecule containing four atoms, with these atoms at the corners of a tetrahedron. Forces between molecules are relatively weak.

39. This substance does not combine with oxygen when heated strongly in air.

40. This substance has a molecular structure with the weakest forces between molecules.

41. This substance burns in air to form a solid oxide.

42. This substance has a structure which closely resembles the structure of silicon.

Questions 43–54 are concerned with the fluorides whose melting points and boiling points are given in the following table.

	NaF	MgF_2	AlF_2	SiF_4	PF_5	SF_6
Melting point, °C	990	1400	1040	−80	−90	−56
Boiling point, °C	1700	2240	1290	−70	−85	−64

Five possible structures for substances are given below:

A giant structure of ions
B freely moving ions
C giant molecule (giant structure of atoms)
D closely packed molecules
E widely spaced molecules

Choose from these possible structures the ONE which in each case is most likely to be the structure of:

43. NaF at room temperature.

44. NaF at 1000 °C.

45. MgF_2 at room temperature.

46. MgF_2 at 1000 °C.

47. AlF_3 at room temperature.

48. AlF_3 at −100 °C.

49. SiF_4 at room temperature.

50. SiF_4 at −100 °C.

51. PF_5 at room temperature.

52. PF_5 at −80 °C.

53. SF_6 at room temperature.

54. SF_6 at −70 °C.

55. Questions 55 and 56 are concerned with the boiling point and molar heat of vaporization of various elements and compounds.

Substance	*Formula*	*B.p., °C*	*Molar heat of vaporization at the boiling point, kJ/mole*
Fluorine	F_2	−188	1.5
Chlorine	Cl_2	−35	4.9
Bromine	Br_2	58	7.3
Sodium chloride	NaCl	1470	41
Hydrogen chloride	HCl	−85	3.9
Potassium chloride	KCl	1400	39
Hydrogen bromide	HBr	−67	4.2
Sodium bromide	NaBr	1390	38
Silver chloride	AgCl	1560	43
Caesium chloride	CsCl	1300	38
Water	H_2O	100	9.8
Hydrogen sulphide	H_2S	−62	4.5
Argon	Ar	−186	1.9
Nitrogen	N_2	−196	1.3
Sulphur dioxide	SO_2	−10	6.0
Sodium	Na	890	25
Magnesium	Mg	1110	31
Lithium	Li	1330	35

(a) Plot a graph of molar heat of vaporization (vertical axis) against boiling point. Your temperature scale should run from −300 °C to +1600 °C. Put the formula of each substance close to its point on the graph.
(b) Put circles of a certain colour around points on the graph which you are certain correspond to substances with molecular structures (separate molecules).
(c) Put circles of another colour around points on the graph which you are certain correspond to substances with a giant structure of atoms or ions. Indicate on the graph paper the colours you are using to distinguish molecular and giant structures.
(d) Use the graph to predict the structures of those remaining substances about which you are uncertain. Can you find any information in reference books to check your predictions?
(e) Extrapolate the curve you have drawn in (a) to meet the temperature axis. Is there any significance about the temperature at which this happens?

(f) Can you represent the relationship between boiling point and molar heat of vaporization by a simple equation?

56. Include the following substances on the graph you have drawn in question 55.

Substance	*Assumed formula*	*B.p., °C*	*Molar heat of vaporization at the boiling point, kJ/mole*
Iodine	I	183	5.0
Sulphur	S	445	2.5
Phosphorus	P	280	3.1
Selenium	Se	685	3.3
Mercury	Hg	356	14
Xenon	Xe	−109	3.0
Iron(III) chloride	$FeCl_3$	315	6.0
Aluminium bromide	$AlBr_3$	263	5.5
Tin tetrachloride	$SnCl_4$	114	8.3

(a) By what factor would the molar heat of vaporization of iodine have to be multiplied to make the iodine point lie close to the curve drawn for the data in question 55? What is the significance of this factor?
(b) Predict the molecular formulae of sulphur, phosphorus, selenium, mercury, xenon, iron(III) chloride, aluminium bromide and tin tetrachloride. Attempt to check your predictions using a reference book.

57. The melting and boiling points of certain hydrocarbons are given in the following table.

Hydrocarbon		*M.p., °C*	*B.p., °C*
Methane,	CH_4	−182	−160
Ethane,	C_2H_6	−183	−88
Propane,	C_3H_8	−188	−41
Butane,	C_4H_{10}	−135	+1
Pentane,	C_5H_{12}	−129	38
Hexane,	C_6H_{14}	−94	69
Octane,	C_8H_{18}	−58	126
Nonane,	C_9H_{20}	−53	151

(a) Plot a graph of boiling point (vertical axis) against number of carbon atoms in the molecule of the hydrocarbon. Use this graph to predict the boiling points of heptane, C_7H_{16}, and decane, $C_{10}H_{22}$.
(b) Plot a graph of melting point (vertical axis) against number of carbon atoms in the molecule of the hydrocarbon. Use this graph to predict the melting points of heptane, C_7H_{16}, and decane, $C_{10}H_{22}$.
(c) Which of these hydrocarbons will be liquid at room temperature?
(d) Which of these hydrocarbons will be composed of widely separated molecules at room temperature?

Chapter 11

1. Calculate the mass of 1 mole of molecules of each of the following substances:
(a) hydrogen, H_2
(b) carbon dioxide, CO_2
(c) chlorine, Cl_2
(d) oxygen, O_2
(e) ammonia, NH_3
(f) argon, Ar
(g) hydrogen chloride, HCl
(h) sulphur dioxide, SO_2

2. Calculate the number of moles of molecules present in each of the following masses:
(a) 1 g of hydrogen, H_2
(b) 88 g of carbon dioxide, CO_2
(c) 142 g of chlorine, Cl_2
(d) 32 g of oxygen, O_2
(e) 1.7 g of ammonia, NH_3
(f) 8 g of argon, Ar
(g) 73 g of hydrogen chloride, HCl
(h) 16 g of sulphur dioxide, SO_2

3. The following table gives information about the volume occupied by 1 g of various gases at room temperature and pressure.

Gas	*Volume occupied by 1 g in cm^3*
Chlorine, Cl_2	334
Carbon dioxide, CO_2	540
Argon, Ar	600
Hydrogen chloride, HCl	653
Oxygen, O_2	755
Ammonia, NH_3	1390

Plot a graph of volume occupied by 1 g of gas (vertical axis) against molar mass of gas. Use this graph to answer the questions which follow.

(a) Predict the volumes of 1 g of methane, CH_4, and 1 g of hydrogen fluoride, HF.
(b) 1 g of a gaseous compound formed between sulphur and oxygen occupies 375 cm^3 at room temperature and pressure. Predict the molar mass and hence the formula of the compound.
(c) A gaseous compound formed between boron and hydrogen was thought to have the formula BH_3. It was found that 1 g of this compound occupied 860 cm^3. What does this suggest about the formula of the compound?
(d) When ethanol (a liquid of molecular formula C_2H_5OH) is vaporized and passed over a hot catalyst, a gas is formed. It is found that 1 g of this gas also occupies 860 cm^3. What information does this give about the action of the hot catalyst on ethanol vapour?

4. A glass syringe contains 50 cm^3 of air at 19 °C and 1 atm pressure. Calculate the volume of air in the syringe at

(a) 19 °C and 2 atm
(b) 19 °C and 0.5 atm
(c) 100 °C and 1 atm
(d) 0 °C and 1 atm
(e) 100 °C and 2 atm

5. A glass syringe containing 100 cm^3 of air at room temperature and pressure was sealed and clamped in a vertical position with the plunger uppermost. Masses of 2 kg were gradually loaded onto the plunger to increase the pressure of the air in the syringe. A load of 2 kg was known from the diameter of the syringe piston to be equivalent to an extra pressure of 0.25 atm. The volume of air in the syringe was recorded at various pressures. The results are shown below.

Pressure of air in syringe in atm	*Volume of air in syringe in cm^3*
1.0	100
1.25	80
1.5	66
1.75	56
2.0	49
2.25	40
2.5	0

Are these results in agreement with Boyle's law? Explain your answer clearly.

6. 75 cm^3 of air was sealed in a glass syringe at 20 °C. The syringe was then placed in a large beaker of water which was gradually increased in temperature to 100 °C. The volume of air in the syringe was recorded at various temperatures. The results are shown below.

Temperature of air in °C	*Volume of air in cm^3*
20	75
40	80
60	85
80	90
100	95

(a) Plot a graph of volume of air (vertical axis) against temperature. The temperature scale should run from −300 °C to +100 °C.
(b) What would the volume of air have become if the syringe had been cooled in an ice/water mixture?
(c) These readings suggest that the volume of air in the syringe might become zero at a certain low temperature. Use your graph to predict this temperature. What would happen to the air in the syringe before this temperature could be reached?
(d) Are the experimental results in agreement with Charles' law? Explain your answer clearly.

7. 100 cm^3 of a certain gaseous hydrocarbon (a compound composed of carbon and hydrogen only) was found to need 300 cm^3 of oxygen for complete combustion to carbon dioxide and water. The carbon dioxide produced by this combustion had a volume of 200 cm^3.

(a) If there are *n* molecules of the hydrocarbon in 100 cm^3, how many molecules of oxygen are there in 300 cm^3 of oxygen?
(b) If there are *n* molecules of the hydrocarbon in 100 cm^3, how many molecules of carbon dioxide are there in 200 cm^3 of carbon dioxide?
(c) Use your answers in (a) and (b) to help decide the formula of the hydrocarbon. Show clearly how you have obtained your answer.
(d) Draw a diagram to show what happens to each atom's outer electrons when a molecule of the hydrocarbon is formed from atoms of its elements.
(e) Predict and sketch the shape of a molecule of the hydrocarbon. Explain how you have made this prediction.

8. 200 cm^3 of a gaseous hydrocarbon was burnt completely using 600 cm^3 of oxygen (this was an excess of oxygen). The volume of gas produced by this combustion was 500 cm^3 which was reduced to 100 cm^3 after removing carbon dioxide with alkali. The liquid water formed by the combustion was of negligible volume.

(a) Use this information to decide the formula of the hydrocarbon, showing clearly how you have obtained your answer.
(b) 1000 cm^3 of the hydrocarbon has a mass of 1.08 g (measurement at room temperature and pressure). Is this what would be expected from the formula you have decided in (a)?

9. (a) Calculate the mass of one mole of molecules of each of the following gaseous compounds. The measurements are at room temperature and pressure where one mole of gas occupies 24 000 cm^3.

Elements in compound	*Mass of 1000 cm^3 of gas*
Carbon and oxygen	1.17 g
Hydrogen and bromine	3.37 g
Hydrogen and sulphur	1.42 g
Carbon and hydrogen	2.41 g
Carbon, hydrogen and chlorine	2.10 g
Carbon, oxygen and chlorine	4.13 g

(b) Suggest a formula for each compound which fits the information given.
(c) Select three of the compounds and draw diagrams to show what happens to each atom's outer electrons when a molecule of the compound is formed from atoms of its elements.

10. The following table gives information about the percentage composition by mass of various compounds containing carbon, hydrogen and oxygen.

Compound	*% C*	*% H*	*% O*	*Mass of 1 mole of molecules in grammes*
A	52.2	13.0	34.8	46
B	40.0	6.7	53.3	30
C	26.7	2.2	71.1	90
D	40.0	6.7	53.3	60
E	38.7	9.7	51.6	62
F	40.6	5.1	54.3	118
G	40.0	6.7	53.3	180

(a) Use the percentage composition by mass to calculate the empirical formulae of each compound.
(b) Use all the information in the table to calculate the molecular formula of each compound.

11. Ammonia burns in oxygen according to the following equation.

$$4NH_3(g) + 3O_2(g) \rightarrow 2N_2(g) + 6H_2O(l)$$

What volume of oxygen is needed for the complete combustion of 800 cm^3 of ammonia, and what volume of nitrogen will be produced? All volumes are measured at room temperature and pressure.

12. Ammonia reacts with excess chlorine according to the following equation.

$$2NH_3(g) + 3Cl_2(g) \rightarrow N_2(g) + 6HCl(g)$$

If 40 cm^3 of ammonia is mixed with 80 cm^3 of chlorine, what volume of chlorine remains unreacted? What volumes of nitrogen and hydrogen chloride will be formed? All volumes are measured at room temperature and pressure.

13. A large family house with natural gas central heating consumed 1 000 000 m^3 (i.e. 10^6 m^3) of gas during one year. Natural gas is almost pure methane, CH_4, which burns in air according to the equation

$$CH_4(g) + 2O_2(g) \rightarrow CO_2(g) + 2H_2O(l)$$

(a) What volume of oxygen is needed for the complete combustion of the natural gas consumed by this family in one year?
(b) What volume of air is needed for the complete combustion of the natural gas consumed by this family in one year? (Air is 20 per cent oxygen.)
(c) The number of families on the Earth is about 1 000 000 000 (i.e. 10^9). If each family had its own house with this same consumption of natural gas, what is the total volume of air which would be consumed in one year?
(d) The total volume of air in the atmosphere is about 10^{18} m^3 at room temperature and pressure. For how long would the Earth's air supply last if it was consumed at the rate calculated in part (c)?

14. This question is concerned with an investigation of the equation for the reaction between ammonia gas, NH_3, and hydrogen bromide gas, HBr. 50 cm^3 of ammonia was mixed with 60 cm^3 of hydrogen bromide. When reaction was complete there was found to be 10 cm^3 of hydrogen bromide remaining. The only product of the reaction was ammonium bromide, a white solid of formula NH_4Br.

(a) What volume of hydrogen bromide reacts with 50 cm^3 of ammonia?
(b) How many moles of hydrogen bromide react with one mole of ammonia?
(c) Write the equation for the reaction between hydrogen bromide and ammonia.

15. When ammonia gas, NH_3, is heated it decomposes into nitrogen, N_2, and hydrogen, H_2. 100 cm^3 of ammonia were decomposed in this way and were found to produce 50 cm^3 of nitrogen and 150 cm^3 of hydrogen (all volumes measured at room temperature and pressure). Write the equation for the decomposition of ammonia, showing clearly how you have used the information given.

16. The following table gives the densities of several gaseous elements at room temperature and pressure.

Element	*Relative atomic mass*	*Density, g/litre*
H	1	0.083
He	4	0.165
N	14	
O	16	1.33 one form
O	16	2.00 another form
F	19	1.58
Ne	20	0.83
Cl	35.5	2.99
Ar	40	1.66
Ar	40	1.66
Xe		5.45

(a) Calculate the volume of 1 mole of atoms of each element for which sufficient data is given in the table. Comment on any patterns which emerge.

(b) Use the information given in the table to predict
(i) the density of nitrogen
(ii) the relative atomic mass of xenon
(iii) the molecular formulae of the two forms of oxygen

17. Many gaseous elements and compounds can be obtained in very pure form for special purposes. The cost of several of these gases is given in the following table.

Substance	*Molecular formula*	*Cost, £/kg*
Carbon dioxide	CO_2	350
Carbon monoxide	CO	550
Helium	He	3800
Hydrogen	H_2	7600
Neon	Ne	760
Oxygen	O_2	480

(a) Plot a graph of cost (vertical axis) against mass of 1 mole of molecules for these six gases, and use the curve to predict the cost of 1 kg of very pure ethene (ethylene, C_2H_4) and very pure argon (Ar).
(b) The cost of a very pure sample of a gaseous compound formed between carbon and hydrogen is £960/kg. Suggest a possible value for the mass of 1 mole of molecules of this compound. What is the formula of this compound likely to be?
(c) Calculate the cost of 1 mole of molecules of each of the gases in the table. Can you offer any explanation for the rather unexpected result you obtain?
(d) Predict the cost of 1 kg of xenon (Xe). Compare this predicted value with the actual cost of £1100 and attempt to explain any difference.

Chapter 12

1. Calculate the mass of each of the following:
(a) 1 mole of bromine atoms
(b) 1 mole of bromine molecules, Br_2
(c) 1 mole of hydrogen bromide, HBr
(d) 1 mole of sodium bromide, NaBr
(e) 1 mole of potassium nitrate, KNO_3
(f) 0.1 mole of magnesium sulphate, $MgSO_4$
(g) 0.25 mole of calcium hydroxide, $Ca(OH)_2$
(h) 0.01 mole of silicon dioxide, SiO_2

2. (a) How many moles of NaBr are there in 1030 g of sodium bromide?
(b) How many moles of KNO_3 are there in 10.1 g of potassium nitrate?
(c) How many moles of $MgSO_4$ are there in 24 g of magnesium sulphate?
(d) How many moles of $Ca(OH)_2$ are there in 37 g of calcium hydroxide?
(e) How many moles of SiO_2 are there in 20 g of silicon dioxide?

3. Calculate the number of moles of each compound present in the given volumes of the following solutions.
(a) 1000 cm^3 (1 litre) of molar NaBr
(b) 2000 cm^3 of 2 molar NaBr
(c) 500 cm^3 of 0.5 molar NaBr
(d) 20 cm^3 of 2 molar KNO_3
(e) 10 cm^3 of 0.1 molar $MgSO_4$
(f) 25 cm^3 of 0.001 molar $Ca(OH)_2$

4. Calculate the molarity of each of the following solutions:
(a) sodium bromide containing 103 g of NaBr in 1000 cm^3 of solution
(b) sodium bromide containing 206 g of NaBr in 2000 cm^3 of solution
(c) sodium bromide containing 1.03 g of NaBr in 50 cm^3 of solution
(d) potassium nitrate containing 10.1 g of KNO_3 in 200 cm^3 of solution
(e) magnesium sulphate containing 24 g of $MgSO_4$ in 100 cm^3 of solution
(f) potassium hydroxide containing 0.029 g of KOH in 10 cm^3 of solution

5. Calculate the number of moles of each reactant and each product in the reactions described below. Use this information to write an equation for each reaction, showing clearly how you have used the information. All gas volumes are measured at room temperature and pressure where one mole of molecules occupies 24 000 cm^3.
(a) 6.62 g of lead(II) nitrate, $Pb(NO_3)_2$, were heated. 4.46 g of lead(II) oxide, PbO, remained after heating. The other products of the reaction were 1.84 g of liquid dinitrogen tetroxide, N_2O_4, and 480 cm^3 of oxygen gas.
(b) 2400 cm^3 of ammonia, NH_3, were passed over 21.6 g of heated copper(I) oxide, Cu_2O. The reaction produced 1200 cm^3 of nitrogen gas, N_2, together with 19.2 g of copper and 2.7 g of water.
(c) 13 g of zinc reacted with 200 cm^3 of molar copper(II) sulphate, $CuSO_4$. The reaction produced 200 cm^3 of molar zinc sulphate, $ZnSO_4$, and 12.8 g of copper were precipitated.
(d) 75 cm^3 of 2 molar sodium hydroxide, NaOH, reacted with 50 cm^3 of molar phosphoric acid, H_3PO_4. The reaction produced 16.4 g of sodium phosphate, Na_3PO_4.

6. Various masses of potassium chlorate, $KClO_3$, were heated in boiling tubes of known mass. A

gas was produced which was collected in each case in a gas syringe connected to the boiling tube. The volume of gas was recorded by the syringe reading and the mass of gas was found by the decrease in mass of the boiling tube and contents. The other product of the reaction was a white solid. The experimental results are shown below.

Mass of $KClO_3$, grammes	*Mass of gas formed, grammes*	*Volume of gas formed, cm^3*
0.060	0.024	18
0.121	0.047	35
0.153	0.060	45
0.186	0.073	56
0.244	0.095	71
0.309	0.121	92

(a) Plot a graph of volume of gas (vertical axis) against mass of gas. Use the graph to estimate the mass of 24000 cm^3 of the gas and hence identify this gas. What test would you carry out on the gas to confirm that you had identified it correctly?
(b) Plot a graph of volume of gas (vertical axis) against mass of potassium chlorate used. Estimate from this graph the volume of gas which would be produced on heating 1 mole of $KClO_3$. Use this information to write an equation for the action of heat on potassium chlorate. You must show clearly how you have used the information.

7. A student took a flask containing about 50 cm^3 of 4M hydrochloric acid, HCl. He added a small lump of marble (calcium carbonate, $CaCO_3$) to the acid and waited until the gas produced by the reaction (carbon dioxide, CO_2) was no longer evolved. He then prepared a plug of cotton wool to fit tightly in the top of the flask.

The student wanted to investigate the equation for the reaction between calcium carbonate and hydrochloric acid. He first found the mass of the flask, acid and cotton wool plug. Then he added a known mass of calcium carbonate and replaced the plug. When reaction was over and no more calcium carbonate remained, he found the mass of the complete apparatus. Finally he removed the plug for a few minutes, replaced it and found the mass of the complete apparatus again. The readings taken in the experiment are shown below.

Mass of flask, acid and plug	= 151.31 g
Total mass of flask, acid, plug and calcium carbonate before reaction	= 152.63 g
Total mass of apparatus after reaction	= 152.22 g
Total mass of apparatus after removing plug for a few minutes at end of reaction	= 152.05 g

(a) Suggest the reason for adding the small lump of marble to the acid at the very beginning of the experiment.
(b) What mass of calcium carbonate was used in the main part of the experiment?
(c) How many moles of calcium carbonate were used in the main part of the experiment?
(d) It is necessary in this experiment to take sufficient acid to react completely with the calcium carbonate. Did the student take sufficient acid? Explain how you have arrived at your answer.
(e) Why is the total mass lower after the reaction than before the reaction?
(f) Suggest the reason for using the cotton wool plug.
(g) Why did the total mass decrease after removing the plug for a few minutes?
(h) What mass of carbon dioxide was formed in the reaction?
(i) How many moles of carbon dioxide were formed in the reaction?
(j) What information do your answers to (c) and (i) give about the equation for the reaction? You must explain how you have used your answers to (c) and (i).

8. The experiment described in question 7 was repeated using sodium carbonate, Na_2CO_3, in place of calcium carbonate. It was found that 10.6 g of sodium carbonate produced 4.4 g of carbon dioxide. The reaction consumed exactly 50 cm^3 of 4 M HCl, and 11.7 g of sodium chloride, NaCl, were obtained by evaporation when reaction was over.

(a) How many moles of sodium carbonate were used in the reaction?
(b) How many moles of hydrochloric acid were consumed by the sodium carbonate?
(c) How many moles of carbon dioxide were formed in the reaction?
(d) How many moles of sodium chloride, NaCl, were formed in the reaction?
(e) Use your answers in parts (a) to (d) to write an equation for the reaction between sodium carbonate and hydrochloric acid. You must show clearly how you have used your earlier answers.

9. A student had been provided with aqueous solutions of sodium hydroxide (2 molar) and hydrobromic acid, HBr (4 molar). He was asked to investigate the equation for the reaction between these two substances. He carried out the reaction and obtained the following readings.

Volume of 4 M HBr (in cm^3) added to 20 cm^3 of 2M NaOH	*Temperature of mixture in °C*
4	26.8
6	30.0
8	33.2
10	36.0
12	35.2
14	34.4
20	30.8

(a) Sketch the apparatus that the student might have used to obtain these results, and describe briefly how the experiment might have been carried out.
(b) Draw a graph of temperature of mixture (vertical axis) against volume of acid added.
(c) Use your graph to estimate the temperature of the mixture when 17 cm^3 of acid had been added.
(d) Both solutions were at room temperature at the start of the experiment. Use your graph to estimate room temperature.
(e) The temperature of the mixture increased during the first additions of acid. Why did the temperature increase?
(f) Suggest two reasons why the temperature decreased during the later part of the experiment.
(g) Use your graph to estimate the volume of 4 M HBr which would just neutralize 20 cm^3 of 2 M NaOH.
(h) How many moles of HBr are present in your answer to (g)?
(i) How many moles of NaOH are present in 20 cm^3 of 2 M NaOH?
(j) Use your answers in parts (h) and (i) to write an equation for the reaction. You must explain clearly how you have used your answers in parts (h) and (i).

10. Use the following information to calculate the molarity of the first named solution in each section.

(a) 20 cm^3 of aqueous sodium hydroxide is neutralized by 10 cm^3 of 2 M hydrochloric acid. The equation for the reaction is

$$NaOH(aq) + HCl(aq) \rightarrow NaCl(aq) + H_2O(l)$$

(b) 40 cm^3 of aqueous magnesium chloride reacts with 50 cm^3 of 0.1 M aqueous silver nitrate. The equation for the reaction is

$$MgCl_2(aq) + 2AgNO_3(aq) \rightarrow 2AgCl(s) + Mg(NO_3)_2(aq)$$

(c) 50 cm^3 of aqueous nitric acid is neutralized by 25 cm^3 of 0.02 M aqueous barium hydroxide. The equation for the reaction is

$$2HNO_3(aq) + Ba(OH)_2(aq) \rightarrow Ba(NO_3)_2(aq) + 2H_2O(l)$$

(d) 20 cm^3 of aqueous aluminium sulphate reacts with 30 cm^3 of 0.4 M aqueous barium chloride. The equation for the reaction is

$$Al_2(SO_4)_3(aq) + 3BaCl_2 \rightarrow 3BaSO_4(s) + 2AlCl_3(aq)$$

11. Which of the reactions in question 10 would produce a precipitate?

Chapters 13 and 14

1. How much heat energy is lost to the surroundings, or gained from the surroundings (you must say which) during the following changes?

(a) 100 cm^3 of water cools from 35 °C to 25 °C.
(b) 50 cm^3 of water cools from 100 °C to 20 °C.
(c) 200 cm^3 of water heats up from 0 °C to 15 °C.
(d) 80 cm^3 of water heats up from 5 °C to 20 °C.

2. 6.5 g of zinc was added to 100 cm^3 of hydrochloric acid at 20 °C. The zinc dissolved completely and the highest temperature reached by the solution was 50 °C.

(a) What was the temperature rise in the reaction?
(b) Was an excess of acid used in the reaction? Explain how you have arrived at your answer.
(c) How much heat energy would be lost to the surroundings when the reaction mixture cooled back to room temperature?
(d) How many moles of zinc were used in the reaction?
(e) How much heat energy would be lost to the surroundings if one mole of zinc reacted with excess hydrochloric acid in a similar experiment?
(f) The reaction produces zinc chloride, $ZnCl_2$, in aqueous solution and hydrogen gas. Write an equation for the reaction and include the heat change in this equation.
(g) Draw an energy level diagram for the reaction.

3. The experiment illustrated in Fig. 14.1 was carried out using methanol, CH_3OH, in place of ethanol. It was found that the temperature of 100 cm^3 of water increased from 18 °C to 48 °C on burning 1 g of methanol. Use this information to calculate the heat of combustion of methanol. Would you expect the value for the heat of combustion of methanol given in a data book to be higher or lower than this experimental value? Explain your answer clearly.

4. Information about heats of reaction is included in the following equations.

$$C(s) + O_2(g) \rightarrow CO_2(g) \quad \Delta H = -405 \text{ kJ/mole}$$

$$2CO(g) + O_2(g) \rightarrow 2\,CO_2(g) \quad \Delta H = -570 \text{ kJ/mole}$$

$2NH_3(g) \rightarrow N_2(g) + 3H_2(g)$
$\Delta H = +92$ kJ/mole

$2H_2(g) + O_2(g) \rightarrow 2H_2O(g)$
$\Delta H = -485$ kJ/mole

$2Mg(s) + O_2(g) \rightarrow 2MgO(s)$
$\Delta H = -1220$ kJ/mole

(a) Which of these reactions are exothermic?
(b) Draw an energy level diagram for the most exothermic of these reactions.
(c) Give the formulae of the substances in the equations which may be suitable gaseous fuels.
(d) What is the energy change when 6 g of carbon burns completely to carbon dioxide?
(e) 1 g of the substance appearing first in each equation is available for reaction. In which case will the most heat be evolved when this 1 g is completely consumed?

5. The following table gives information about the heats of combustion of the first four members of a series of compounds called alcohols.

Compound	*Heat of combustion in kJ/mole*
Methanol, CH_3OH	710
Ethanol, C_2H_5OH	1370
Propanol, C_3H_7OH	2020
Butanol, C_4H_9OH	2670

(a) Plot a graph of heat of combustion (vertical axis) against number of carbon atoms in the molecule.
(b) The next member of the alcohol series contains five carbon atoms. Predict its formula. Use your graph to predict its heat of combustion.

6. A student took 12.5 g of copper sulphate crystals, $CuSO_4,5H_2O$, and dissolved these in 100 cm^3 of water in an expanded polystyrene mug. The temperature of the water dropped from 20 °C to 18.8 °C.

(a) How much heat energy would be absorbed from the surroundings when the solution warmed up again to the original temperature?
(b) How many moles of copper sulphate, $CuSO_4,5H_2O$, were used in the experiment?
(c) What is the heat energy change when 1 mole of copper sulphate, $CuSO_4,5H_2O$, dissolves in water? Draw an energy level diagram for this change.

7. When 50 cm^3 of 2 M nitric acid, HNO_3, reacted with 50 cm^3 of 2 M potassium hydroxide, KOH, in an expanded polystyrene mug the temperature of the solutions changed from 20 °C to 32 °C.

(a) Is this reaction endothermic or is it exothermic?
(b) What is the heat energy change in the reaction described?
(c) How many moles of nitric acid, HNO_3, are present in 50 cm^3 of 2 M solution?
(d) How many moles of potassium hydroxide, KOH, are present in 50 cm^3 of 2 M solution?
(e) What is the heat energy change when 1 mole of nitric acid reacts with 1 mole of potassium hydroxide?
(f) Write an equation for this reaction including the heat energy change in the equation.
(g) Draw an energy level diagram for the reaction.

8. It is possible to construct simple cells similar to the Daniell cell to produce electrical energy from other electron transfer reactions. The maximum voltages of four cells are given below.

Equation for electron transfer reaction	*Number of moles of electrons transferred for 1 mole of reaction*	*Maximum voltage of cell in volts*	*Heat of reaction in kJ/mole (all exothermic)*
$Mg(s) + Cu^{2+}(aq) \rightarrow Mg^{2+}(aq) + Cu(s)$	2	2.71	526
$2Al(s) + 3Cu^{2+}(aq) \rightarrow 2Al^{3+}(aq) + 3Cu(s)$	6	2.00	1242
$Fe(s) + Cu^{2+}(aq) \rightarrow Fe^{2+}(aq) + Cu(s)$	2	0.78	152
$Pb(s) + Cu^{2+}(aq) \rightarrow Pb^{2+}(aq) + Cu(s)$	2	0.47	63

(a) Calculate the maximum electrical energy that can be obtained from 1 mole of each reaction.
(b) Compare your answers in (a) with the values for the heat of reaction. Would you expect each answer in (a) to be the same as the heat of reaction? Explain your answer briefly.

Chapter 15

1. Two pieces of fresh calcium of equal mass were used in separate experiments. In one experiment a piece was added to excess water. In the other experiment a piece was added to excess dilute hydrochloric acid. The volume of hydrogen produced was measured at intervals. The results of the experiments are shown below.

Time from start of reaction in seconds	*Total volume of hydrogen produced using water*	*Total volume of hydrogen produced using HCl*
10	6	24
20	12	50
30	20	80
50	40	116
60	54	120
80	88	120
90	109	120
100	119	120
110	120	120

(a) Plot a graph of volume of hydrogen produced (vertical axis) against time.
(b) Use your graph to predict the volumes of hydrogen produced in each experiment after 40 and 70 seconds.
(c) Which reaction was most rapid at the start?
(d) At what time (or times) were the rates of both reactions the same?
(e) Sketch on your graph the curve that would be expected if the same mass of calcium was reacted with water which was colder than the water used in the original experiment.
(f) Predict the volume of hydrogen which would have been formed in 60 seconds if the same mass of calcium had reacted with an excess of more concentrated hydrochloric acid.
(g) The mass of calcium used in each experiment was 0.2 g. What mass of calcium remained unreacted in each experiment at 65 seconds after the start of the reaction?
(h) How many moles of calcium atoms were used in each experiment, and how many moles of hydrogen molecules, H_2, were formed in each case?
(i) Use your answers in (h) to help write equations for the two reactions, explaining clearly how you have used these answers.

2. Dilute sulphuric acid reacts fairly slowly with lumps of zinc to produce hydrogen and aqueous zinc sulphate. Draw a diagram of the apparatus you would use to investigate how rapidly the hydrogen is produced in this reaction.

If the zinc is always completely covered by the acid, how would you expect the rate of production of hydrogen at the start of the reaction to be influenced by the following changes?

(a) using a greater mass of zinc
(b) using more acid of the same concentration
(c) using acid of a higher concentration
(d) increasing the temperature of the reaction mixture
(e) adding a suitable catalyst
(f) using a larger reaction vessel
(g using equal volumes of acid and water
(h) using the same mass of powdered zinc
(i) adding a piece of ice to the acid before reaction

3. This question is concerned with a student's investigation of the rate of reaction of magnesium ribbon with various acids. He used a 10 cm length of ribbon each time and added this in separate experiments to 20 cm^3 (an excess) of molar solutions of the following acids:

Hydrochloric acid, HCl
Acetic acid (ethanoic acid), CH_3CO_2H
Chloracetic acid (chloroethanoic acid), CH_2ClCO_2H

The student's readings are shown below.

Time in seconds after adding Mg to acid	*Total volume of hydrogen collected in cm³*		
	HCl	*CH_3CO_2H*	*CH_2ClCO_2H*
20	20	4	14
40	40	8	27
60	60	12	40
80	79	19	54
100	95	29	67
120	100	42	80
140	100	57	90

(a) Plot a graph of volume of hydrogen collected (vertical axis) against time, using the same sheet of graph paper for all three reactions.
(b) Estimate the volume of hydrogen which would have been collected from each reaction after 3 minutes.
(c) How long after the start of the experiment did the rate of reaction of magnesium with dilute hydrochloric acid begin to decrease? Suggest a reason for this decrease in rate.
(d) The rate of reaction of magnesium with acetic acid began to increase after a certain time. Suggest a reason for this increase in rate.
(e) Which acid appears from the results of this experiment to be the most highly ionized, and which acid appears to be the least ionized? Explain briefly how you have made your decision.
(f) What further test could you make on the acids to confirm that your choice in (e) is correct?
(g) In a further experiment, a 10 cm length of magnesium ribbon was put into excess pure acetic acid. Only 3 cm^3 of hydrogen was collected in 120 seconds. Suggest a reason for the very slow rate of reaction with pure acetic acid.

Chapter 16

1. Aqueous hydrogen sulphide reacts with aqueous zinc ions to form a precipitate of zinc sulphide. The reaction is reversible.

$$H_2S(aq) + Zn^{2+}(aq) \rightleftharpoons ZnS(s) + 2H^+(aq) \quad \Delta H = -9 \text{ kJ/mole}$$

Which three of the following changes are likely to alter the equilibrium position in this reaction so that less zinc sulphide is formed?

(a) addition of extra zinc ions to increase the concentration of $Zn^{2+}(aq)$
(b) addition of extra hydrogen sulphide to increase the concentration of $H_2S(aq)$
(c) addition of extra hydrogen ions to increase the concentration of $H^+(aq)$
(d) addition of an oxidizing agent which removes hydrogen sulphide

(e) addition of OH^- ions to convert H^+ ions to H_2O
(f) addition of a suitable catalyst for the reaction
(g) increase in temperature of the reaction mixture

2. Gaseous hydrogen chloride dissolves in water and forms an acidic solution by reaction with the water. The reaction is reversible.

$$HCl(g) + H_2O(l) \rightleftharpoons H_3O^+(aq) + Cl^-(aq) \quad \Delta H = -74 \text{ kJ/mole}$$

The ion H_3O^+(aq) represents a possible formula for the hydrogen ion, H^+, in water.

Which of the following changes are likely to alter the equilibrium position in this reaction so that more hydrogen chloride gas dissolves?

(a) increasing the concentration of HCl(g) by increasing its pressure
(b) removing H_3O^+(aq) ions by neutralization with OH^-(aq) ions
(c) removing Cl^-(aq) ions by addition of Ag^+(aq) ions to form insoluble AgCl
(d) increasing the temperature of the reaction mixture
(e) Adding a suitable catalyst for the reaction

3. Iodine vapour at high temperature dissociates reversibly into gaseous iodine atoms.

$$I_2(g) \rightleftharpoons 2I(g) \quad \Delta H = +151 \text{ kJ/mole}$$

The equilibrium position in the reaction may be affected by changing conditions in the following ways:

(a) equilibrium moves to the right
(b) equilibrium moves to the left
(c) equilibrium position unchanged

Select from (a) to (c) the one which in each case is most likely to take place when the reaction conditions are changed in the following ways:

(i) the temperature is increased
(ii) the pressure is increased
(iii) a catalyst for the decomposition is added

4. Pure water is slightly ionized. The reaction is reversible and the equation for the reaction can be written in the following way. H_3O^+(aq) represents a possible formula for the hydrogen ion, H^+, in water.

$$2H_2O(l) \rightleftharpoons H_3O^+(aq) + OH^-(aq) \quad \Delta H = +58 \text{ kJ/mole}$$

The equilibrium position in the reaction can be affected in the following ways by changing the conditions:

(a) more water ionizes
(b) less water ionizes
(c) no change in ionization of water

Select from (a) to (c) the one which in each case is most likely to take place when the reaction conditions are changed in the following ways:

(i) the temperature of the water is increased
(ii) a suitable catalyst for the reaction is added
(iii) extra OH^-(aq) ions are added to the water
(iv) extra H_3O^+(aq) ions are added to the water

5. Acetic acid (ethanoic acid) ionizes slightly in water. It is said to be a weak acid. The ionization is reversible and can be represented by the following equation.

$$CH_3CO_2H(l) + H_2O(l) \rightleftharpoons CH_3CO_2^-(aq) + H_3O^+(aq) \quad \Delta H = +78 \text{ kJ/mole}$$

The ion $CH_3CO_2^-$(aq) is the acetate (ethanoate) ion. The ion H_3O^+(aq) represents a possible formula for the hydrogen ion, H^+, in water.

Various attempts to alter the ionization of acetic acid may have the following effects:

(a) more acetic acid will ionize
(b) less acetic acid will ionize
(c) acetic acid will be unchanged in ionization

Select from (a) to (c) the effect on the ionization of acetic acid which is likely to be produced by each of the following:

(i) addition of more water
(ii) addition of extra acetate ions
(iii) addition of OH^-(aq) ions to remove H_3O^+(aq) ions by neutralization
(iv) addition of a suitable catalyst for the reaction
(v) increase in temperature of the reaction mixture

6. The production of ammonia by the Haber process involves the following reversible reaction:

$$N_2(g) + 3H_2(g) \rightleftharpoons 2NH_3(g) \quad \Delta H = -92 \text{ kJ/mole}$$

(a) Four ways in which the conditions of reaction can be changed are:

(i) increase in temperature
(ii) decrease in temperature
(iii) increase in pressure
(iv) decrease in pressure

Select from (i) to (iv) the two which are most likely to make the equilibrium position move from left to right and so increase the conversion of nitrogen and hydrogen to ammonia.
(b) How will the conversion of nitrogen and

hydrogen to ammonia be altered if the ammonia is removed from the equilibrium mixture as rapidly as it is formed?
(c) What effect will a suitable catalyst have on the equilibrium position in the reaction?
(d) What effect will a suitable catalyst have on the rate at which equilibrium is reached in the reaction?

Chapter 17

1. Which of the following equations represent electron transfer reactions:

(a) $Zn(s) + SnCl_2(aq) \rightarrow ZnCl_2(aq) + Sn(s)$
(b) $2Li(s) + O_2(g) \rightarrow 2LiO(s)$
(c) $Mg(s) + CuCl_2(aq) \rightarrow MgCl_2(aq) + Cu(s)$
(d) $2Na(s) + O_2(g) \rightarrow 2NaO(s)$
(e) $4Na(s) + O_2(g) \rightarrow 2Na_2O(s)$
(f) $LiOH(aq) + HCl(aq) \rightarrow LiCl(aq) + H_2O(l)$
(g) $Ba(s) + Cl_2(g) \rightarrow BaCl_2(s)$
(h) $Ba(OH)_2(aq) + H_2SO_4(aq) \rightarrow BaSO_4(s) + 2H_2O(l)$
(i) $Sn(s) + 2HCl(aq) \rightarrow SnCl_2(aq) + H_2(g)$
(j) $Pb(NO_3)_2(aq) + 2KI(aq) \rightarrow PbI_2(s) + 2KNO_3(aq)$

2. Draw up a table showing electron donors (reducing agents) and electron acceptors (oxidizing agents) in the following reactions:

(a) $Fe(s) + Cu^{2+}(aq) \rightarrow Fe^{2+}(aq) + Cu(s)$
(b) $Mg(s) + Sn^{2+}(aq) \rightarrow Mg^{2+}(aq) + Sn(s)$
(c) $Cu(s) + 2Ag^{+}(aq) \rightarrow Cu^{2+}(aq) + 2Ag(s)$
(d) $2Na(s) + S(s) \rightarrow 2Na^{+}S^{2-}(s)$
(e) $2Mg(s) + O_2(g) \rightarrow 2(Mg^{2+}O^{2-})(s)$

3. Write ionic half-reactions to represent the conversion of:

(a) Fe atoms to $Fe^{2+}(aq)$ ions
(b) $Fe^{2+}(aq)$ ions to Fe atoms
(c) Cu atoms to $Cu^{2+}(aq)$ ions
(d) $Fe^{3+}(aq)$ ions to $Fe^{2+}(aq)$ ions
(e) $Cl^{-}(aq)$ ions to $Cl_2(g)$ molecules
(f) $O_2(g)$ molecules to $OH^{-}(aq)$ ions
(g) $H^{+}(aq)$ ions to $H_2(g)$ molecules
(h) $Mg^{2+}(aq)$ ions to Mg atoms
(i) $S^{2-}(aq)$ ions to S atoms
(j) $Br_2(l)$ molecules to $Br^{-}(aq)$ ions

4. (a) Point out one important way in which base metals differ from noble metals.
(b) Select the base metals from the following list:

sodium silver magnesium zinc platinum gold potassium lithium lead

5. Which of the following elements donate electrons readily and act as reducing agents?

sodium sulphur oxygen magnesium chlorine potassium calcium bromine

6. Which of the following ions donate electrons readily and act as reducing agents?

Mg^{2+} Br^{-} Na^{+} Al^{3+} I^{-}

7. How many electrons are there in the outer shell of atoms of each of the following elements?

lithium oxygen fluorine neon sodium magnesium aluminium sulphur chlorine argon potassium calcium bromine krypton

8. How many electrons are there in the outer shell of each of the following ions?

Li^{+} O^{2-} F^{-} Na^{+} Mg^{2+} Al^{3+} S^{2-} Cl^{-} K^{+} Ca^{2+} Br^{-}

Chapter 18

1. Use Table 18.1 to answer the following questions.

(a) How much heat energy is lost to the surroundings when 0.15 mole $Cu^{2+}(aq)$ ions are discharged by Zn(s) atoms?
(b) How much heat energy is lost to the surroundings when 2 moles Pb(s) atoms discharge $Cu^{2+}(aq)$ ions?
(c) Which of the following liberates the greatest amount of heat energy to the surroundings?

(i) discharge of 3 moles $Cu^{2+}(aq)$ by Al(s) atoms,
(ii) discharge of 4 moles $Cu^{2+}(aq)$ by Fe(s) atoms,
(iii) discharge of 5 moles $Cu^{2+}(aq)$ by Pb(s) atoms.

2. Which of the following is the least exothermic half-reaction?

(a) $Al(s) \rightarrow Al^{3+}(aq) + 3e$
(b) $Pb(s) \rightarrow Pb^{2+}(aq) + 2e$
(c) $Mg(s) \rightarrow Mg^{2+}(aq) + 2e$
(d) $Fe(s) \rightarrow Fe^{2+}(aq) + 2e$

3. Use Table 18.2 to find which of these cells has the lowest voltage:

(a) magnesium against copper
(b) lead against copper
(c) aluminium against copper
(d) zinc against copper

4. (a) Arrange the following metals in order of their readiness to lose electrons when they are made electrodes in cells. Start the list with the metal losing electrons most easily.

Mg Zn Pb Al Cu

(b) Where would you put Na if you added it to the list?
(c) Where would you put Au if you added it to the list?

5. In each of the following equations the arrow representing direction of reaction is missing. Most of the equations are not balanced. Write the equations the correct way round, add the arrow and balance if necessary. Use Table 18.3 if you need to.

E.g. $Fe(s)+Mg^{2+}(aq)$ $Fe^{3+}(aq)+Mg(s)$ should be rewritten:

$$2Fe^{3+}(aq)+3Mg(s) \rightarrow 2Fe(s)+2Mg^{2+}(aq)$$

(a) $Al(s)+Pb^{2+}(aq)$ $Al^{3+}(aq)+Pb(s)$
(b) $Mg(s)+Cu^{2+}(aq)$ $Mg^{2+}(aq)+Cu(s)$
(c) $Na^{+}(aq)+H_2(g)$ $Na(s)+H^{+}(aq)$
(d) $Ag^{+}(aq)+Cu(s)$ $Ag(s)+Cu^{2+}(aq)$
(e) $Cl^{-}(aq)+Br_2(aq)$ $Cl_2(aq)+Br^{-}(aq)$
(f) $Li^{+}(aq)+H_2(g)$ $Li(s)+H^{+}(aq)$
(g) $Fe(s)+H^{+}(aq)$ $Fe^{2+}(aq)+H_2(g)$
(h) $Br^{-}(aq)+I_2(aq)$ $Br_2(aq)+I^{-}(aq)$
(i) $Ag^{+}(aq)+Pb(s)$ $Ag(s)+Pb^{2+}(aq)$
(j) $Mg(s)+H^{+}(aq)$ $Mg^{2+}(aq)+H_2(g)$

6. (a) Which of the metals in the following list can displace hydrogen from acids?

Mg Al Pb Cu Fe Zn Ag Na

(b) Which metal does so most readily?
(c) Write balanced equations for reactions between each metal and hydrochloric acid (if reaction does take place).

7. The following reactions take place from left to right as shown. Break up each equation into two ionic half equations. Explain clearly by reference to the ECS (Table 18.3) why each reaction would be expected to take place from left to right.

(a) $Ca(s)+2H^{+}(aq) \rightarrow Ca^{2+}(aq)+H_2(g)$
(b) $2Al(s)+6H^{+}(aq) \rightarrow 3Al^{3+}(aq)+3H_2(g)$
(c) $Zn(s)+Cu^{2+}(aq) \rightarrow Zn^{2+}(aq)+Cu(s)$
(d) $Zn(s)+Cl_2(aq) \rightarrow Zn^{2+}(aq)+2Cl^{-}(aq)$
(e) $Mg(s)+Br_2(aq) \rightarrow Mg^{2+}(aq)+2Br^{-}(aq)$
(f) $Cl_2(aq)+2I^{-}(aq) \rightarrow 2Cl^{-}(aq)+I_2(aq)$

8. Use the ECS (Table 18.3) to predict what will happen (if anything) in the following situations:

(a) zinc is added to an aqueous solution of lead ions, $Pb^{2+}(aq)$
(b) copper is added to an aqueous solution of lead ions, $Pb^{2+}(aq)$
(c) zinc is added to an aqueous solution of sodium ions, $Na^{+}(aq)$
(d) aluminium is added to an aqueous solution of iodine molecules, $I_2(aq)$
(e) magnesium is added to an aqueous solution of bromide ions, $Br^{-}(aq)$
(f) fluorine gas is bubbled through aqueous solutions of

(i) chloride ions, $Cl^{-}(aq)$
(ii) bromide ions, $Br^{-}(aq)$
(iii) iodide ions, $I^{-}(aq)$

9. Twelve pairs of students constructed simple cells using a beaker containing potassium dichromate dissolved in dilute hydrochloric acid into which was fixed a carbon rod connected to one terminal of a bulb holder. The other connection to the bulb holder was made using a wire whose free end was attached by a clip to a piece of metal. When this metal was dipped into the solution in the beaker, the bulb was observed to light in several cases. The students noted how the brightness of the bulb varied with the metal used.

Metal	*Estimate of bulb brightness*
Iron nail	Fairly bright
Aluminium foil	Nothing
Aluminium foil after dipping in solution of mercury(II) chloride	Very bright
Lead sheet	Dim
Magnesium	Bulb burned out in four cases
Copper	Very dim
Calcium	Bulb burned out in ten cases
Zinc	Bright

(a) Draw a diagram of the circuit used in this experiment.
(b) What information does this experiment give about the position of the metals in the ECS?
(c) Can you suggest any reason for the difference between the two observations on aluminium?
(d) Suggest how you would modify this experiment to obtain more information about the order of the metals in the activity series.

Chapter 19

1. Which of the following liquids would be expected to conduct electricity with decomposition at the electrodes?

(a) molten lead(II) bromide
(b) molten sodium iodide
(c) tetrachloromethane, CCl_4
(d) mercury
(e) silicon tetrachloride

2. (a) What would be liberated at the positive electrode in each of your answers in question 1?
(b) What would be liberated at the negative electrode in each of your answers in question 1?

3. Which of the following aqueous solutions would be expected to be good conductors of electricity?
   (a) sodium chloride
   (b) magnesium chloride
   (c) ethanol
   (d) sulphuric acid
   (e) potassium bromide
   (f) sugar
   (g) calcium iodide
   (h) soap

4. Give the symbols and charges of the ions present in each of the following molten electrolytes:
   (a) lead(II) chloride
   (b) sodium chloride
   (c) magnesium bromide
   (d) calcium iodide
   (e) lithium oxide

5. (a) Write ionic equations to show what happens at an inert positive electrode on electrolysis of each of the molten electrolytes in question 4.
   (b) Write ionic equations to show what happens at an inert negative electrode on electrolysis of each of the molten electrolytes in question 4.

6. Calculate the masses of:
   (a) Cu(s) from Cu^{2+}(aq) deposited by a current of 0.1 ampere flowing for 30 minutes
   (b) Ag(s) from Ag^{+}(aq) deposited by a current of 1.5 ampere flowing for 40 minutes
   (c) H_2(g) from H^{+}(aq) liberated by a current of 0.2 ampere flowing for 10 minutes

7. Calculate the number of coulombs passing when a current of:
   (a) 5 A flows for 1 hour
   (b) 0.1 A flows for 30 min
   (c) 1.5 A flows for 200 sec
   (d) 1000 A flows for 24 hours
   (e) 30 000 A flows for 6 days

8. What mass of aluminium would be deposited by a current of 100 000 A flowing for 12 hours?

9. How many Faradays are needed to deposit 230 g sodium from molten sodium chloride?

10. What currents in amperes would be necessary to pass
    (a) 965 coulombs in 30 min
    (b) 10 000 coulombs in 6 hours
    (c) 3600 coulombs in 1 hour
    (d) 9.65×10^{10} coulombs in 7 days
    (e) 1000 Faradays in 15 min

11. The output of an aluminium plant was 250 metric tons per week. Calculate:
    (a) the number of coulombs required in a week
    (b) the number of Faradays
    (c) the average amperage rate, assuming full-time production

12. A current of 0.5 A is passed through copper (Cu^{2+}(aq)), water (H^{+}(aq)), and silver (Ag^{+}(aq)) voltameters in series for 45 minutes. Calculate:
    (a) the mass of copper deposited
    (b) the mass of silver deposited
    (c) the mass of hydrogen liberated
    (d) the volume of the hydrogen, measured at room temperature and pressure

13. A steady current was passed through copper, Cu^{2+}(aq), and silver, Ag^{+}(aq), voltameters in series for 1 hour. The loss in mass of the copper anode was 0.45 g. Calculate
    (a) the gain in mass of the copper cathode
    (b) the gain in mass of the silver cathode
    (c) the current in amperes
    (d) how long this current would have to be passed for 2.5 g silver to be deposited on the silver cathode

14. (a) What FOUR ions are present in aqueous sodium chloride?
    (b) Which of these ions is most likely to be discharged at an inert negative electrode during electrolysis of the solution?
    (c) Which ion is most likely to be deposited at an inert positive electrode during electrolysis of the solution?
    (d) Explain how you made your prediction in (b) and (c).
    (e) Write ionic equations to show what happens at each inert electrode during this electrolysis.

15. (a) What FOUR ions are present in aqueous copper(II) sulphate?
    (b) Which of these ions is most likely to be discharged at an inert negative electrode during electrolysis of the solution?
    (c) Which ion is most likely to be deposited at an inert positive electrode during electrolysis of the solution?
    (d) Explain how you made your prediction in (b) and (c).
    (e) Write ionic equations to show what happens at each inert electrode during this electrolysis.

Chapters 20 and 21

1. Draw up a table of three columns in which elements are listed as *Extracted electrolytically*, *Extracted chemically, e.g. smelted*, *Extracted by*

physical processes, e.g. distillation. Enter each of the elements below in the appropriate column.

Al Ar Br Ca Cl Cu F He H I
Fe Kr Li Pb Mg Ne N O P K
Na S Xe Zn

2. Below are given a list of manufacturing and raw materials and a list of final products. Select, as far as you can, the materials from the left-hand list which are required in the production of substances in the right-hand list.

List of manufacturing and raw materials	*List of final products*
Haematite	Aluminium
Brine	Argon
Sodium hydrogen sulphite	Bromine
Bauxite	Calcium
Common salt	Chlorine
Galena	Copper
Sodium hydroxide	Fluorine
Coke	Helium
Chlorapatite	Hydrogen
Siderite	Iodine
Cryolite	Iron
Silica	Krypton
Magnetite	Lithium
Mercury	Lead
Iron pyrites	Magnesium
Chalcopyrite	Neon
Beech twigs	Nitrogen
Limestone	Oxygen
Zinc blende	Phosphorus
Air	Potassium
Petroleum	Sodium
Fluorapatite	Sulphur
Sea water	Xenon
Caliche	Zinc
Hydrogen chloride gas	

3. Describe the extraction of any one important solid. Why is it important in your opinion?

4. Describe the extraction of any one important gas. Why is it important in your opinion?

5. Sodium carbonate only occurs naturally in the so-called soda lakes, which have no outlet to the sea. It is not found in other areas on account of its high solubility in water. It is an important chemical and is used in glassmaking, to soften water, and in making sodium hydrogen carbonate. Use a library or other reference source to find out details of the manufacture of sodium carbonate. Technical details of plant design are not important. You should concentrate on the chemical principles underlying the manufacturing process. Chapter 16 on Dynamic Equilibrium will be found useful.

6. Draw up a table of the sources and methods of extraction of the Group 7 elements. In the last column list two uses for each element.

7. Sodium hydroxide was at one time manufactured from sodium carbonate and slaked lime (calcium hydroxide):

$$Na_2CO_3(aq) + Ca(OH)_2(aq) \rightarrow 2NaOH(aq) + CaCO_3(s)$$

The $CaCO_3$ was separated by filtration and the NaOH obtained by evaporating the filtered solution. Nowadays sodium hydroxide is made by electrolysing an aqueous solution of sodium chloride. Can you suggest why the older, simpler method has been displaced by the modern method?

8. Metals high in the ECS are usually extracted by electrolysis of their molten chlorides. Outline the process for extracting aluminium and explain why it cannot be extracted by electrolysis of aluminium chloride.

9. (a) Calculate the current in amperes required to produce 100 metric tons of aluminium in 24 hours.
(b) How many metric tons of magnesium would be produced by this current in 24 hours?
(c) Calculate the masses of chlorine produced in (a) and (b).

10. (a) List the chemical reactions involved in the extraction of iron from an ore such as haematite, Fe_2O_3, in the blast furnace.
(b) Which of these are oxidation/reduction reactions?
(c) Which reaction has a salt as its product?
(d) Which reaction involves thermal decomposition of a salt?

11. (a) What is meant by the term 'steel'?
(b) Give essential details of the conversion of iron into steel.
(c) More than 50 per cent of the world's steel is produced in Open Hearth furnaces at present (1973). In the more industrially advanced countries the L–D and similar processes are gaining ground. Can you suggest a reason for this?

Chapter 22

In each of the first 20 questions an element is described briefly. You must decide in each case the element's group number in the Periodic Table. None of these elements is a transition metal; they are all normal elements. The symbols A to T are not the usual symbols for these elements.

1. Element A reacts vigorously with water to form hydrogen and an aqueous solution of the hydroxide with formula AOH.

2. Element B forms a gaseous compound with hydrogen of formula BH_4.

3. Element C reacts readily with sodium to form an ionic compound of formula NaC.

4. Element D forms an oxide of formula DO which dissolves in water to give an alkaline solution.

5. Element E is composed of molecules of formula E_2. The element combines with phosphorus to form two covalent compounds of formula PE_3 and PE_5.

6. Element F reacts with dilute acids to give hydrogen but the element does not react with water. The oxide and hydroxide of F are insoluble in water.

7. Element G does not react with water or with dilute acids. The element forms two oxides of formula GO_2 and GO_3.

8. Element H exists in two polymorphic forms. The element forms covalent chlorides of formula HCl_3 and HCl_5.

9. Element I tarnishes rapidly in air to form the oxide I_2O. This oxide dissolves in water to give an alkaline solution.

10. Element J forms a gaseous compound with hydrogen of formula JH_3.

11. Element K reacts with cold water to give hydrogen. The carbonate and sulphate of K are insoluble in water. The oxide and hydroxide of K are slightly soluble in water to form an alkaline solution.

12. Element L is very unreactive and forms only a small number of compounds.

13. Element M forms a compound with silver of formula AgM. This compound is insoluble in water.

14. Element N forms two chlorides of formula NCl_2 and NCl_4. NCl_2 has a fairly high melting and boiling point, while NCl_4 is a liquid at room temperature. The element forms two oxides of formula NO and NO_2.

15. Element O forms a chloride of formula OCl. The molten chloride is a good conductor of electricity.

16. Element P forms a covalent compound with hydrogen of formula H_2P.

17. Element Q forms acidic oxides of empirical formula Q_2O_3 and Q_2O_5.

18. Element R forms a chloride of formula RCl_2 which is a solid with a very high melting and boiling point.

19. The compound of element S with hydrogen is a gas with formula HS. This gas is very soluble in water to give a strongly acidic solution.

20. Element T reacts vigorously when heated in chlorine to form a compound of formula TCl_3. The chloride is composed of ions.

21. Element U is a hard solid which reacts readily with water to form hydrogen and an alkaline solution of the sparingly soluble hydroxide, $U(OH)_2$. Use this information to predict the answers to the following questions.

    (a) Is U likely to be a metal or is it likely to be a non-metal?
    (b) What is likely to be observed when U is added to dilute hydrochloric acid?
    (c) What is likely to be observed when U is heated in air?
    (d) Predict the formula and solubility in water of the carbonate of U.
    (e) Predict the formula and type of bonding present in the compound of U with chlorine.
    (f) Can you make any further predictions about the properties of U and its compounds?

22. Element V is composed of covalent molecules of formula V_2. The element forms a yellow compound with silver of formula AgV which is insoluble in water. The element forms a gaseous compound with hydrogen of formula HV. Use this information to predict the answers to the following questions.

    (a) Is V likely to be a metal or is it likely to be a non-metal?
    (b) Predict the formula and physical properties of the compound that V forms with sodium.
    (c) What is likely to be observed when chlorine is bubbled through an aqueous solution of the compound that V forms with sodium?
    (d) Predict the approximate pH of an aqueous solution of HV.
    (e) Predict what will be observed when phosphorus is added to a sample of V.

(f) Can you make any further predictions about the properties of V and its compounds?

23. Element W is a soft metal which tarnishes readily in air. The hydroxide of W has the formula WOH and is extremely soluble in water. Use this information to predict the answers to the following questions.

(a) Is W likely to form compounds with many non-metals?
(b) Is W likely to form compounds with many metals?
(c) Predict the formula and type of bonding present in the compound of W with chlorine.
(d) Predict the approximate pH of an aqueous solution of WOH.
(e) What is likely to be observed when W is added to cold water?
(f) Can you make any further predictions about the properties of W and its compounds?

Chapter 23

1. (a) Draw a graph of boiling point in °C (vertical axis) against atomic number for the elements Na to Ar.
(b) Find out the boiling points of the elements in the period Li to Ne and plot them on the same graph.
(c) Predict boiling points for the elements K, Ca, Br and Kr and afterwards check to see how near your answers are to the correct values.

2. (a) Write the formulae of oxides of elements in the period Na to Ar.
(b) Which of these oxides are acidic?
(c) Which oxides would you expect to be acidic in the period K to Kr?
(d) Write expected formulae of oxides in the period Li to Ne.

3. List the ionic compounds for which you have data in the period Na to Ar.

4. Write equations to represent the action of water on the chlorides of elements in the period Na to Ar.

5. (a) Write electron configurations for the elements in the period Na to Ar.
(b) Write electron configurations for the elements Li to Ne.
(c) Write electron configurations for the elements Ca, Br and Kr.

6. The volumes occupied by 1 mole of atoms (1 gramme-atom) at r.t.p. (room temperature and pressure) of Cl and Ar are 12 000 cm^3 and 24 000 cm^3. Explain the difference between these two values.

7. Write electron configurations for elements 11, 13, 15 and 18.

8. Which elements would have the following electron configurations:

(a) 2.8.2 (b) 2.8.4 (c) 2.8.6 (d) 2.2
(e) 2.4 (f) 2.6 (g) 2.8.8.2 (h) 2.8.8.4
(i) 2.8.18.2 (j) 2.8.18.4

9. Use data in Chapter 23 to identity the elements from the following descriptions:

(a) has an acidic oxide; exists in two crystalline forms (polymorphs); heat of formation of oxide is greater than 1000 kJ/mole
(b) forms a peroxide; compounds are ionic; all atoms of same relative atomic mass (no isotopes)
(c) reacts with water slowly; forms more than one oxide
(d) forms negative ions; does not react with water; forms a chloride in which 1 mole of atoms of the element combine with 2 moles of chlorine atoms
(e) does not react with water; does not burn; obtained from air

Chapter 24

Select elements from the following list which fit the description given in each question.

Sodium Magnesium Iron Copper

1. These two elements have a small number of electrons in the outer shell. The shell before the outer one is not full.

2. This element always forms an ion with two positive charges in its compounds.

3. This element is below hydrogen in the ECS.

4. These two elements have very high boiling points due to strong bonding between their atoms.

5. These two elements have low densities.

6. This element can use either one electron or two electrons in compound formation.

7. This element reacts with steam when red-hot, but has no reaction at all with pure cold water.

8. This element reacts very slowly with cold water and rapidly with dilute hydrochloric acid. Hydrogen is produced in both cases.

9. This element does not react with water or with dilute hydrochloric acid. It does, however, react with dilute nitric acid.

10. This element reacts with chlorine but not with hydrogen chloride.

11. This element forms a chloride, $ElCl_2$, with hydrogen chloride and a chloride, $ElCl_3$, with chlorine.

12. This element forms a chloride, ElCl, which can be prepared by reducing its chloride, $ElCl_2$.

13. The hydroxide of this element is very soluble in water.

14. This element can form a green and a brown hydroxide.

15. The oxides of these two elements are completely insoluble in water.

16. This element forms a hydroxide of formula $El(OH)_2$ which is white.

17. This element forms an ion El^{2+}. When aqueous OH^- ions are added to an aqueous solution containing El^{2+} ions a blue precipitate is formed.

18. This element forms a red and a black oxide.

19. This element can be used as a catalyst in the production of ammonia from nitrogen and hydrogen.

20. These two elements and their compounds would be expected to increase the rate of many chemical reactions when present in only small amounts.

Chapter 25

1. Write molecular formulae for the following compounds:

   (a) methane
   (b) ethane
   (c) ethene
   (d) chloroethane
   (e) trichloromethane
   (f) ethanol
   (g) sodium ethoxide
   (h) acetic acid
   (i) ethyl acetate
   (j) aminoacetic acid (glycine)

2. Give the name and structural formula of one example of each of the following types of compound:

   (a) an alcohol
   (b) an alkane
   (c) an alkene
   (d) an alkyne
   (e) a dibromoalkane
   (f) a chloroalkane
   (g) an amino acid
   (h) an ester
   (i) a polymer

3. Use Table 25.1 to draw a graph of boiling point (vertical axis) against number of carbon atoms per molecule. From the graph predict the boiling points of heptane, octane, nonane and decane, which contain 7, 8, 9 and 10 carbon atoms per molecule. Predict the structural formula of each of these compounds.

4. Use Table 25.2 to draw a graph of boiling point (vertical axis) against number of carbon atoms per molecule. From the graph predict the boiling points of heptene, octene, nonene and decene. Predict the structural formula of each of these compounds.

5. Draw structural formulae to represent the possible isomers of the compounds with formula $C_4H_8Cl_2$. If you can make plasticine ball-and-stick models the task will be much easier: there are nine isomers with this formula.

6. There are fifteen isomers of the alkene C_6H_{12}. How many isomers are there of the alkene C_5H_{10}?

7. How could you make the ester ethyl acetate using only ethanol, sodium dichromate, concentrated sulphuric acid and the necessary laboratory glassware, etc? Write equations for all reactions.

8. Write molecular formulae for

   (a) an alcohol containing three carbon atoms
   (b) an alcohol containing five carbon atoms
   (c) Write two structural formulae for three-carbon alcohols.
   (d) Name the compounds which would be formed if each three-carbon alcohol were reacted with PCl_3.

9. What is meant by the term polymerization? Illustrate your answer by reference to addition and condensation polymers.

10. What is meant by the term fractional distillation? What is meant by the term cracking? Illustrate your answers by reference to petroleum.

11. (a) Give one way in which soaps and detergents resemble each other.
    (b) Give one way in which soaps and detergents differ from each other.

12. An organic compound reacts with PCl_3 (or PCl_5) to form a compound C_3H_7Cl.
(a) What is the organic compound?
(b) How would the organic compound react with sodium?

13. An organic compound reacts readily with bromine to form a compound $CH_3.CHBr.CH_2Br$
(a) What is the organic compound?
(b) How would the organic compound react with alkaline potassium permanganate?

14. The boiling points in °C of some members of the paraffin series are given in the following table:

C_2H_6 ethane	−89
C_3H_8 propane	−42
C_6H_{14} hexane	+69
C_8H_{18} octane	+126
C_9H_{20} nonane	+151

The cracking of a paraffin which has a boiling point of +174 °C gives, among the products, compounds with boiling points of −250 °C, −162 °C and +98 °C. None of these compounds is capable of being polymerized and they have no effect on bromine water.
(a) Suggest what the original paraffin might have been.
(b) Suggest what the three products might have been. You must show how you arrive at your conclusion.
(c) Some paraffins are gases, some liquids and some solids. Which of the liquid paraffins has the lowest mass for one mole of molecules?

Chapter 26

1. Name a substance which colours the Bunsen flame
(a) yellow/orange
(b) brick red
(c) lilac or pink
(d) blue/green
(e) scarlet

2. Name a substance which is
(a) white; soluble in hot water but insoluble in cold
(b) pale blue in aqueous solution and turns royal blue when ammonia is added
(c) pale green and insoluble in water
(d) white; dissolves in excess aqueous sodium hydroxide solution but not in excess aqueous ammonia
(e) yellow when hot but white when cool

3. You are given three tubes containing aqueous solutions of sodium salts A, B and C. When barium nitrate solution is added to each there is a white precipitate. On adding dilute nitric acid the precipitates in B and C dissolve, giving gases D and E respectively. The precipitate in A does not dissolve. Gas D has a choking smell and decolourizes potassium permanganate. Gas E makes limewater cloudy. Identify A, B, C, D and E.

4. Identify each gas from the following descriptions:
(a) reddish-brown; does not bleach litmus
(b) alkaline; gives blue/white smoke with HCl(g)
(c) makes limewater cloudy
(d) turns potassium dichromate paper from orange to green
(e) relights glowing splint

5. A substance Q gave a negative result for the flame test. When aqueous sodium hydroxide solution was added to a solution of Q a pale green precipitate was produced. Dilute nitric acid had no effect on Q. When a solution of lead nitrate was added to Q there was a white precipitate which did not disappear on warming.
(a) Identify Q.
(b) What would you expect to observe when aqueous barium nitrate was added to an aqueous solution of Q?
(c) A little Q was warmed with concentrated nitric acid, a powerful oxidizing agent (electron acceptor). The contents of the tube were then diluted and aqueous sodium hydroxide solution was added in excess of the amount required to neutralize the acid. What would you expect to observe when the sodium hydroxide was added?

6. A coloured substance R gave a gas S and steam on strong heating. The gas made limewater cloudy. The residue in the tube after heating was a black solid T. T was soluble in dilute nitric acid to give solution U. When aqueous ammonia was added to solution U a blue/green precipitate was produced which dissolved to give a royal blue solution when excess ammonia was added.
(a) Identify R, S, T and U.
(b) Write equations for the reactions

$$R \rightarrow T + S$$

and

$$T + \text{nitric acid} \rightarrow$$

7. How would you distinguish between each member of the following pairs of substances:
(a) sodium sulphate and sodium carbonate
(b) magnesium chloride and lithium nitrate
(c) potassium carbonate and sodium bicarbonate
(d) ammonium sulphate and iron(II) sulphate.
(e) copper(II) bromide and lithium iodide.

Full details of tests for both substances must be given in each case.

Index

ACETYLENE (ETHYNE), 98, 228
Acid, acetic, 19, 98, 231
  carbonic, 18, 30
  chloric, 101
  hydrobromic, 17, 30, 101
  hydrochloric, 17, 19, 30, 101
  hydrofluoric, 101
  hydriodic, 101
  nitric, 19, 30
  periodic, 99
  phosphoric, 19, 30, 99
  stearic, 31, 33
  sulphuric, 18, 19, 30, 99
  sulphurous, 18, 19, 30, 100
Acids, attack on metals, 17, 157–8, 166, 216
  formulae, 29
  heat of neutralization, 133–4
  organic, 231
  reaction with, alkalis, 20
    metals, 17, 157–8, 166, 216
  strong, 19
  weak, 19
Active electrodes, 174
Addition polymer, 232
Adipoyl chloride, 233
Air, composition of, 15, 16
  used in, contact process, 221
    copper extraction, 191
    iron extraction, 186–7
    nitric acid production, 222
    nitrogen and oxygen production, 192
    partial oxidation of hydrocarbons, 220
    steel production, 188–90
Alcohol (ethanol), 4, 10, 31, 97
Alcohols, 230–1
Alkali and pH, 18
  reaction with acid, 20
Alkanes, 225–6
Alkenes, 227–8
Alkynes, 228
Allotropes, 103
Aluminium, bromide, 82, 99
  chloride, 70, 80
  data on element and compounds, 206–12
  electrons in atom, 64
  extraction by electrolysis, 181–2
  fluoride, 82
  group elements, general properties, 198–9
  iodide, 82
  oxide, 30
    electrolysis of molten compound, 181–2
    purification of, 182
  uses, 181
Aminoacetic acid, 231–2
Amino acids, 231–2
Ammonia, 31, 96
  oxidation to nitric acid, 222
  production, 220
  reaction with hydrogen chloride, 112
  solution in analysis, 237
  test for, 240
Ammonium ions, test for, 237
Amphoteric oxides, 182, 199–200
  and hydroxides, 199–200
Analysis, 236–40
Anions, 84
  tests for, 238–9
Anode, reactions at in electrolysis, 172–4
Antimony, 103
Argon, 15, 31
  data on element, 206–12
  Group elements, general properties, 203
Arsenic, 15, 103, 105
Atomic mass, scale, 26, 27
  nucleus, 61
  number, 27, 62, 64
  structure, 60
  theory, 23, 25
  volume of elements in period sodium to argon, 206–7
Atoms, size of, 37
Avogadro constant, 40

BARIUM, nitrate in analysis, 239
  test for, 236–7
Bases, 17–21
Beryllium, chloride, 80, 91
  general properties, 197–8
Bicarbonate, test for, 238–9
Bismuth chloride, 81
Blast furnace, extraction of, iron, 186–8
    lead, 191
    zinc, 192
Boiling points of, chlorides, 247, 249
  elements, 44
  elements in period sodium to argon, 208
Bond breaking and bond forming, 134–6
Bonding, covalent, 88, 96, 98
  in various covalent molecules, 86 ff.
  ionic, 70 ff.
  metallic, 102
  non-ionic, 86 ff.
Boron, 15, 105
  general properties, 199
  in giant networks, 105
  trichloride, 91
  trifluoride, 96
Boyle's law, 108, 151
Brass, 192
Brine, electrolysis of, 183–4
Bromide, ions, reaction with chlorine, 167, 184
  test for, 238–9
Bromides, ionic, 82
Bromine, 14, 15, 31
  displacing halogens from aqueous solution, 167
  extraction of, 184–5
  reaction with water, 168
  reversible reaction with sodium hydroxide, 153–7
  test for, 240
  trichloride, 101
  uses, 184

Butane, 97
Butene, 227
Butyne, 228

CALCIUM, 14, 18
  carbonate, 30, 114
  chloride, electrolysis of molten compound, 178–9
  extraction by electrolysis, 178–9
  hydroxide, 30
  nitrate, 30
  oxide, 30
  reaction with water, 168
  test for, 236–7
  uses, 179
Calorimeter, 161–2
Carbon, 14, 15
  chemistry of, 224–35
  compared with other elements in Group 4, 200
  diamond polymorph (allotrope), 36, 102
  graphite polymorph (allotrope), 36, 102
Carbon dioxide, 4, 18, 25, 31
  in extraction of iron, 186–8
  test for, 240
Carbon monoxide, 25, 31
  in extraction of iron, 187
  in extraction of zinc, 192
  in production of hydrogen, 220
Carbon tetrachloride, 31, 91, 226
Carbonate, 18
  and hydrogencarbonate (bicarbonate), 18
  ion, 98
  test for, 238–9
Catalysts, industrial, 219–22
  influence, on rate of reaction, 142–4
    in reversible reaction, 154
Catalytic activity of transition metals, 219
Cathode, reactions at in electrolysis, 172–4
Cations, 84
  tests for, 236–7
Cell, electrical, 123
Cellulose, 233–4
Change, chemical, 1
Charge flow during electrolysis, 174–6
Charles' law, 108, 151
Chlorate ion, 101
Chloric acid, 101
Chloride, discharge of ion in electrolysis, 173
  test for, 238–9
Chlorides, covalent, 54, 73, 90
  formulae, 70
  ionic, 54, 73, 77, 90
  melting and boiling points, 247, 249
  of elements in period sodium to argon, 211–2
  physical properties, 71, 77
Chlorine, 14, 15, 31
  data on element and compounds, 206–12
  displacing halogens from aqueous solution, 167
  extraction, 183–4
  group elements, general properties, 201–3
  liberated in electrolysis, 173
  reaction, with, bromide ions, 167, 184
    elements in period sodium to argon, 70, 209
    normal and transition metals, 217
    water, 168
  uses, 184
Chloroethane, 228, 231
Chloromethane, 226
Chromate ions, movement in electrolysis, 170–1
Coke in extraction of, iron, 186–7
  hydrogen, 194
  lead, 191
  phosphorus, 194–5
  zinc, 192
Colour of transition metal compounds, 218
Combustion, heat of, 128–9
Concentration, influence on, equilibrium position, 153–4
  rate of reaction, 138–40
Condensation polymer, 233
Conductivity, electrical, 72
Conductors of electricity, 8, 10, 12
Configuration, electronic, 65–7, 81, 158
Conservation of mass, 6
Constant composition, law of, 24
Contact process, 221
Copper, 28
  compared with sodium, magnesium and iron, 213–9
  extraction, 190–1
  oxide, 28, 30
  purification, 174, 191
  reaction with iron, 156–7, 161–4
  sulphate, 3, 30
    electrolysis of solution, 11, 13, 170, 174–5, 191
    heat of solution, 130–3
    reaction with zinc, 121
  test for, 236–7
  uses, 190
Copper(I), preparation of chloride, 217
  oxide, 28, 30
    preparation, 217–8
Copper(II), carbonate, 3, 5
  ions, discharge in electrolysis, 172
    movement in electrolysis, 170–1
Coulomb, 174
Covalence, 86 ff., 96, 98
Covalent chlorides, 54, 73, 90
Cracking, 194, 224–5
Cryolite, 181
Cyclohexane, 224

DANIELL CELL, 123
Decomposition, electrolytic, 9
  thermal, 6
Density, of elements in period sodium to argon, 208
  periodic fluctuation in, 45, 48
Detergents, 234
Diagonal, the, 57
Diamond, 36, 102
Dibromoethane, 228
Dichloromethane, 226
Diffusion, 39
Dinitrogen tetroxide, 31
  preparation, 147
  reversible dissociation, 149–51
Discharge of ions in electrolysis, 170–6
Dissociation, as endothermic change, 149
  influence, of pressure, 152–3
    of temperature, 150–1
  of dinitrogen tetroxide, 149–51
Distillation, 21
  fractional of petroleum, 224
Dodecane, 224–5
Dynamic equilibrium, 146–54

ELECTRIC FURNACE for steel production, 188
Electrical energy change and heat energy change, 163–4
Electricity, conductors of, 8
  non-conductors of, 8
Electrochemical series, 161–9
  chemical properties, 167–8
  discharge of ions in electrolysis, 172–3
  displacement of, one halogen by another, 167
    one metal by another, 166
  metals reacting with acids, 166
  reaction of, halogens with water, 168
    metals with water, 168
  table, 165, 168
Electrode, 11
  active and inert, 174

comparison, 164
hydrogen, 164–5
standard, 164–5
Electrolysis, 9, 170–6
of aqueous copper(II) sulphate, 11, 13, 170, 174–5, 191
aqueous sodium chloride, 173
charge flow during, 174–6
competition between reactions, 170–4
of dilute sulphuric acid, 174
Faraday's laws, 175–6
in extraction of, aluminium, 181
calcium, 178
chlorine, 183
fluorine, 185
magnesium, 180
sodium, 179
zinc, 192
in purification of, copper, 174, 191
lead, 191
movement of ions in, 170–1
of molten lead bromide, 172
used to extract elements, 178–85
Electrolytes, 9, 11, 172
Electron, acceptors and donors, 157
transfer, 123 ff., 155–60
and energy changes, 161–4
in reaction between metals and acids, 157–8
investigated in calorimeter, 161–2
investigated in electric cell, 156–7, 163–4
Electronic configuration of, atoms, 65, 67
ions, 158–9
noble gases, 81
Electronic structure, 65, 67, 81, 158–9
Electrons, 60
donated to electron acceptor, 157
outer, 63, 66, 74
provided by electron donor, 157
Electrostatic attraction, 75
Elements, abundance, 14, 247
densities, 45, 48, 208
extraction by chemical methods, 186–95
extraction by electrolysis, 178–85
gaseous, densities of, 256
Group properties of, 196–204
heat capacity, specific, 248
heat of vaporization, molar, 253–4
isolation, date of, 246
metallic, 9, 53, 213 ff.
non-metallic, 9, 53
normal, 43
price, 247
properties, chemical, 52 ff.
physical, 42 ff.
Elimination reaction, 233
Endothermic change, 1, 10, 130, 151
End point, 118
Energy changes, 121 ff.
combustion, 128–9
in bond forming and breaking, 134–6
in reaction between metals and copper(II), 161–4
neutralization, 133–4
solution, 129–33
Energy, conservation of, 125
electrical, 123
maximum electrical, 124
ionization, 65
levels, 123
of vaporization, 72, 73, 75, 87, 253–4
Enthalpy, 124
Equations, 114 ff.
Equilibrium, dynamic, 146–7
influence of catalysts, 154
concentration, 153–4
pressure, 152–4
temperature, 150–1, 154
position, influence of, pressure, 220–2
temperature, 220–2
Esters, 231, 234
Ethane, 31, 97
Ethanediol, 228, 233
Ethanol, 31, 97
heat of combustion of, 128–9
reactions of, 230–1
Ethene (ethylene), 31, 97, 98, 225, 227–8, 231–2
Ethyl acetate, 98, 231
Ethyne, 98, 228
Exothermic change, 1
Extraction of elements in period sodium to argon, 212

FARADAY, 175
Faraday's laws, 175–6
Fats, 234
Fermentation, 230
Fertilizers, phosphate, 194
Flame test, 236
Fluorides, ionic, 81
melting and boiling points, 250, 253
Fluorine, 15, 31
electrons in atom, 64
extraction, 185
reaction with water, 168, 202
uses, 185
Formula, empirical, 29, 111
molecular, 111
Fractional distillation of petroleum, 224
Frasch process for sulphur extraction, 192–3

GALVANIZING, 192
Gases, volume occupied by, 50, 107
Germanium, compared with other elements in Group 4, 200
extraction of, 194
Giant structures, 75, 82, 101, 104
Glucose, 31, 233–4
Glycerol, 234
Glycine, 231–2
Glycol, 228, 233
Graphite, 36, 42, 102
Group, 57
1 (sodium group), chlorides, 80
electronic configuration of ions, 159
electron transfer to non-metals, 159–60
2 (magnesium group), chlorides, 80
electronic configuration of ions, 159
electron transfer to non-metals, 159–60
3 (aluminium group), 198
chlorides, 80
4 (silicon group), chlorides, 81
5 (phosphorus group), chlorides, 81
6 (sulphur group), chlorides, 81
electronic configuration of ions, 159
electron transfer to metals, 159–60
7 (chlorine group), chloride, 81
electronic configuration of ions, 159
electron transfer to metals, 159–60
Group properties of normal elements, 196–203
Group 1, 196–7
Group 2, 197–8
Group 3, 198–9
Group 4, 199–200
Group 5, 200–1
Group 6, 201
Group 7, 201–3
Group 0, 203
Gunmetal, 192

HABER PROCESS, 220
influence of, pressure on equilibrium position, 151

temperature on equilibrium position, 153
Half-reaction, 157, 159–60, 162
Halides, Groups 1 and 2, 91
Group 3, 91
Group 4, 91
Group 5, 92
Group 6, 92
Group 7, 93
Group 0, 94
Halogens, 68, 167, 168, 201–3
displacement related to ECS, 167
group properties, 201–3
reaction with, hydrogen, 202–3
water related to ECS, 168
Heat energy change and electrical energy change, 163–4
Heat of, combustion, 128–9
neutralization, 133–4
reaction, 122
solution, 129–33
vaporization, 72, 73, 75, 87, 253–4
Helium, 15, 31
Hexamethylene diamine, 233
Hexane, 31
Hexene, 227
Homologous series, 226
Hydrides, 54
Hydrobromic acid, 17, 30
Hydrocarbons, 225–8
in production of hydrogen, 220
melting and boiling points, 254
partial oxidation, 220
Hydrochloric acid, 17, 19, 30
heat of neutralization of, 133–4
reaction with magnesium investigated, 138–43
Hydrogen, 14, 17, 31, 87
bonding in, 87
bromide, 31, 101
chloride, 31, 101, 112
reaction with normal metals, 217
reaction with transition metals, 217
electrode, 164–5
electrons in atom, 64
fluoride, 101
iodide, 101
ions, discharge in electrolysis, 172–4
in neutralization reactions, 134, 154
peroxide, rate of decomposition, 143–4
production from, hydrocarbons, 220
petroleum, 194
water, 194
reaction with halogens, 202
resemblance to halogens, 203
sulphide, 100
test for, 240
uses, 194
Hydrogencarbonate, 18
test for, 238–9
reaction with transition metals, 217
test for, 240
Hydrolysis, 234
Hydroxides, amphoteric, 199–200
Hydroxide ions in neutralization reactions, 134
reaction with bromine, 153–4

INDICATORS, 19–20
Indium chlorides, 80–1, 99
Inert electrodes, 174
Iodine, 15
dissolving in water, 146–8
extraction, 195
reaction with water, 168
test for, 240
uses, 195
Iodine, heptafluoride, 93, 99
pentafluoride, 93
trichloride, 94–5
Iodide ions reacting with chlorine and bromine, 167
Iodides, ionic, 82
test for, 238–9
Ionic, chlorides, 54, 73, 77, 90
fluorides, 81
half-reaction, 157, 159–60, 162
Ionization energies, 65
of elements in period sodium to argon, 208
Ions, 53
coloured, 170–1
discharge in electrolysis, 170–6
electronic configuration, 98, 100–1, 158–9
formed by elements in period sodium to argon, 208
movement in electrolysis, 170–1
produced by electron loss and gain, 155–6
transition metal, 170–1
Iron, 1, 4, 14, 18
as catalyst in Haber process, 220
compared with sodium, magnesium and copper, 213–9
conversion to steel, 188–90
extraction, 186–8
ore, 186–8
reaction with copper(II), 156–7, 161–4
test for, 237
Irreversible reactions, 146
Isomerism, 229–30
Isotopes, 63
of elements in period sodium to argon, 206

KINETIC MODEL, and rate of reaction, 140, 142–3
and dynamic equilibrium, 148
Krypton, 15

LATTICE, ionic, 86
Lead, as catalyst, 143–4
bromide, 10, 30
bromide in electrolysis, 172
chloride, 81
dioxide, 30
extraction and purification, 191
monoxide, 3, 30
nitrate, 3, 5, 30
oxide, reduction by coke, 191
oxides as catalysts, 143–4
sulphide in extraction of lead, 191
test for, 236–7
tetroxide (red lead), 2–3, 5–6, 30
uses, 191
Limestone in extraction of, iron, 187–8
lead, 191
steel, 188
Linz-Donawitz process for steel production, 188–9
Lithium, chloride, 78
electronic configuration, 67
hydroxide, 30
reaction with water, 168
test for, 236–7
Lone pairs of electrons, 101

MAGNESIUM, 4, 6, 14, 18
bromide, 82
chloride, 70, 180–1
compared with sodium, iron and copper, 213–9
data on elements and compounds, 206–12
extraction by electrolysis, 180–1
group elements, general properties, 197–8
iodide, 82
oxide, 4, 6, 30
rate of reaction with acid, 138–43
reaction with water, 168
test for, 237

uses, 180
Manganese(IV) oxide as catalyst, 143–4
Mass, conservation of, 6
Melting points, of chlorides, 72–3
of elements, 44, 208
Mendeléef, 42
Mercury, in manufacture of chlorine and sodium hydroxide, 183–4
Metal, ions discharged in electrolysis, 172
oxides as catalysts, 143–4
Metals, 9, 53, 56, 70
base and noble, 158
displacement from aqueous solution, 166
electron transfer in reaction with acids, 157–8, 166
energy changes in reaction with copper(II), 161–4
extraction, 178–82, 186–92
reaction with, acids, 17, 157–8, 166, 216
halogens, 202
hydrogen, 203
water, 17–18, 53, 168
Methane, 31, 96, 226
as source of hydrogen, 220
Methanol, 97
Methyl cyanide, 98
Molar solution, 115
Mole, definition, 40, 107
of atoms, 28
of electrons, 175
of gas, 107
Molecules, formula, 30
shape, 88
size, 34
Motion, molecular, 39
Movement of ions in electrolysis, 170–1

NATURAL GAS, as source of hydrogen, 220
Negative ions, discharge in electrolysis, 172
Neon, 15, 31
electronic configuration, 67
Neutralization, 20
heat of, 128, 133–4
of H^+ ions by OH^- ions, 134, 154
Neutrons, 63
Nickel, 14
Nitric acid, in analysis, 238–9
production, 222
Nitrate, test for, 238–9
Nitrite, test for, 238–9
Nitrogen, 15, 31
obtained from air, 192
production for Haber process, 220
uses, 192
dioxide, 5, 31
reaction with water, 222
test for, 240
monoxide, reaction with air, 222
used to prepare nitrogen dioxide, 147, 149
trichloride, 92
Noble, gases, 14, 30, 55, 73
general properties, 203
production, 195
metals, 166
Non-conductors of electricity, 8
Non-electrolytes, 10
Non-metals, 9, 53, 56, 70
as electron acceptors, 159
extraction, 183–5, 192–5
Non-metal ions, discharge in electrolysis, 172
Nucleus, atomic, 61
Nylon, 233

OIL-DROP EXPERIMENT, 33
Open Hearth process for steel production, 188–90
Oxidation, 14, 17
Oxides, 23
acidic, 53, 57
amphoteric, 182, 199–200
basic, 53, 57
ionic, 83
melting and boiling points, 250
metallic, 23, 53, 57
non-ionic, 83
non-metallic, 23, 53, 57
of elements in period sodium to argon, 210–11
Oxidizing agents, 157, 168
Oxygen, 5, 15, 31
dichloride, 93
obtained from air, 14, 192
reaction with elements in period sodium to argon, 209
transport of, 14
test for, 240
uses, 14, 188–90, 192

PENTANE, 224, 229
Pentene, 227
Pentyene, 228
Peptide bonds, 232–3
Periodic, Law, 42
Table, 55
Periods, 56
Petrol, 2
Petroleum, cracking, 194, 224–5
pH scale, 18
Phosphate fertilizer, 194
Phosphine, 100
Phosphoric acid, 99
Phosphorus, red, 2, 5, 15, 103
white, 2, 5, 15, 31, 103
data on element and compounds, 206–12
extraction, 194–5
group elements, general properties, 200–1
molecule, 104
pentachloride, 31, 70, 92, 231
pentoxide, 5, 30
trichloride, 31, 54, 70, 89, 231
uses, 194
Platinum, as catalyst in nitric acid production, 222
Polymers and polymerization, 232–4
Polymorphs, 103
Polypeptides, 232–3
Polythene, 102, 232
Positive ions, discharge in electrolysis, 172
Potassium, 14, 18
chlorate, 6, 30
chloride, 30
electronic configuration, 65
extraction by electrolysis, 180
hydroxide, 30
nitrate, 6
permanganate, 3, 5–6, 30, 39
reaction with water, 168
test for, 236–7
Pressure, influence on equilibrium position in reversible reactions, 152–3, 220–2
Propane, 97, 110, 225–6
Propanetriol, 234
Propene, 227
Propyne, 228
Proteins, 231–3
Protons, 63

RADIOACTIVITY, 60
Radon, 15
Rate of reaction, 138–45
influence of, catalysts, 142–4
concentration, 138–40
surface area, 140–2
temperature, 140

Reactivity series, 167
Reducing agents, 157
  connection with ECS, 168
Reforming, steam, 220
Reversible reactions, 146–54
Rust, 1, 4
Rutherford, 62

SALT BRIDGE, 163
Salts, 20
Saturated compounds, 226–7
Selenium, 15, 102, 104
Series, electrochemical, 161–9
  homologous, 226
  reactivity, 167
Silicon, 14–15, 105
  compared with other elements in Group 4, 200
  data on elements and compounds, 206–12
  dioxide, 30, 102, 187–8
  group elements, general properties, 199–200
  hydride, 99
  tetrabromide, 82
  tetrachloride, 31, 70, 89
  tetraiodide, 82
Silver, halides, 202
  nitrate in analysis, 239
  sulphate, 30
Soap, 234
Sodium, 14, 18
  bromide, 30, 82
  carbonate, production, 266
  chloride, 30, 70
    electrolysis, 179–80, 183–4
  compared with magnesium, iron and copper, 213–9
  data on elements and compounds, 206–12
  extraction by electrolysis, 179–80
  fluoride, 82
  group elements, general properties, 196–7
  hydroxide, 30
    heat of neutralization, 133–4
    in analysis, 237
    production of, 183–4
    reversible reaction with bromine, 153
  iodate, conversion to iodine, 195
  iodide, 82
  nitrate, 30
  reaction with ethanol, 231
  reaction with water, 168
  stearate, 234
  sulphate, 30
  test for, 236–7
  uses, 180
Solution, volumetric, 115
  in titration, 118
  heat of, 129
Solvay cell, 183–4
Spectator ions, 153
Standard electrode potential, 164–5
Starch, 233–4
Steam reforming, 220
Steel, 188–90
Strontium, reaction with water, 197
Structure, electronic, 65–7, 81, 158
  ionic compounds, 70
  non-ionic compounds, 86
Sulphate, test for, 238–9
Sulphide, test for, 238–9
Sulphite, test for, 238–9
Sulphur, 2, 4, 15, 31, 104
  data on element and compounds, 206–12
  dichloride, 70, 89
  dioxide, 4, 31, 100
    production, 221
    reaction with air, 221
    test for, 240
  extraction, 192–3
  group elements, general properties, 201
  heptafluoride, 99
  hexafluoride, 92, 99
  trioxide, 99
    conversion to sulphuric acid, 221
    production, 221
  uses, 193
Sulphuric acid, 18–19, 30, 99
  in analysis, 238
  production, 221
Sulphurous acid, 18–19, 30, 100
Surface area, influence on rate of reaction, 140–2

TELLURIUM, 15, 105
Temperature influence on, dissociation of dinitrogen tetroxide, 150–1
  equilibrium position, 150–1
  rate of reaction, 140
Terylene, 233
Tetrachloromethane, 31, 91, 226
Thallium, 81
Tin, 64, 103
  compared with other elements in Group 4, 200
  chloride, 81, 100
Titration, 118
Transition metals, bonding in, 215
  catalytic activity, 218
  colour of ions, 218
  compared with normal elements, 213–19
  electronic, configuration, 214
    configuration of ions, 158–9
  formulae of compounds, 215
  preparation of compounds, 217–8
  reaction with, acids, 216
    chlorine, 217
    hydrogen chloride, 217
    water, 216
  reactivity, 215–17
  solubility of hydroxides, 218
Trichloromethane, 226
Trilead tetroxide, 2–3, 5–6, 30

UNSATURATED COMPOUNDS, 227

VANADIUM(V) OXIDE, 221
Volume, atomic, 206–7
  molar, 206–7
Volumetric analysis, 118–20

WATER, bonding in, 96
  composition, 17, 31
  of crystallization, 17
  properties, 3
  reaction with, elements in period sodium to argon, 209
    halogens related to ECS, 168
    metals related to ECS, 168
    normal and transition metals, 216
  solvent for iodine, 146–8

XENON, 15
  fluorides, 94–5

ZINC, 121
  extraction, 192
  reaction with copper(II) sulphate, 121
  sulphate, 121
  test for, 237
  uses, 192

LOGARITHMS

	0	1	2	3	4	5	6	7	8	9	1	2	3	4	5	6	7	8	9
10	0000	0043	0086	0128	0170	0212	0253	0294	0334	0374	4	8	12	17	21	25	29	33	37
11	0414	0453	0492	0531	0569	0607	0645	0682	0719	0755	4	8	11	15	19	23	26	30	34
12	0792	0828	0864	0899	0934	0969	1004	1038	1072	1106	3	7	10	14	17	21	24	28	31
13	1139	1173	1206	1239	1271	1303	1335	1367	1399	1430	3	6	10	13	16	19	23	26	29
14	1461	1492	1523	1553	1584	1614	1644	1673	1703	1732	3	6	9	12	15	18	21	24	27
15	1761	1790	1818	1847	1875	1903	1931	1959	1987	2014	3	6	8	11	14	17	20	22	25
16	2041	2068	2095	2122	2148	2175	2201	2227	2253	2279	3	5	8	11	13	16	18	21	24
17	2304	2330	2355	2380	2405	2430	2455	2480	2504	2529	2	5	7	10	12	15	17	20	22
18	2553	2577	2601	2625	2648	2672	2695	2718	2742	2765	2	5	7	9	12	14	16	19	21
19	2788	2810	2833	2856	2878	2900	2923	2945	2967	2989	2	4	7	9	11	13	16	18	20
20	3010	3032	3054	3075	3096	3118	3139	3160	3181	3201	2	4	6	8	11	13	15	17	19
21	3222	3243	3263	3284	3304	3324	3345	3365	3385	3404	2	4	6	8	10	12	14	16	18
22	3424	3444	3464	3483	3502	3522	3541	3560	3579	3598	2	4	6	8	10	12	14	15	17
23	3617	3636	3655	3674	3692	3711	3729	3747	3766	3784	2	4	6	7	9	11	13	15	17
24	3802	3820	3838	3856	3874	3892	3909	3927	3945	3962	2	4	5	7	9	11	12	14	16
25	3979	3997	4014	4031	4048	4065	4082	4099	4116	4133	2	3	5	7	9	10	12	14	15
26	4150	4166	4183	4200	4216	4232	4249	4265	4281	4298	2	3	5	7	8	10	11	13	15
27	4314	4330	4346	4362	4378	4393	4409	4425	4440	4456	2	3	5	6	8	9	11	13	14
28	4472	4487	4502	4518	4533	4548	4564	4579	4594	4609	2	3	5	6	8	9	11	12	14
29	4624	4639	4654	4669	4683	4698	4713	4728	4742	4757	1	3	4	6	7	9	10	12	13
30	4771	4786	4800	4814	4829	4843	4857	4871	4886	4900	1	3	4	6	7	9	10	11	13
31	4914	4928	4942	4955	4969	4983	4997	5011	5024	5038	1	3	4	6	7	8	10	11	12
32	5051	5065	5079	5092	5105	5119	5132	5145	5159	5172	1	3	4	5	7	8	9	11	12
33	5185	5198	5211	5224	5237	5250	5263	5276	5289	5302	1	3	4	5	6	8	9	10	12
34	5315	5328	5340	5353	5366	5378	5391	5403	5416	5428	1	3	4	5	6	8	9	10	11
35	5441	5453	5465	5478	5490	5502	5514	5527	5539	5551	1	2	4	5	6	7	9	10	11
36	5563	5575	5587	5599	5611	5623	5635	5647	5658	5670	1	2	4	5	6	7	8	10	11
37	5682	5694	5705	5717	5729	5740	5752	5763	5775	5786	1	2	3	5	6	7	8	9	10
38	5798	5809	5821	5832	5843	5855	5866	5877	5888	5899	1	2	3	5	6	7	8	9	10
39	5911	5922	5933	5944	5955	5966	5977	5988	5999	6010	1	2	3	4	5	7	8	9	10
40	6021	6031	6042	6053	6064	6075	6085	6096	6107	6117	1	2	3	4	5	6	8	9	10
41	6128	6138	6149	6160	6170	6180	6191	6201	6212	6222	1	2	3	4	5	6	7	8	9
42	6232	6243	6253	6263	6274	6284	6294	6304	6314	6325	1	2	3	4	5	6	7	8	9
43	6335	6345	6355	6365	6375	6385	6395	6405	6415	6425	1	2	3	4	5	6	7	8	9
44	6435	6444	6454	6464	6474	6484	6493	6503	6513	6522	1	2	3	4	5	6	7	8	9
45	6532	6542	6551	6561	6571	6580	6590	6599	6609	6618	1	2	3	4	5	6	7	8	9
46	6628	6637	6646	6656	6665	6675	6684	6693	6702	6712	1	2	3	4	5	6	7	7	8
47	6721	6730	6739	6749	6758	6767	6776	6785	6794	6803	1	2	3	4	5	5	6	7	8
48	6812	6821	6830	6839	6848	6857	6866	6875	6884	6893	1	2	3	4	4	5	6	7	8
49	6902	6911	6920	6928	6937	6946	6955	6964	6972	6981	1	2	3	4	4	5	6	7	8
50	6990	6998	7007	7016	7024	7033	7042	7050	7059	7067	1	2	3	3	4	5	6	7	8
51	7076	7084	7093	7101	7110	7118	7126	7135	7143	7152	1	2	3	3	4	5	6	7	8
52	7160	7168	7177	7185	7193	7202	7210	7218	7226	7235	1	2	2	3	4	5	6	7	7
53	7243	7251	7259	7267	7275	7284	7292	7300	7308	7316	1	2	2	3	4	5	6	6	7
54	7324	7332	7340	7348	7356	7364	7372	7380	7388	7396	1	2	2	3	4	5	6	6	7
	0	**1**	**2**	**3**	**4**	**5**	**6**	**7**	**8**	**9**	**1**	**2**	**3**	**4**	**5**	**6**	**7**	**8**	**9**